CONCRETE REPAIR, REHABILITATION AND RETROFITTING IV

PROCEEDINGS OF THE 4TH INTERNATIONAL CONFERENCE ON CONCRETE REPAIR, REHABILITATION AND RETROFITTING (ICCRRR 2015), LEIPZIG, GERMANY, 5–7 OCTOBER 2015

Concrete Repair, Rehabilitation and Retrofitting IV

Editors

F. Dehn

Leipzig University/MFPA Leipzig, Germany

H.-D. Beushausen, M.G. Alexander & P. Moyo

University of Cape Town, South Africa

CRC Press
Taylor & Francis Group
Boca Raton London New York Leiden

CRC Press is an imprint of the
Taylor & Francis Group, an **informa** business

A BALKEMA BOOK

CRC Press/Balkema is an imprint of the Taylor & Francis Group, an informa business

Typeset by V Publishing Solutions Pvt Ltd., Chennai, India
Printed and bound in Great Britain by CPI Group (UK) Ltd, Croydon, CR0 4YY

Published by: CRC Press/Balkema
P.O. Box 11320, 2301 EH Leiden, The Netherlands
e-mail: Pub.NL@taylorandfrancis.com
www.crcpress.com – www.taylorandfrancis.com

ISBN: 978-1-138-02843-2 (Hbk + CD-ROM)
ISBN: 978-1-315-67764-4 (eBook PDF)

Concrete Repair, Rehabilitation and Retrofitting IV – Dehn et al. (Eds)
© 2016 Taylor & Francis Group, London, ISBN 978-1-138-02843-2

Table of contents

Concrete Repair, Rehabilitation and Retrofitting IV – Dehn et al. (Eds)
© 2016 Taylor & Francis Group, London, ISBN 978-1-138-02843-2

Preface

These conference proceedings contain papers presented at the Fourth International Conference on Concrete Repair, Rehabilitation and Retrofitting (ICCRRR 2015), Leipzig, Germany, October 2015, which can be grouped under six main themes:

- Concrete durability aspects
- Condition assessment of concrete structures
- Modern materials technology
- Concrete repair, rehabilitation and retrofitting
- Performance and health monitoring
- Education, research and specifications.

A large number of papers discusses performance and assessment of innovative materials for durable concrete construction. Interesting fields, some quite new, are covered, such as self-healing techniques, high performance concretes, and strain hardening composites. The number of papers submitted on the topic of modelling and prediction of durability confirms the positive international developments towards performance-based methods for durability design and specification. Another fact that is evident from the paper submissions is that large advances have recently been made in the fields of condition assessment of concrete structures. The papers in the proceedings cover interesting new techniques for the assessment of reinforcement corrosion and its interpretation.

The majority of papers discusses recent developments in concrete repair, rehabilitation and retrofitting techniques. An important research area lies in the field of specifications for repair materials and systems. Here, an integrated approach is needed, linking assessment techniques and service life modelling to appropriate repair methods. A number of papers deals with these important issues, confirming that the industry is on the right track towards efficient and durable repairs. Based on research reports and case studies, latest developments on repair strategies and materials are presented, ranging from surface protection techniques to full-scale repairs. Numerous papers were submitted on the topic of strengthening and retrofitting, highlighting the need to cope with increasing loads and deteriorating structures and showcasing latest developments in strengthening systems.

The Fourth International Conference follows the highly successful previous International Conferences on Concrete Repair, Rehabilitation and Retrofitting. This conference is a collaborative venture between researchers from the South African Research Programme in Concrete Materials (based at the Universities of Cape Town and The Witwatersrand) and the Material Science Group at Leipzig University and The Leipzig Institute for Materials Research and Testing (MFPA) in Germany. The organization and implementation of the conference continues to embody a strong South African-German link, reflected in the excellent support given to the conference by researchers and practitioners from these two countries. However, the range of presenters at the conference continues to indicate its truly international nature. This continues to fulfill an aim of these conferences, to strengthen relationships not only between Africa and Europe but also between countries and regions from all over the world.

The backdrop, in industry and the state of national infrastructures, continues to be highly challenging and demanding. The facts remain that much of our concrete infrastructure deteriorates at unacceptable rates, that we need appropriate tools and techniques to undertake the vast task of sound repair, maintenance and rehabilitation of such infrastructure, and that all this must be undertaken with due cognizance of the limited budgets available for such work. New ways need to be found to extend the useful life of concrete structures cost-effectively. Confidence in concrete as a viable construction material into the 21st century needs to be retained and sustained, particularly considering the environmental challenges that industry and society now face.

This fourth conference also continues to seek to extend a sound base of theory and practice in repair and rehabilitation, through both theoretical and experimental studies, and through good case study literature. It also seems to the organizers that two key aspects need to be addressed currently: that of developing sound and easily applied standard practices for repair, possibly codified, and the need to seriously study the service performance of repaired structures and the repair system. In fact, without substantial effort at implementing the latter goal, much of the effort in repair and rehabilitation may prove to be less than economic or satisfactory.

All papers submitted to ICCRRR 2015 were subjected to a review process, and the proceedings contain only those papers that were accepted following this process. The review of manuscripts was undertaken by identified leading experts, acting independently on one or more assigned manuscripts. This invaluable assistance, which has greatly enhanced the quality of the proceedings, is gratefully acknowledged.

Special acknowledgements are due to the following organizations:

- Deutscher Beton- und Bautechnik-Verein e.V. (DBV)
- Deutscher Ausschuss für Stahlbeton e.V. (DAfStb)
- Deutsche Bauchemie e.V.
- International Union of Laboratories and Experts in Construction Materials, Systems and Structures (RILEM)
- International Federation for Structural Concrete (*fib*).

Finally, the editors wish to thank the authors for their efforts at producing and delivering papers of a high standard. We trust that the proceedings will be a valued reference for many working in this important field and that they will form a suitable base for discussion and provide suggestions for future development and research.

F. Dehn
H.-D. Beushausen
M.G. Alexander
P. Moyo
Editors

Concrete Repair, Rehabilitation and Retrofitting IV – Dehn et al. (Eds)
© 2016 Taylor & Francis Group, London, ISBN 978-1-138-02843-2

ICCRRR committees

LOCAL ORGANIZING COMMITTEE

F. Dehn, *Co-chairman, Leipzig University/MFPA Leipzig, Germany*
H.-D. Beushausen, *Co-chairman, University of Cape Town, South Africa*
M.G. Alexander, *Co-chairman, University of Cape Town, South Africa*
P. Moyo, *Co-chairman, University of Cape Town, South Africa*
C. Klimke, *Secretary, MFPA Leipzig, Germany*
K. Karsten, *Secretary, MFPA Leipzig, Germany*

INTERNATIONAL SCIENTIFIC AND TECHNICAL ADVISORY BOARD

Professor H. Abdelgader, *Tripoli University, Libya*
Dr. H.-P. Andrä, *LAP Consult, Stuttgart, Germany*
Professor Y. Ballim, *University of the Witwatersrand, South Africa*
Professor A. Bentur, *Israel Institute of Technology, Israel*
Professor K. Bergmeister, *BOKU Vienna, Austria*
Professor B. Bissonnette, *CRIB Laval, Canada*
Professor D. Bjegovic, *University of Zagreb, Croatia*
Professor B. Boshoff, *University of Stellenbosch, South Africa*
Professor W. Brameshuber, *RWTH University Aachen, Germany*
Mr. G. Breitschaft, *DIBt, Germany*
Professor E. Brühwiler, *EPFL Lausanne, Switzerland*
Dr. P. Castro Borges, *CINVESTAV-Mérida, Mexico*
Dr. G. Concu, *University of Cagliari, Italy*
Professor L. Courard, *University of Liège, Belgium*
Professor M. Curbach, *Technical University of Dresden, Germany*
Professor L. Czarnecki, *Warsaw University of Technology, Poland*
Professor N. de Belie, *Ghent University, Belgium*
Professor G. de Schutter, *Ghent University, Belgium*
Professor M. Di Prisco, *Politecnico di Milano, Italy*
Professor D. Fowler, *University of Texas at Austin, USA*
Professor O. Fischer, *Technical University of Munich, Germany*
Professor A. Garbacz, *Warsaw University of Technology, Poland*
Professor C. Gehlen, *Technical University of Munich, Germany*
Professor M. Grantham, *Concrete Solutions, UK*
Professor C. Grosse, *Technical University of Munich, Germany*
Dr. P. Haardt, *Federal Highway Research Institute (BASt), Germany*
Professor K. Imamoto, *Tokyo University of Science, Japan*
Professor M. Iqbal Khan, *King Saud University, Kingdom of Saudi Arabia*
Professor C.K. Leung, *UST, Hong Kong*
Professor E. Kearsley, *University of Pretoria, South Africa*
Dr. A. Klemm, *Glasgow Caledonian University, UK*
Professor E. Koenders, *TU Darmstadt, Germany*

Concrete durability aspects

Concrete Repair, Rehabilitation and Retrofitting IV – Dehn et al. (Eds)
© 2016 Taylor & Francis Group, London, ISBN 978-1-138-02843-2

Chloride ingress testing of concrete

D. Dunne & C. Christodoulou
AECOM Europe, Birmingham, UK

M.D. Newlands
Concrete Technology Unit, University of Dundee, UK

P. McKenna
CH2M Hill, Glasgow, UK

C.I. Goodier
School of Civil and Building Engineering, Loughborough University, Loughborough, UK

ABSTRACT

Concrete is recognised as a durable material that can provide long-term protection to embedded carbon steel reinforcement. However, effective protection can only be achieved if the mechanisms of deterioration and the durability properties of the protective material involved are fully understood. A well-known phenomenon is the ingress of chloride ions into concrete, which takes place via the solution-filled pores. These ions, combined with optimum contents of air and moisture, result in reinforcement corrosion, and subsequently, loss in functionality of a reinforced concrete element. Currently, available literature suggests that measurement of chloride ion ingress into concrete can be misleading. Reasons for erroneous measurements include: (i) continuing formation of hydration products resulting in pore refinement/pore blocking and; (ii) chloride diffusion coefficients being commonly applied as the "effective" diffusivity, which does not take into account the effect of chemical binding on the chloride ion transport process. This paper reports the findings of an investigation of different chloride test methods for measuring chloride ingress of concrete, which are currently used. This was undertaken to establish their suitability to measure this intrinsic concrete durability property. Five test methods were selected, covering the two main test method types that currently finding application: (i) electrically accelerated short-term; and (ii) naturally accelerated long-term chloride test methods. Three concrete types of CEM I, PC/FA (20–55%) and PC/GGBS (25–75%) at varying water-cement (w/c) ratios of 0.35, 0.50 and 0.65 were utilised during the investigation. This work has established that whilst all the test methods demonstrated capacities to measure chloride ingress into concrete, operator accuracy is of significant importance.

INTRODUCTION

The durability of concrete structures exposed to chloride-containing solutions has being widely studied (Dunne, 2010). Nonetheless, chloride induced corrosion continues to be recognized as a major reason for the loss of durability in reinforced concrete structures. The predominant sources of these chlorides into concrete emanate from exposure to marine environments and from the application of de-icing salt on roads, parking and bridge structures. Whilst recognizing transport mechanisms, such as, capillary suction (absorption) and/or permeability, it is generally considered that long-term, diffusion is the predominant mechanism, by which chlorides are transported through concrete.

As a result, determining chloride ion diffusivity of concrete is recognised as one of the most important parameters when assessing the potential durability of concrete (Lu et al., 1995; Thomas and Matthews, 2004). Consequently, whilst all physical processes are subject to variability, it is essential that the measurement techniques which are applied to characterise this intrinsic concrete property are robust and provide a representation of the likely trends, which may be expected for a particular concrete in service.

EXPERIMENTAL

Materials & mix proportions

A single sourced CEM I 42.5 N cement, conforming to BS EN 197-1 was utilised (BSI, 2000). In addition the influences of two cement addition materials were examined in this work: (i) Fly Ash (FA); (ii) Ground Granulated Blast furnace Slag (GGBS). Natural sand and gravel conforming to BS EN 12620 was used as aggregate in all mixes.

Concrete mixes were proportioned for three water-cement ratios of 0.35, 0.50 and 0.65, a total cement content of 330 kg/m^3 and single total water content of 165 kg/m^3.

The chloride test methodologies which were assessed, included:

- Short-term Rapid chloride test methods.
  1. Rapid chloride permeability (Method A)
  2. Split and spray (Method B).
  3. Rapid chloride migration (Method C)
- Short-term Rapid chloride test methods.
  1. Apparent chloride diffusion (Method D)
  2. Effective chloride diffusion (Method E).

FINAL EVALUATION OF TEST METHODS

In determining the overall accuracy of a test method ISO 5725-1 (BS ISO, 1994) makes reference to both the repeatability and trueness of a test method. This repeatability term is expressed as 'precision', which is a measure of the variability observed between repeated measurements of a test method. The suitability of a test method in performing a particular measurement is in many cases, subjective to the individual assessor.

In response to this a measured approach was applied in arriving at a consensus, as to the five chloride test methods suitability in measuring chloride ingress of concretes. The methods were assessed with respect to their simplicity, application of their measured chloride parameter, length of test and the relevance of the measured parameter of these chloride test methods to actual working life performance.

From comparison it was established, that Method C was the most consistently scoring method. On the other hand, Method D scored highest on the more critical criteria of application of measured parameter to modelling and relevance to actual chloride penetration. These are in contrast to Methods A, B and E with ratings of 14, 12 and 9 respectively. Therefore, Method D is recommended as the method most suitable for measuring chloride ingress of concrete. In particular, the measurements taken with this method will reflect expected concrete performance in working life, whilst being more appropriate to long-term durability modelling application.

The study identified that all the methods demonstrated capacities to characterise chloride ingress. Variations in magnitudes of the measured results do exist and this intensifies with increasing w/c. Nonetheless, this study has allowed each method to be ranked in the order of the most suitable for measuring chloride ingress. Method C and D were selected as being the most suitable and thereafter Methods A and B respectively, whilst Method E is deemed the fifth ranking method.

CONCLUSION

When determining the chloride ingress of concrete the technique applied should be robust and reliable in its measurement. Findings from this study have demonstrated that each method has the potential for use in measuring the chloride ingress of concretes.

Thereafter, each method demonstrated sensitivities to changes in w/c, albeit Method A's sensitivity reduced at the w/c of 0.65, periods of moist curing/test duration and to different cement contents and types.

Taking account of all the investigated parameters, Methods C and D are recommended as being the methods with the greatest potential for accurately measuring chloride ingress of concrete. Due to insensitivity to change in w/c and concrete composition, Method A is recommended for quality control purposes only.

Concrete Repair, Rehabilitation and Retrofitting IV – Dehn et al. (Eds)
© 2016 Taylor & Francis Group, London, ISBN 978-1-138-02843-2

Concrete corrosion in an Austrian sewer system

C. Grengg, A. Baldermann & M. Dietzel
Institute of Applied Geosciences, Graz University of Technology, Graz, Austria

F. Mittermayr
Institute of Technology and Testing of Building Materials, Graz University of Technology, Graz, Austria

M.E. Böttcher
Leibniz Institute for Baltic Sea Research (IOW), Warnemünde, Germany

A. Leis
Joanneum Research, RESOURCES—Institute for Water, Energy and Sustainability, Graz, Austria

ABSTRACT

This study comprises the application of a multi proxy approach, where a strongly deteriorated Austrian sewer system was intensively investigated. Understanding the underlying reaction mechanisms leading to the deterioration by microbial induced sulfuric acid attack on concrete structures is highly complex and often not fully understood. The aim of this study is to contribute to a deeper understanding by introducing a novel approach that comprises a range of mineralogical methods, as well as hydro-geochemical analyses, analyses of gases, hydro-geochemical modelling, and microbiological analyses. Results revealed an extremely fast propagating Microbial Induced Concrete Corrosion (MICC), with corrosion rates of up to 1 cm/y. Expressed pore fluids contained sulfate concentrations of up to 104 g/l at low pH of between 0.7 and 3.1. Sulfuric acid produced triggered the dissolution of the cementitious matrix and the carbonatic additives, as well as massive formations of gypsum, anhydrite and bassanite. Microprobe analyses revealed sequences of element distributions within the corrosion fronts, controlled by the suggested pH gradient of 13 to <1 within the pore fluids (Figure 1).

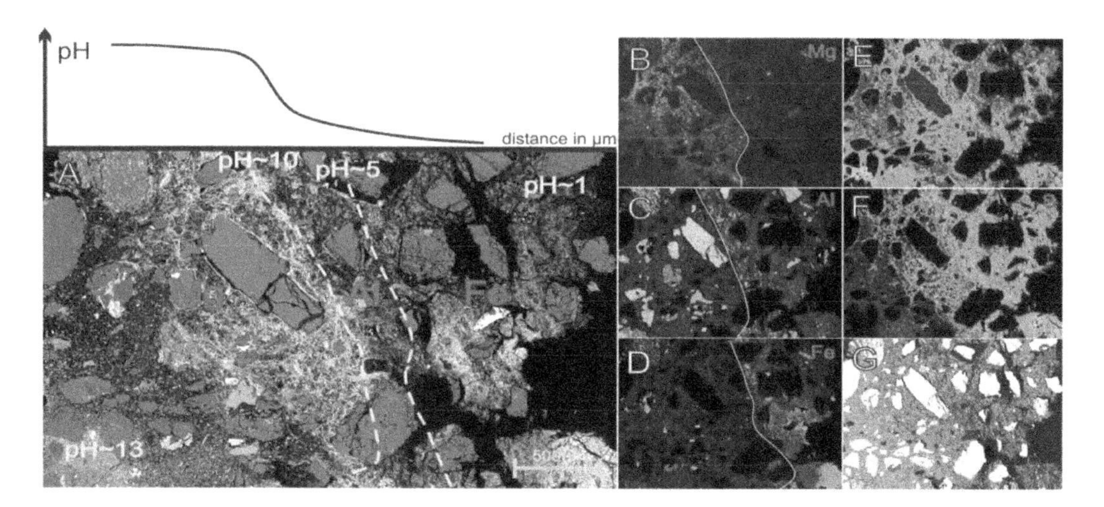

Figure 1. Displaying the pH depending dissolution and re-precipitation of Mg, Al, and Fe within a progressive corrosion front from the right to the left side (A). Mg accumulations indicate pH >10 (B) while Al is enriched in areas with pH between 5 and 10 (C). Fe enrichment can be explained by the ability of *Acidithiobacillus ferrooxidans* to reduce Fe (III) to Fe (II) and subsequent oversaturation of distinct iron bearing phases within the interstitial solutions causing precipitation (D). Additionally, concentrations of sulfur (S), calcium (Ca) and silica (Si), within the propagating corrosion front, emanating from top right to bottom left, are presented (E-G). Notice the opposing trend of Si and S enrichment.

Concrete Repair, Rehabilitation and Retrofitting IV – Dehn et al. (Eds)
© 2016 Taylor & Francis Group, London, ISBN 978-1-138-02843-2

One-dimensional scanning of water transport in hardened cement paste during freeze-thaw attack by NMR imaging

Z. Djuric, M. Haist & H.S. Müller
Institute of Concrete Structures and Building Materials, Karlsruhe Institute of Technology (KIT), Germany

J. Sester & E.H. Hardy
Institute for Mechanical Process Engineering and Mechanics, Karlsruhe Institute of Technology (KIT), Germany

ABSTRACT

The frost resistance of concrete is highly influenced by the water transport processes in the matrix of the hardened cement paste. Herein the water content of the pores in relation to the pore volume (the so-called degree of saturation) plays an important role. Thus, every freeze-thaw attack provokes an increase of the water content in concrete (the so called frost suction) that can significantly exceed the saturation which is normally reached by capillary suction (Setzer 2001). Upon reaching a critical value of saturation structural damage within the concrete, i. e. the formation of cracks, will occur after one freezing attack (Fagerlund 2004). Hence, when the water saturation behaviour of the hardened cement paste during the frost exposure is understood and the critical degree of saturation is known, it will be possible to predict the time of failure. Relating to laboratory investigations, the critical degree of saturation can be easily determined (Fagerlund 1977). However, the challenge is primarily the quantification of the frost suction process as a function of environmental conditions and material quality considering the underlying physical mechanisms. Therefore, it is essential to observe continuously and spatially resolved the water transport in concrete during the freezing and thawing process. A suitable measurement method for the determination of moisture profiles is the Nuclear Magnetic Resonance imaging (NMR). For this analysis method, a special set-up has been developed which allows a non-destructive, highly spatially resolved in situ detection of the water uptake during the frost exposure.

Thus, using this NMR technique the time dependent progress of the water uptake in hardened cement paste during the frost attack has been investigated (see Figure 1) and compared to a pure capillary suction. For the examination a special calibration method was developed that allows the conversion of the measured NMR signal intensity to the degree of saturation.

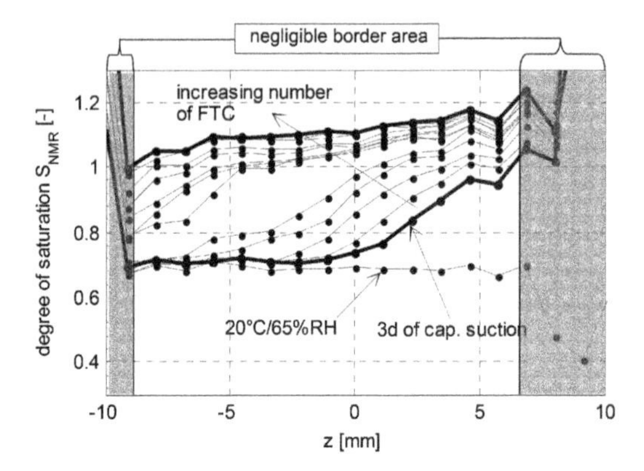

Figure 1. Averaged degree of saturation profiles distributed over the longitudinal axis of hardened cement paste samples with a water/cement ratio of w/c = 0.4 during the frost suction (SPI with 70 μs echo time).

As can be seen from Figure 1, the increase of the saturation degree in the sample in dependence of the freeze-thaw cycles could be closely followed with a high longitudinal resolution of 1.14 mm.

Furthermore an extended NMR set-up for the in situ one-sided temperature control has been developed and presented. With this setup, it is possible to observe the freezing and thawing processes and associated therewith the water transport during the freeze-thaw attack without interruption.

REFERENCES

Fagerlund, G. 1977. The critical degree of saturation method of assessing the freeze/thaw resistance of concrete. In Materials and Structures 10, 58, 217–229.

Fagerlund, G. 2004. A Service Life Model for International Frost Damage in Concrete. Lund Institute of Technology, Division of Building Materials.

Setzer, M. J. 2001. Micro-Ice-Lens Formation in Porous Solid. In J. of Colloid and Interface Science 243, 193–201.

Concrete Repair, Rehabilitation and Retrofitting IV – Dehn et al. (Eds)
© 2016 Taylor & Francis Group, London, ISBN 978-1-138-02843-2

Effects of electrochemical chloride extraction on microstructure of various cement paste systems

Nguyen Thi Hai Yen, Yokota Hiroshi & Hashimoto Katsufumi
Division of Engineering and Policy for Sustainable Environment, Hokkaido University, Sapporo, Japan

ABSTRACT

The adverse effects of the long term application of electrochemical chloride extraction on binding capacity of hydrated cement products are inneg-ligible. In this study, the influences of electro-chemical chloride extraction in the microstructural properties of hydrated cement products of cement pastes were investigated in paste specimens by using a modified migration cell that was conducted with the applied current density of 4.5 A/m^2 and synthesized pore solutions used as electrolytes for 8 weeks. Three types of cement pastes were used with the water/binder ratio of 0.4; that were ordinary Portland cement, fly ash cement and blast furnace slag cement pastes. Chloride source was supplied by adding 1.5% sodium chloride. After extraction, the acid-soluble chloride content significantly reduced. Portlandite content increased proportionally with the time of extraction in ordinary Portland cement and fly ash cement pastes. Furthermore, the alteration and decomposition of C-S-H were inevitable with different rates.

INTRODUCTION

The resources of the earth are depleted day by day, including the materials for concrete, the most popular material used in construction. Therefore, the maintenance and prolongation the lifetime of an existing structure are necessary not only due to the budget issues, but also the sustainability problems for the effective using of earth materials and carbon dioxide emission. Electrochemical realkalization and Electrochemical Chloride Extraction (ECE in this paper) arose as options to solve the issue that, when the source of corrosion is extracted or eliminate out the zone around steel reinforcement, it is not clear if the corrosion process of reinforcement stops, the passive layer is re-passivated or the deterioration progress could be slowdown and the performance and lifetime of the structures prolonged.

Theoretically, they are both potential methods to stop the further corrosion process and mitigate the deterioration progress of reinforced concrete structures when they are exposed to severe environments. Although they have the same principle, the results from many studies have shown out that the efficiency of fore-mentioned methods is different. Meanwhile electrochemical realkalization has been accepted widely [1–2], the effectiveness and efficiency of ECE are remaining suspicion although there were many research have been done. The feasible results of ECE on mitigating the further corrosion progress of reinforcement when it was conducted in several existing reinforced concrete structures; however, have not stopped the arguments, not only about the effectiveness and the possibility of this method on preventing the deterioration progress but also the technical term [1–14]. Moreover, the use of substitution materials has been applied widely for last some decades, but there were very few studies of the impacts of ECE on blended cement.

Therefore, in this study the impacts of ECE in chemical properties and C-S-H, the main component that affects to the binding capacity of the paste on various types of cements including Ordinary Portland cement (OPC), Fly Ash cement (FA) and Blast Furnace Slag cement (BFS) pastes were studied.

CONCLUSIONS

Electrochemical chloride extraction is an efficient method to release chloride ions from the pastes, especially water-soluble chloride ions. It can remove approximately 50% of acid-soluble chloride ions.

There was no significant differences in reduction of acid-soluble chloride content of the ordinary Portland cement paste and the fly ash cement paste, 53% and 48%, respectively. However, the reduction of chloride content in the blast furnace slag cement paste was rather low, only approximately

30%. It was interesting that the reduction of binding chloride in the fly ash cement paste was greater than that in the ordinary Portland cement and the blast furnace slag cement pastes.

Portlandite content in the ordinary Portland cement and fly ash cement pastes gradually increased during extraction. Nevertheless, in the blast furnace slag cement paste it slightly reduced after increasing in the first week of extraction.

The alteration or even the decomposition of C-S-H gel phases in these three popular types of binder was inevitable due to the long application of electrical current. However, it seems that C-S-H gel phase in the blast furnace slag cement paste showed a more stable than that in the ordinary Portland cement and the fly ash cement pastes when they suffered the same quantity of electric current passed through.

ACKNOWLEDGMENT

This work has been supported by JSPS KAKENHI Grant Number 26630208.

Concrete Repair, Rehabilitation and Retrofitting IV – Dehn et al. (Eds)
© 2016 Taylor & Francis Group, London, ISBN 978-1-138-02843-2

Study on possibility of estimation of chloride content in coastal reinforced concrete structures using electromagnetic waves

J. Nojima, M. Uchida & T. Mizobuchi
HOSEI University, Tokyo, Japan

ABSTRACT

Reinforcement corrosion caused by the presence of chloride ions around the reinforcing bars has been identified as one of the major causes of deterioration of concrete structures. The chlorides could find their way to concrete either as part of constituent materials when sea sand is used, or, by gradual permeation and diffusion as in the case of marine structures, or, cases where deicing salts are used to melt away snow on highways, etc. However, a definite understanding about any corrosion of reinforcement is very difficult unless corrosion induced cracks appear on the surface. In order to detect the deterioration caused by the chloride ions in an early stage, it is implemented generally to investigate chloride contents within concrete by carrying out chemical analysis using cores drawn from the RC structure. But, the method for carrying out to the chemical analysis using drawing cores not only gives damage partially to the concrete structure, but also is only to obtain result of analysis at the position of the drawn cores. In addition, drawing cores to estimate the chloride contents in concrete could not make it possible to investigate the changes in chloride contents over time at exactly the same place.

In this study, as an evaluation method of estimation of chloride content using the electromagnetic wave, the relationship between content of chloride ions and the conductivity, which directly influences the attenuation of the reflected waveform of the electromagnetic wave based on the attenuation equation derived from the Maxwell's wave equation, was investigated in the laboratory and the applicability was verified in existing structures.

As the results, it is possible to estimate the conductivity of the concrete using the equation (1) from the temperature, the relative humidity, the dielectric constant, the output amplitude value of the electromagnetic wave radar and the distance of cover concrete in the investigation objective portion.

$$\sigma_{ca} = \frac{10^3}{2\sqrt{\varepsilon_r} \cdot c} \cdot \eta \cdot \ln\left(\frac{|A_w|}{\zeta \cdot \delta_c |A_0|} \right), \quad \eta = -\frac{1}{189 \cdot \xi_t \cdot \kappa_w} \tag{1}$$

Where, c = the cover concrete (mm).

It is possible to estimate the content of chloride ions in the investigation objective portion by obtaining the relationship between the content of chloride ions and the conductivity of concrete calculated from equation (1).

In order to verify this proposed equation, the estimation of content of chloride ions using the electromagnetic waves was carried out in the six power plants from 2010. Until now, drawn cores were collected from more than 60 positions as well as the measurements of electromagnetic wave were carried out around the drawn cores.

As the results of verification, the content of chloride ions increased with the increase in the conductivity of concrete, though the calculations of conductivity of concrete were varied widely. The correlation coefficient was 0.790. The relationship between the conductivity of concrete and the content of chloride ions is shown in the following.

$$W_{acl} = 67.5\sigma_c - 1.88 \tag{2}$$

Where, W_{acl} = content of chloride ions in field survey (kg/m^3).

From the equation (2), it was possible to estimate the average content of chloride ions in cover concrete in the non-destruction using environmental conditions and measurement results by electromagnetic wave in the field survey.

Concrete Repair, Rehabilitation and Retrofitting IV – Dehn et al. (Eds)
© 2016 Taylor & Francis Group, London, ISBN 978-1-138-02843-2

Lithium migration in mortar specimens with embedded cathode

L.M.S. Souza & O. Çopuroğlu
Faculty of Civil Engineering and Geosciences/Materials and Environment, TU Delft, Delft, The Netherlands

R.B. Polder
Faculty of Civil Engineering and Geosciences/Materials and Environment, TU Delft, Delft, The Netherlands
TNO Technical Sciences/Structural Reliability, Delft, The Netherlands

INTRODUCTION

Concrete structures worldwide are affected by a degradation mechanism known as Alkali-Silica Reaction (ASR). Hydroxyl and alkali (sodium and potassium) ions react with siliceous components present in reactive aggregate, forming a hygroscopic gel. As the ASR gel absorbs water from the pore solution, it expands. As the process progresses, it leads to deleterious expansion and cracking of the concrete element. Even though there are numerous preventive methods against ASR, if the reaction is detected in an existing structure, currently, there are no definite treatments.

The addition of lithium ions to the reactive concrete mixture leads to the formation of a non-expansive ASR gel (McCoy & Cadwell 1951, Feng et al. 2010). In hardened concrete, those ions need to be transported into de cementious matrix and migration is the most effective technique. (Thomas et al. 2007).

It is known that the use of the reinforcement as electrode during electrochemical techniques might bring deleterious effects, such as loss of bond strength between rebar and changes in the pore structure, depending on the level of current (Bertolini et al. 2013). In this paper, preliminary results on the experimental study of lithium migration with embedded cathode is presented.

EXPERIMENTAL

Material and specimens preparation

Mortar cylinders (diameter of 98 mm and 60 mm of height) were cast with a titanium mesh at 10 mm from one of the surfaces. The mortar mixture had a water to cement ratio of 0.5 and sand to cement proportion of 3:1. Portland cement type CEM I 42.5 N, standard sand with D_{max} of 2.00 mm (according to EN 196 1:2005) and deionized water were used. The specimens were cured in the fog room for 36 days.

Experimental procedure

Lithium migration test was performed in a set-up as shown in Figure 1. Each specimen was positioned between two chambers—one with LiOH solution (4.9 M) and another that remained empty, working only as support for the specimen. The chamber with electrolyte received a titanium electrode that worked as anode. The embedded titanium mesh in the specimens was the cathode, i.e., it attracted cations, such as lithium ions, when electric potential was applied. Two types of cell were used. They were identical, except for the presence of ventilation holes in the type I.

Specimens were tested during four or eight weeks, under 15 or 25 V. Current and anolyte temperature were continuously monitored. Cell electrical resistance was measured with a multimeter at 120 Hz, in resistance mode. Resistivity was calculated from resistance values. Anolyte pH was measured with pH test strips and samples were collected throughout the experiment and analysed by Inductively Coupled Plasma (ICP), in order to obtain the concentration of sodium, potassium and lithium and calcium.

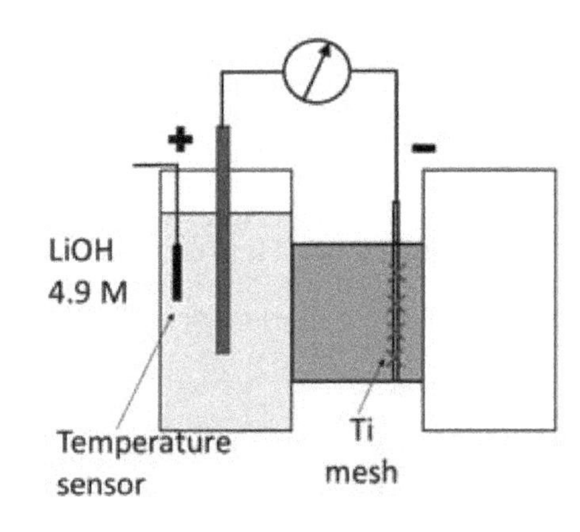

Figure 1. Diagram of the experimental set-up.

Table 1. Charge through the specimens during experiment.

Cells	Charge kC
LiOH 25 V 4 w (I)	123
LiOH 25 V 4 w (II)	159
LiOH 25 V 8 w (I)	186
LiOH 25 V 8 w (II)	253
LiOH 15 V 4 w (I)	88
LiOH 15 V 4 w (II)	103
LiOH 15 V 8 w (I)	138
LiOH 15V 8 w (II)	172

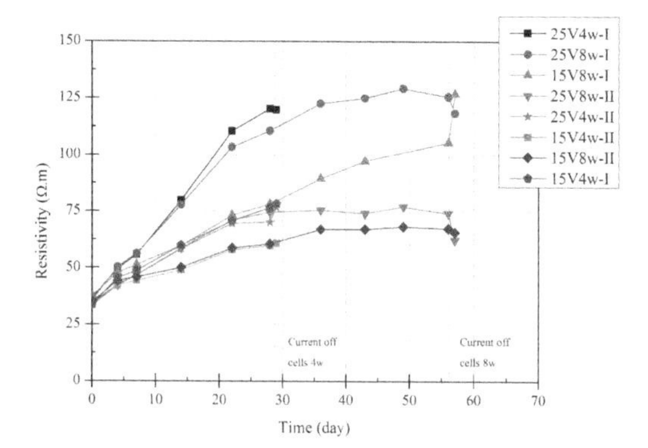

Figure 2. Resistivity variations during test.

Results and discussion

Table 1 brings the passing charge values throughout the test. As expected, the cells under higher voltage for longer exhibited higher charge, that is, higher ionic transport. In addition, when comparing the cells under the same level of voltage, it can be noted that cells in the set-up type II presented higher current density than those in the set-up type I- and the difference is more pronounced when the applied voltage is higher. This happened most likely due to the presence of ventilation holes in ring around the specimen in the cells type I, which lead to lower temperatures.

The cell electrical resistivity values during the experiment are shown in Figure 2. All cells exhibited increasing resistivity until the end of the experiment. At the end, the current was turned off and resistance was once more measured, after 24 hours. Interestingly, the resistivities did not return to the initial values, indicating that the variations during the experiment were not a mere artefact due to temperature. This indicates that there were, in fact, irreversible changes in the microstructure of the specimens and/or in the pore solution composition. One possibility is that the production of H_2 on the titanium mesh in the specimens, due to cathodic reaction lead to development of pressure that resulted in local debonding. Further investigation is planned in order to determine the nature of those changes.

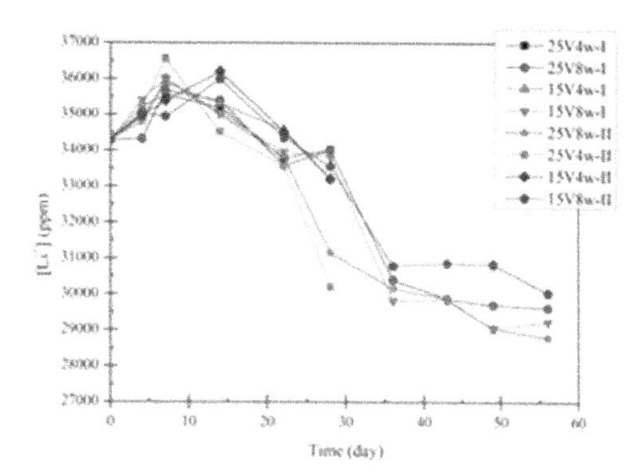

Figure 3. Lithium concentrate ion in anolyte.

Figure 3 shows the lithium concentration anolytes during the experiment. As expected, more lithium ions were removed from anolyte when higher voltage was applied.

CONCLUDING REMARKS

Preliminary results indicate that by applying higher voltage it is possible to drive more lithium ions into the specimen, as expected. However, that voltage should be limited. Studies have shown that high current density levels (above 10 A/m^2) might lead to changes of the microstructure of concrete and debonding of the reinforcing steel used as electrode (Bertolini et al, 2013). Even though it is not yet possible to determine the cause of the irreversible increase in resistivity during the experiment, those results indicate that the magnitude of this effect depends on the applied voltage. Therefore, when using this technique as treatment against ASR, it is necessary to find the current density level that would lead to the necessary lithium penetration without causing damage to the concrete element.

This work is still in progress. Other tests, such as obtaining lithium profile in the specimens, are still needed.

Concrete Repair, Rehabilitation and Retrofitting IV – Dehn et al. (Eds)
© 2016 Taylor & Francis Group, London, ISBN 978-1-138-02843-2

Analysis and visualization of water uptake in cracked and healed mortar by water absorption tests and X-ray radiography

B. Van Belleghem & N. De Belie
Magnel Laboratory for Concrete Research, Department of Structural Engineering, Faculty of Engineering and Architecture, Ghent University, Zwijnaarde, Belgium
Strategic Initiative Materials (SIM), Zwijnaarde, Belgium

J. Dewanckele & V. Cnudde
Centre for X-Ray Tomography, Department of Geology and Soil Science, Faculty of Sciences, Ghent University, Ghent, Belgium

ABSTRACT

Penetration of moisture is one of the most important damage mechanisms that could lead to the deterioration of building materials such as mortar and concrete. The capillary water absorption of sound mortar and concrete has already been studied in depth (Hall 1989, McCarter et al. 1992), but considering concrete in a perfect, uncracked state is usually not realistic. In most cases concrete is cracked and this can significantly accelerate ingress of water into the matrix. A possible way to repair the cracks consists of autonomous crack healing by means of encapsulated healing agents (Van Tittelboom 2011).

In this research, standardized artificial cracks in mortar samples (40 mm × 40 mm × 160 mm) were made by thin metal plates (thickness = 300 μm). The plates were introduced into the mortar upon casting and removed before complete hardening. In order to evaluate the effect of crack healing, glass capsules filled with a polyurethane pre-polymer were embedded inside the mortar specimens. The capsules ran through the thin metal plates, so that they broke at the moment that the plates were removed (and the crack was formed). Gravimetrical water absorption tests were performed on the samples according to the standard NBN EN 13057 in order to evaluate the ability of autonomous crack healing to reduce water absorption.

The gravimetrical absorption tests provide information about the total water content that is absorbed by the samples. In order to understand the process of water transportation, it is necessary to also visualize the water profile inside the samples. X-ray radiography was used for visualization of the time dependent water movement into cracked or healed mortar samples. The radiation experiments were performed on HECTOR, a 240 kV micro-CT setup optimized for research (Masschaele et al. 2013).

Mortar specimens with one, two and four cracks were subjected to a gravimetrical water absorption test. The water absorption per unit of surface area was then plotted in function of the square root of time. From the results it was clear that the water absorption of all specimens (cracked and uncracked) is increasing nearly linear with the square root of time during the first 8 hours of the test. This is in agreement to what is found in literature (Hall 1989, Martys & Ferraris 1997).

During the first hour of the absorption test, the water absorption per unit surface area of the specimens with one and multiple cracks was the same. After the first hour, the prisms with multiple cracks began to take up less water per exposed surface area than the prisms with one crack. The reason for this is the fact that the water in the prisms with one crack can spread horizontally from both crack faces. When two cracks are present in the sample, the surface area subjected to water is doubled and the water can only spread freely from one crack face of each crack. The horizontal water movement in the mortar zone between two adjacent cracks is limited because of the overlap of the water spread.

When the water absorption of specimens with one or two cracks was compared to the water uptake of samples with one or two healed cracks, it could be concluded that the polyurethane was very well able to seal the cracks in most of the samples. After 24 hours, the mean total water absorption of the healed specimens with both one and two cracks was almost half the amount of the mean water absorption of the unhealed specimens (Fig. 1). The mean absorption of the healed specimens was even comparable to that of the uncracked mortar samples.

The distribution of the water in the specimens in function of time was visualized by a series of 2D radiographs. The specimens were placed in individual containers and were illuminated a first time in the dry state. Next, the containers were filled with water and X-ray radiographs were taken again at regular time intervals. By subtracting the X-ray images in the wet state (during the absorption test)

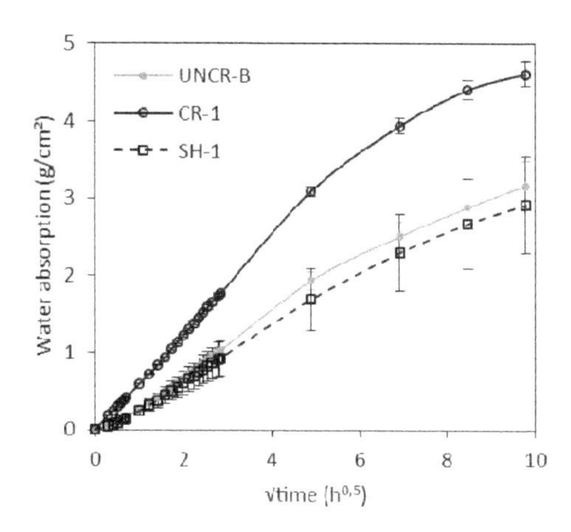

Figure 1. Comparison of water absorption between healed (SH-1) and unhealed (CR-1) specimens with one crack and uncracked samples (UNCR-B). Error bars represent the standard deviations of the mean values of 3 specimens.

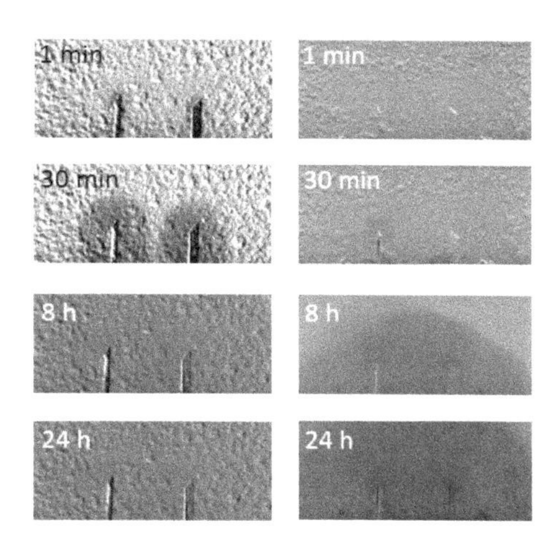

Figure 2. Visualization of the water uptake in a cracked and healed specimen with two adjacent cracks.

from the dry state image the water profile in the mortar specimens could be clearly visualized.

The X-ray images of the specimens with multiple cracks showed that the overlap of water spread between two adjacent cracks occurs somewhere between 30 and 50 minutes of exposure time. The approximation of one hour, based on the results of the gravimetrical absorption test clearly agrees well to that. When the X-ray images of the specimen with two healed cracks were analyzed, it was clear that there was very few water uptake through the cracks compared to the specimen with two unhealed cracks (Fig. 2). This means that the cracks were very well sealed by the polyurethane. One of the cracks did not turn dark on the inside on the X-ray images. This indicates that the crack was completely filled with polyurethane, because no water could enter. The other crack on the other hand did turn dark after 30 minutes, which indicates that this crack was not completely filled with polyurethane. However, the water entering the mortar through the crack faces was very limited, so the crack faces were probably covered with PU for the biggest part. This clearly limited the horizontal water spread inside the mortar sample. After 8 hours, the water front in the healed specimen was comparable to that of the uncracked specimen.

From the results of this study it can concluded that the polyurethane used in this research is a very appropriate healing agent for obtaining autonomous crack healing in cementitious materials. X-ray visualization of the water uptake as a function of time shows that the healed cracks are either completely filled with PU and block the water ingress into the crack completely or the PU may not fill up the whole crack but seal the crack faces. In both cases the water ingress into the mortar is significantly reduced.

ACKNOWLEDGEMENTS

This research under the program SHE (Engineered Self-Healing materials) (project ISHECO: Impact of Self-Healing Engineered Materials on steel Corrosion of reinforced concrete) was funded by SIM (Strategic Initiative Materials in Flanders) and IWT (Agency for Innovation by Science and Technology). The financial support from the foundations for this study is gratefully acknowledged.

REFERENCES

Hall, C. 1989. Water sorptivity of mortars and concretes: a review. *Magazine of Concrete Research* 14(147): 51–61.

Martis, N.S. & Ferraris, C.F. 1997. Capillary transport in mortars and concretes. *Cement and Concrete Research* 27(5): 747–760.

Masschaele, B., Dierick, M., Van Loo, D., Boone, M.N., Brabant, L., Pauwels, E., Cnudde, V., Van Hoorebeke L. 2013. HECTOR: a 240 kV micro-CT setup optimized for research. *Journal of physics: Conference Series* 463: UNSP 012012.

McCarter, W.J., Ezirim, H. & Emerson, M. 1992. Absorption of water and chloride into concrete. *Magazine of Concrete Research* 44(158): 31–37.

Van Tittelboom, K., De Belie, N., Van Loo, D. & Jacobs, J. 2011. Self-healing efficiency of cementitious materials containing tubular capsules filled with healing agent. *Cement & Concrete Composites* 33(4): 497–505.

Concrete Repair, Rehabilitation and Retrofitting IV – Dehn et al. (Eds)
© 2016 Taylor & Francis Group, London, ISBN 978-1-138-02843-2

Experimental study on the long-term leaching properties of CSG materials

Wei Feng, Zhongwei Liu, Jinsheng Jia & Fengling Ma
State Key Laboratory of Simulation and Regulation of Water Cycle in River Basin, China Institute of Water Resources and Hydropower Research, Beijing, China

ABSTRACT

Cemented Sand and Gravel (CSG) is a new–building dam material produced by mixing a small amount of cementitious materials, sand and gravel without screening or washing on site, paving and vibration compaction. Advantages of projects built by CSG are as follows: ① Low cement content and low temperature rising by hydration heat, leading to simple temperature controlling measures. ② Low grading demand of aggregates. Sands and gravels dug from the dam riverbed and excavation. ③ Low cost. The low cement and low aggregate demand reduce the cost. ④ Low environmental impacts. The CSG dams are economical, fast, safe and environment-friendly. So they have broad application potential. In recent years, CSG dams have been used as permanent buildings in Japan, Turkey, Greece, France, Philippines and other countries.

CSG material, as a new dam construction material, needs low cement content (40~60 kg/m^3, total dosage of cementitious materials is less than 100 kg/m^3), so it has low strength and poor permeability resistance. Even equipped with specialized impervious body, the CSG dam still has the risk of penetration and leaching. With long-term leaching under pressure water, will the Ca^{2+} continue to leach, how the strength attenuation change when osmotic quantity reaches a certain degree. Will the CSG material be complete destroyed by large serious leaching? The above questions are the main focus of this paper.

Based on the pressing concerns in engineering, Leaching properties of the main materials in Cemented Sand and Gravel (CSG) Dam were studied. Following phenomenon would occur in CSG under the continuous water pressure ① As the internal pores were filled with water, the porosity would continuously change due to the auto-compacting with the increasing age; ② $Ca(OH)_2$ in hydration products were leached, and other products might be hydrolyzed. ③ Hydraulic fracturing occurred in the micro-cracks by the pressure seepage water. ④ The strength of CSG was obviously decreased because of the seepage and leaching.

The experimental results showed that

1. With the increase of age, permeability coefficient of CSG was decreased, and it kept at 10^{-10} m/s 120 d later, half of the initial value. In the leaching process, activity of fly ash was stimulated gradually, and active SiO_2 and Al_2O_3 had secondary hydration reaction with $Ca(OH)_2$, the hydration product of cement. After the longest leaching duration at 485 d, the dissolved Ca^{2+} was little, and tended to be stable.

   Internal pH of CSG continued to decrease until reached 10.0–11.5, belonging to low alkaline. The strength was obviously decreased with the Ca^{2+} leaching. The total leaching CaO ratio in cementitious materials was 3.16%, reaching the maximum value .The ratio between average reduction in strength to Ca^{2+} leaching quantity was about 17%, that is, when the total leaching CaO ratio in cementitious materials was 1%, with 17% reduction in strength.
2. After curing for 1 year, the soluble Ca^{2+} of CSG was very few, which had good ability to resist Ca^{2+} leaching. It usually needs more than one year from the buildup of CSG dam to water storage. Even leaching occurs, the leached Ca^{2+} is very few.

Keywords: Cemented Sand and Gravel, dam materials, leaching dissolution, durability

ACKNOWLEDGEMENT

This work was supported by the National Basic Research Program (973 Program) of China (Grant No. 2013CB035903) and by the National Key Technology R&D Program (2013BAB06B02).

Concrete Repair, Rehabilitation and Retrofitting IV – Dehn et al. (Eds)
ISBN 978-1-138-02843-2

Effects of resistivity on corrosion rate measurements obtained from a coulostatic monitoring device

A.N. Scott
Department of Civil and Natural Resources Engineering, University of Canterbury, New Zealand

ABSTRACT

There are a number of factors which can influence the measurement of the corrosion rate of reinforcing steel in concrete structures. This paper reports on the influence of concrete resistivity on the measured corrosion rate of reinforcing steel in cracked concrete and compares corrosion rate measurements obtained from a coulostatic monitoring device with gravimetric results taken after approximately 18 months of corrosion. The theoretical development and principles of the coulostatic method have been presented in a number of sources (Glass et al 1993, Rodriguez and Gonzalez 1994, Hassanein et al 1998).

The corrosion rate of mild steel round bar embedded in concrete was investigated for three mix designs using a coulostatic measuring device. Three different concrete mix designs were chosen to provide a range of resistivity values. Two GP cement mixes were cast at a water/cement (w/c) ratio of 0.45 and 0.6, and one 8% Micro Silica (MS) mix with a w/c ratio of 0.45.

The corrosion rates for GP 0.45, GP 0.6 and MS 0.45 were 0.72, 1.18 and 0.39 $\mu A/cm^2$ respectively. The corrosion rates of the passive uncracked GP 0.45 (u) control sample was measured at 0.2 $\mu A/cm^2$.

The total measured corrosion current over the investigation was calculated for each specimen based on the individual corrosion rate curves. The total charge was converted to an equivalent mass loss through the use of Faraday's law and compared to the actual mass loss calculated as the difference between the initial mass of the steel prior to casting and that measured at the end of the investigation after cleaning.

Table 1 provides a comparison between the average calculated mass loss obtained from the coulostatic corrosion rates with those of the measured gravimetric mass loss. It can be seen that the calculated coulostatic mass loss was approximately 60% greater than that obtained from gravimetric analysis for all the actively corroding specimens.

Table 1. Comparison of coulostatic and gravimetric mass loss.

	Mass loss coulostatic	Mass loss gravimetric	Percent difference
GP 0.45	0.80	0.55	+45
GP 0.6	1.27	0.75	+70
MS 0.45	0.54	0.32	+68
GP 0.45 (u)	0.02	0.03	−13

The difference in mass loss between the calculated and measured non-corroding specimens was negligible.

The difference in average values are within a factor of 2 which is considered an acceptable range of accuracy as suggested by Andrade and Alonso (1996).

The coulostat used in this investigation provided an accurately estimate of the rate of corrosion for reinforcing steel embedded in concrete. The resistivity of the concrete did not have a detrimental effect on the coulostatic corrosion measurements as concretes with both high and low resistivity showed similar variations in corrosion rate when compared to gravimetric measurements.

REFERENCES

Andrade, C. Alonso, C. 1996, Corrosion Rate Monitoring in the Laboratory and On-Site. Corrosion and Building Materials, 10 (5), pp. 315–328.

Glass, G. 1995, An Assessment of the Coulostatic Method Applied to the Corrosion of Steel in Concrete, Corrosion Science, Vol 37, No. 4, pp. 597–605.

Hassanein, A., Glass, G. and Buenfled, N. 1998, The use of Small Electrochemical Perturbations to Assess the Corrosion of Steel in Concrete, NDT&E International, Vol. 31, No. 4, pp. 265–272.

Rodriguez, P. and Gonzalez, J. 1994, Use of the Coulostatic Method for Measuring Corrosion Rates of Embedded Metal in Concrete, Magazine of Concrete Research, Vol. 45, No 167, pp. 91–97.

Concrete Repair, Rehabilitation and Retrofitting IV – Dehn et al. (Eds)
© 2016 Taylor & Francis Group, London, ISBN 978-1-138-02843-2

Study of residual protection following interruption of impressed current cathodic protection in concrete

D.W. Law & S. Bhuiyan
RMIT University, Melbourne, Vic, Australia

ABSTRACT

Impressed Current Cathodic Protection (ICCP) is a well-established technique to rehabilitate and protect corroding steel reinforcement in concrete. The ICCP system works on the basis of shifting steel potentials to more negative values using a DC power supply in order to thermodynamically prevent corrosion from occurring. The negative potential repels chloride ions and allows the generation of hydroxyl ions at the steel-concrete interface which leads to an increase in the pH, helping restore passivity. Once passivity is achieved, the applied current can be reduced to maintenance levels that sustain the passive state. The reduction in the current also assists in preserving the operational lifetime of the cathodic protection system, reducing the likelihood of acid attack at the anode.

Recent studies have demonstrated that the protective effects of ICCP do not cease immediately after the system is interrupted, but can persist for a period of time before corrosion re-initiates. The residual effect is attributed to the beneficial effects of ICCP, where chlorides are repelled from the steel and re-alkalisation of the concrete around it occurs, thereby re-establishing passivity. This paper looks into this phenomenon of this residual effect and investigates how long it lasts and how it varies with current densities and duration of ICCP application.

The test regime included ten specimens. The specimens were made from a 100% Ordinary Portland (OP) cement mix with a nominal strength of 40 MPa. This was chosen to simulate a standard site mix. In order to initiate corrosion, 0.5% and 3% NaCl by weight of cement, was added to the mix. The 0.5% was selected to replicate levels commonly found in reinforced concrete structures, the 3% was selected to accelerate the corrosion process given the time constraints in laboratory trials.

A number of parameters including steel potentials, depolarisation values, corrosion rates and concrete resistance were monitored for salt-contaminated reinforced concrete specimens that were subjected to ICCP and subsequently interrupted. The specimen is 300 × 150 × 100 mm with a ribbed mild steel bar, diameter 16 mm and length and 250 mm, with 40 mm cover. The cathodic protection current was applied via an activated titanium mesh ribbon anode De Nora Type 1—current rating of 5.5 mA/m at 110 mA/m^2). Monitoring of the specimen included an embedded Ag/AgCl electrode (Castle type LD15), a welded-tip thermocouple, a humidity sensor (Honeywell HIH4000-01) and three pairs of resistivity probes.

After de-moulding, specimens were kept in a spray cabinet which was set to spray twice a day (every 12 hours), with each spray cycle lasting 30 minutes. The specimens for the 3% NaCl were left to corrode for 3.5 months before the Cathodic Protection (CP) was applied, the specimens with 0.5% salt were conditioned for 2 months prior to application of the CP. Corrosion was evaluated via the steel potential and corrosion rate measurements using Linear Polarisation Resistance (LPR). The current was applied by connecting the bar to the negative terminal of the DC power source and the anode mesh to the positive terminal, with a multimeter connected in series to the circuit to measure the current. The current was selected to enable a comparison between current density and total charge passed. As such current densities of 20, 60, 180 and 540 mA/m^2 were selected and durations of 1 and 3 months.

Following the application of the CP residual protection was observed in all specimens. The duration of the residual protection increased with time of application. However, the increase in the duration of the residual protection was not linear with time. The duration of residual protection increased with current density, though again this was not a direct relationship between the duration and current density. In general specimens with the higher current density, but shorter CP period provided the longer residual protection. The potential and corrosion data indicated that active corrosion had not been initiated in the

0.5% specimens during the conditioning process, while low level corrosion was occurring in the 3% specimens at the time of the application of the CP. The resistance measurements suggested that the CP had affected the ion concentration in the pore solutions around the bar, possibly due to the production of hydroxyl ions and the removal of chloride ions in the interfacial zone. The relative humidity and temperature in the specimens were consistent with those experienced in exposed concrete structures.

Overall the results show that residual protection can be achieved by the application of ICCP to reinforced concrete structures. The duration and current density of the applied current both affect the duration for which this protection will remain.

Concrete Repair, Rehabilitation and Retrofitting IV – Dehn et al. (Eds)
© 2016 Taylor & Francis Group, London, ISBN 978-1-138-02843-2

Deterioration of service reservoirs constructed in accordance with EN 206

R. Brueckner
Mott MacDonald, Bristol, UK

C. Atkins & P. Lambert
Mott MacDonald, Altrincham, UK

ABSTRACT

Water can be a highly aggressive medium attacking concrete as a result of erosion or chemical attack. In general, concrete has excellent resistance to chemical attack provided an appropriate mix is used and it is well compacted. However, due to its high alkalinity, Portland Cement (PC) based concrete is not particularly resistant against strong acids or compounds which can convert to acids, as noted by Neville.

The most important factors in corrosive attack of concrete are the amount of fluid flowing over the exposed surface of the concrete and the pH. If a significant flow rate is occurring, the attack on the concrete may be considerable even for mildly acidic conditions. However, if one of the corrosion products formed is insoluble it can provide a protective layer on the surface of the concrete.

Where concrete is exposed to natural soils and ground water then EN 206 provides guidance on the exposure class for Chemical Attack-XA. The National Annex of BS EN 206 recommends that XA3, as the most severe exposure class, requires a minimum concrete strength of C35/45 with a maximum w/c-ratio of 0.45 and a minimum cement content of 360 kg/m^3 whereas DIN 1045-2 recommends 320 kg/m^3 of cement.

In the UK concrete is designed in accordance with BS EN 206, BS 8500, BRE Special Digest 1 and other current applicable guidelines but despite compliant specifications, issues with concrete durability have occurred during recent years. This was notably encountered at water service reservoirs where structures had recently been constructed or refurbished. It was observed that concrete faces started to transform into a mushy paste, fully depleted of calcium compounds, within one year after construction. Reservoirs constructed in the 1970s and 1980s have, however, not experienced the reported deterioration when exposed to the same water.

Petrographic analysis has confirmed that the softening of the concrete cover zone was due to ingress of moisture containing dissolved carbon dioxide, causing mild acid attack. The extent of surface softening has not yet been quantified. The modifications to the cement paste are up to a depth of 3 mm, without considering the loss of eroded concrete. The deterioration progress was identified to be represented by three distinct zones:

- Zone 1 (outer zone): Zone of substantial porosity enhancement and calcium depletion.
- Zone 2: Zone of carbonation.
- Zone 3 (inner zone): Zone of dissolved and leached Portlandite.

The water characteristics indicate that the rate of deterioration is dependent on the alkalinity of the water, despite the total hardness generally indicating slightly/moderately hard water according to Thresh et al.

To assess the rate of deterioration, susceptibility of varying concretes and to identify durable concrete mixes, specimens have been produced and immersed into the reservoirs. Upon completion, the investigation will also provide more details on the effects of the calcium content, total hardness, pH and of the aggressive carbon dioxide content on the concrete resistance.

Final conclusions with regards to the recommendations provided in BS EN 206 for the chemical attack classes cannot yet be drawn. It is, however, evident that the current BS EN 206 recommendations are not sufficient and that EN 206 may need to be reviewed or amended.

Concrete Repair, Rehabilitation and Retrofitting IV – Dehn et al. (Eds)
© 2016 Taylor & Francis Group, London, ISBN 978-1-138-02843-2

Modelling of chloride diffusion coefficient in concrete with supplementary cementitious materials

K.N. Shukla & R.G. Pillai
Indian Institute of Technology Madras, Chennai, India

ABSTRACT

Chloride ingress through diffusion is a major deterioration mechanism, making it absolutely necessary to account for it in the Service Life Prediction (SLP) framework. However, in most SLP frameworks, there are limited provisions for the estimation of the chloride diffusion coefficient ($D_{chloride}$), a parameter that has a direct strong influence on the chloride induced corrosion initiation. Developing a probabilistic model for $D_{chloride}$ based on raw material characteristics and mixture proportions can help in making decisions on the selection of cementitious materials during the planning and design phases of projects. This thesis presents the development of probabilistic models for the estimation of $D_{chloride}$ based on 42 data sets collected from literature. First, the predictor variables such w/b, specific surface area of binder (SSA_{binder}), and SiO_2 and CaO contents in the binder were identified. Then, a diagnostic study was conducted to investigate the nature and degree of influence of each independent variable on $D_{chloride}$. Then, probabilistic models that relate $D_{chloride}$ to the significant independent variables were developed.

A database was generated using the data from literature. Using this data, diagnostic plots were prepared for the 2- and 3-way interactions of the identified critical predictor variables and the following model form was obtained.

$$D_{28,predicted} = \theta_0 exp\left\{\left(\frac{w}{b}\right)\times\left[\left(\frac{SSA_{binder}}{SSA_{max}}\right)p_{SiO_2}\right]^{\theta_1}\right\} + \sigma\varepsilon \tag{1}$$

Then, two models were developed for two different SSA_{binder} groups (< and ≥ 345 m²/kg). The Low Specific Surface Area (LSSA) group consists of cases with SSA less than 345 m²/kg (least of all high specific surface area cases) and the High Specific Surface Area (HSSA) group consists of cases with SSA_{binder} greater than or equal to 345 m²/kg.

Table 1. Maximum Likelihood Estimates (MLE) model parameters.

Group	Statistic	θ_0	θ_1
LSSA	MLE	4.95	− 0.60
	Standard deviation	0.26	0.02
	Coefficient of variation	0.05	0.03
HSSA	MLE	0.24	− 0.95
	Standard deviation	0.04	0.02
	Coefficient of variation	0.15	0.02

The generated database of 48 cases was screened for outliers and 6 cases were removed. The model form in Eq (1) was calibrated using the remaining 42 cases and the LSSA and HSSA models were obtained.

LSSA model for $SSA_{binder} < 345\ m^2/kg$

$$D_{28,\,predicted} = 4.95\exp\left\{\frac{w}{b}\left[\left(\frac{SSA_{binder}}{SSA_{max}}\right)p_{SiO_2}\right]^{-0.60}\right\} + \sigma\varepsilon \tag{2a}$$

HSSA model for $SSA_{binder} \geq 345\ m^2/kg$

$$D_{28,\,predicted} = 0.24\exp\left\{\frac{w}{b}\left[\left(\frac{SSA_{binder}}{SSA_{max}}\right)p_{SiO_2}\right]^{-0.95}\right\} + \sigma\varepsilon \tag{2b}$$

Calibration was done through a maximum likelihood estimation MATLAB™ code. This code provides the estimates for each θ_i along with the standard deviation and coefficients of variation for each (Table 1).

These models have Mean Absolute Percentage Error (MAPE) equal to 3.77% and Weighted Absolute Percentage Error (WAPE) equal to 1.59%, indicating reasonably good predictions of $D_{chloride}$ using raw material characteristics of binder and mixture proportions of concrete.

Figure 1 illustrates the goodness of fit of the model. It can clearly be seen that most data points

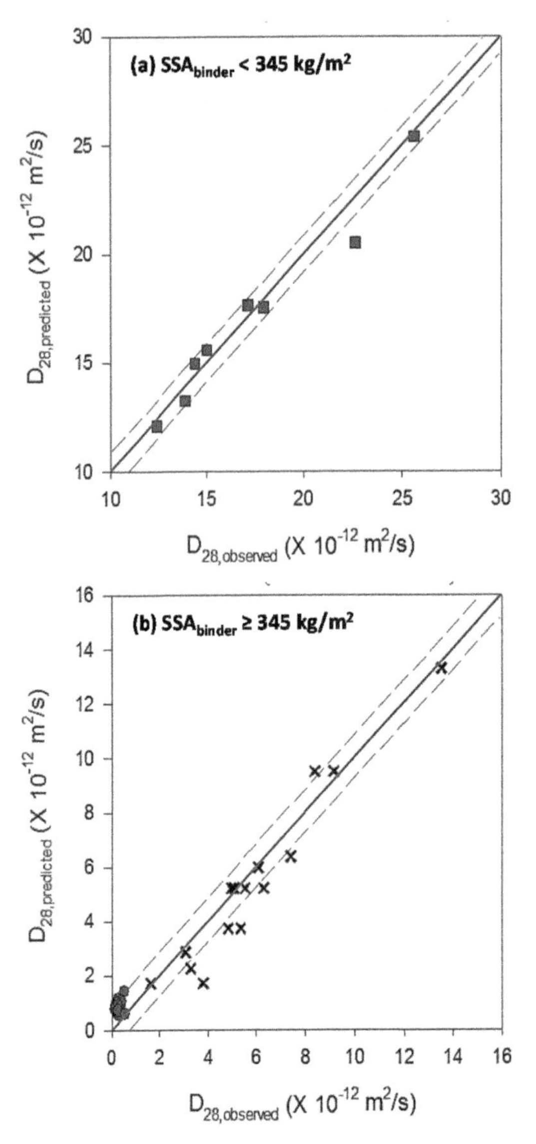

Figure 1. Correlation between the observed and predicted data on chloride diffusion coefficient.

fall inside one standard deviation – indicating reasonably good prediction.

Although significant amount of literature exist on durability performance of various concrete systems, most of the available data sets are incomplete (i.e., data on raw material characteristics and mixture proportions are not available). This emphasizes the need to include data on raw material characteristics, mixture proportions, mixing procedures, porosity, etc. in the scientific publications on strength and durability of concrete systems.

Concrete Repair, Rehabilitation and Retrofitting IV – Dehn et al. (Eds)
© 2016 Taylor & Francis Group, London, ISBN 978-1-138-02843-2

Damage risk and development in concrete pavements caused by an Alkali-Silica Reaction

A. Wiedmann, E. Kotan & H.S. Müller
Department Building Materials, Institute of Concrete Structures and Building Materials (IMB), Karlsruhe Institute of Technology (KIT), Karlsruhe, Germany

ABSTRACT

Concrete pavements underlie climatic and interfering mechanical impacts as well as additional stresses resulting from construction conditions or the construction method. An Alkali-Silica Reaction (ASR), on the other hand, leads to expansion and strain. These two combined load cases may cause cracking. The aim of this work is to investigate the macroscopic behaviour of the different exposures. For this purpose an appropriate numerical model has to be developed, which allows describing the complex interactions. By means of extensive experimental fracture mechanical examinations the temporal change of strength and deformation of concrete due to ASR is investigated. Based on this work, it will be possible to develop suitable measures to reduce the risk of cracking and possibly to prevent any damage in the future.

INTRODUCTION

In recent years, some cases of damage to concrete pavements have occurred in the German motorway network, which are attributed to an ASR. These damages and the related drastic degradation of durability have led to a strong increase in activity in this research area. In Germany, repairing damage caused by an ASR costs millions of euros year over year.

ASR in concrete means a reaction of aggregate-components with alkalis in the pore water. In this reaction, an alkali-silica gel is formed. This gel is able to swell and the increasing volume leads to cracking, finally to a concrete damage (Stark 2013).

The chemical and mineralogical processes that occur in an ASR have already been described in many works. Research concerning a change of material properties and cracking-mechanisms, on the other hand, is quite rare or incomplete (Bödeker 2003).

EXPERIMENTAL INVESTIGATIONS

To describe the realistic resistance of concrete against cracking, the temporal change of strength and deformation properties of pavement concrete under ASR and possibly loading caused by traffic is necessary. The investigation of these aspects is carried out as subproject six of research group DFG 1498 "alkali-silica reaction in concrete members at simultaneous cyclic loadings and external alkali support", which is funded by the DFG since 2011. For this purpose, each individual subgroup produces reference specimens (non-damaged) under the same conditions with centrally provided source materials of pavement concrete (Table 1).

Afterwards, specimens will be mechanically pre-damaged and stored under ASR-provoking conditions. Hereby, it will be distinguished between internal and external alkali supply.

By means of specific experiments, using the example of centric tensile test, (fracture-) mechanical material properties of the selected pavement concrete were determined.

For this purpose, first experiments were performed to identify just the influence of mechanically pre-damage due to a typical fatigue loading on the stress-deformation relation and two mechanical material parameters (net tensile strength $f_{ct,n}$ and fracture energy G_F) of the selected pavement concrete.

Table 1. Composition of the pavement concrete.

Setting	Designation/value
Cement	CEM I 42.5 N
Cement content [kg/m³]	360
w/c-ratio [-]	0.42
Aggregate	granodiorite and Upper Rhine crushed gravel
Grading curve	A/B 22
Air pore content [vol%]	4.0–4.5

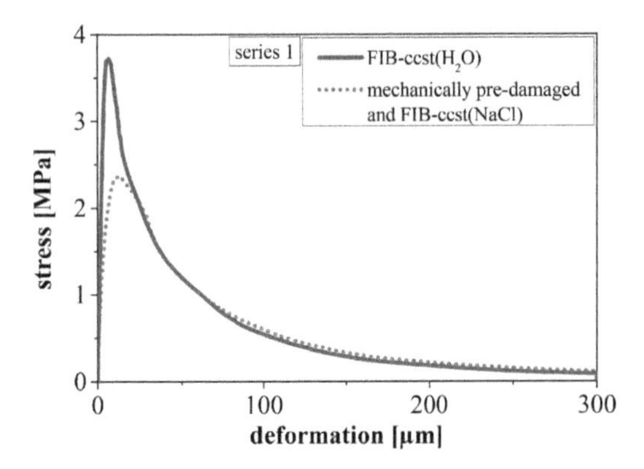

Figure 1. Influence on the mean stress-deformation relation identified in the centric tensile test by using bilateral notched prismatic test specimens (with and without mechanically pre-damage) after FIB cyclic climatic storage test (FIB-ccst).

In addition to this, within the scope of further tests the mechanically induced pre-damage in combination with a deleterious ASR (storage in the FIB cyclic climatic storage test (FIB-ccst)) under external alkali supply on the top side of the test specimen reduces the net tensile strength about 40 per cent, see Figure 1.

For these investigations, the specimens were mechanically pre-damaged at the Institute for Materials Research and Testing (BAM) and stored at the F. A. Finger-Institute for Building Material Science (FIB).

THEORETICAL AND ANALYTICAL INVESTIGATIONS

On one hand, material laws have to be formulated on basis of the obtained experimental observations as part of theoretical-analytical analysis. This is absolutely necessary to be able to implement the time- and load-dependent stresses of concrete in terms of numerical calculations realistically. At this point the focus is on the influence of the internal damage as a result of ASR and the associated impact on all mechanical material properties.

On the other hand, the quantitative description of the complex stress of concrete pavements for the numerical analysis is being carried out. In this connection the quality of the selected stress scenarios affects the accuracy of the obtained numerical results. To describe the climatic conditions such as air temperature and relative humidity, it can be referred to already existing records of the Institute of Concrete Structures and Building Materials. In this case the focus is on the determination of the course of zero-stress temperature also considering the manufacturing period and the construction method. The traffic load will be estimated on the current state as quasi-dynamic load on the basis of appropriate standard.

NUMERICAL INVESTIGATIONS

Using numerical analysis, both the time-dependent material properties in connection with the realistic display of the fracture mechanical behaviour of concrete influenced by the ASR (resistance), as well as the implementation of the complex stress scenarios (impacts) are realised. These eventually serve to analyse and forecast of the formation of macroscopic cracks in the concrete pavement and to derive measures for damage reduction and prevention in the future.

The numerical simulation is made using the finite elements program DIANA. In recent years, for this purpose the numerical calculation model, which was developed at the Institute of Concrete Structures and Building Materials, is refined. The composite system of concrete carriageway consists of concrete pavement, hydraulically bound base course, frost blanket course and corresponds to the standard structure of carriageways with concrete pavement.

Finally, the numerical investigations will be used to determine measures, like sawing joints, or regularities to reduce and prevent cracking in concrete pavements due to an ASR.

CONCLUSIONS AND OUTLOOK

On the one hand, the presented approach of subproject six as part of the DFG research group 1498, allows characterising ASR-damaged pavement concrete and the derivation of corresponding material laws in the future.

On the other hand, it is described which complex stresses in the numerical model of a concrete pavement will be considered. Both are necessary to extend the life cycle and to minimize the damage risk of a concrete carriageway the in the future.

ACKNOWLEGDEMENT

Further information about the DFG research group 1498 "alkali-silica reaction in concrete members at simultaneous cyclic loadings and external alkali support" can be found on the internet at http://www.for1498.sd.ruhr-uni-bochum.de. The authors wish to thank the DFG for funding the project.

Concrete Repair, Rehabilitation and Retrofitting IV – Dehn et al. (Eds)
 ISBN 978-1-138-02843-2

Reinforcement corrosion behavior in bending cracks after short-time chloride exposure

F. Hiemer, S. Keßler & C. Gehlen
Centre for Building Materials, Technische Universität München, Munich, Germany

ABSTRACT

Even if properly designed, crack formation in reinforced concrete structures is inevitable for activating the reinforcements' tensile capacity. Although in cracked concrete the barrier property of the protective concrete cover against corrosion initiating substances, such as chlorides, is considerably reduced (Šavija & Schlangen 2010), it is not yet completely understood whether cracks accelerate only the corrosion initiation process or the overall corrosion process (Yu et al. 2014).

Cracks allow a locally limited but rapid ingress of aggressive agents like chlorides which can lead to an accumulation in the crack area. When exceeding a critical chloride content the crack crossing reinforcement is locally depassivated. This discrete depassivation promotes the formation of macro-cell corrosion, in which the anodic and cathodic sub-processes are spatially separated.

To protect direct driven concrete surfaces such as in multi-storey car parks against early initiated corrosion, German Standards recommend the application of a crack-bridging surface coating. On the one hand, an accumulation of chlorides is reduced, minimizing the risk of reinforcement depassivation. At the same time, the water intake through the concrete surface is hindered, resulting in a reduction and homogenisation of the moisture content in the concrete component. Hence, the electrolytic conductivity decreases, reducing the overall corrosion rate of the macro-cell (Warkus & Raupach 2008).

The planning principles given in DBV (2010) recommend coating the concrete surface after crack formation is finished, even if a chloride exposure cannot be ruled out. It is not expected that a short-time chloride exposure before sealing the surface induces and maintains endangering corrosion. However, this assumption was not yet scientifically secured nor was it sufficiently investigated through practical examinations.

Thus, a research project was initiated, in which reinforced concrete specimens are used to investigate the corrosion process in cracked concrete after a short-time chloride exposure. Subsequently the effect of an applied surface protection system on the corrosion situation can be determined.

In the course of this project, 19 reinforced concrete specimens with bending cracks, like shown in Figure 1, were produced. In order to reproduce the structural situation of a multi-storey car park, the concrete composition was chosen in accordance with DIN 1045-2 (2008) and the requirements resulting for a concrete component in a corresponding exposure situation. The specimens had to be designed allowing the initiation of a defined bending crack orthogonal to a steel bar representing the depassivated anode. Besides other boundary conditions, the positioning of reinforcement elements had to ensure the electrical isolation of anodic and cathodic sub-processes, so all electrochemical measurements could be conducted on each half cell separately.

To simulate the intensity of a chloride attack by de-icing salts in the course of one winter period the specimens were loaded cyclically with 1.5 M% sodium chloride solution for twelve weeks. This consequently led to depassivation of the electrically isolated steel surface crossing the crack, forming macro-cell corrosion. Beginning with the temporary exposure, electrochemical parameters of reinforcement corrosion were recorded in order to draw conclusions on the corrosion state of the reinforcement and its boundary conditions. Afterwards the surface of the specimens was sealed with a two component polyurethane coating. Hence, the influence of an applied surface protection system on the already initiated corrosion process could be investigated.

The first results on the corrosion behavior in cracked concrete after a short-time chloride exposure showed that a depassivation of the crack crossing reinforcement is very likely to occur. Even only after twelve weeks of exposure to the sodium chloride solution the anodic potentials dropped significantly leading to potential differences between anode bar and cathode cages of approximately 250–300 mV. These drops were accompanied by increasing macro-cell corrosion currents of up to 0.065 mA, which afterwards mostly showed a degressive progress. On the one hand this could be lead back to an increasing electrolytic resistance resulting from the continuously hydrating cement paste. On the other hand during the corrosion

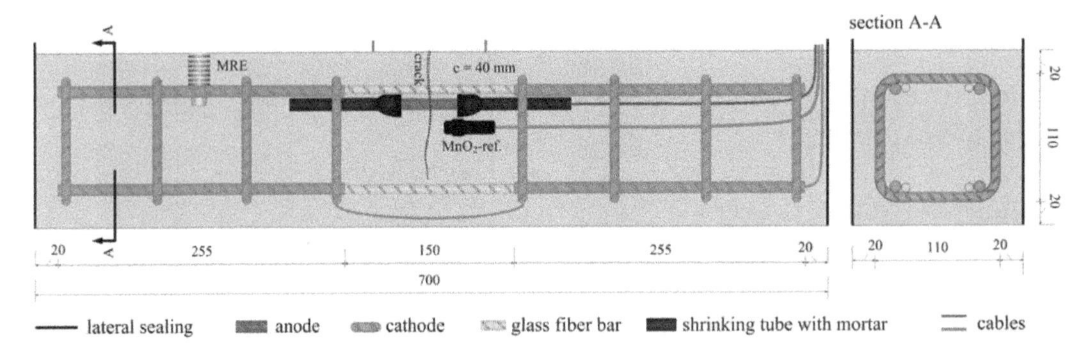

Figure 1. Geometry and reinforcement setup of the specimens, all dimensions in [mm].

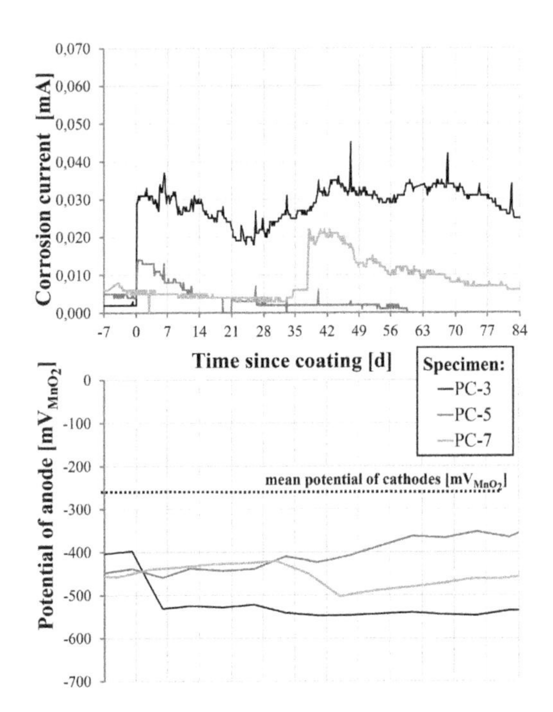

Figure 2. Macro-cell corrosion current (top) and corresponding free corrosion potentials of the anodes (bottom) of three exemplary specimens after the application of the coating.

process oxidation products are formed that can shield the area of active corrosion. This can lead to a shortage of corrosion promoting agents, reducing their overall concentration and therefore increasing the anodic polarization resistance in the corrosion system.

Although the corrosion currents tended to decrease throughout the exposure, not all corrosion processes came to a complete halt, even not in the time after the exposure was ended. Additionally, the application of a crack-bridging surface protection system on the concrete specimens had different effects on their corrosion behavior. In some actively corroding specimens, e.g. specimen PC-5 in Figure 2, the process was halted after the coating was applied. Other specimens (e.g. PC-3 and PC 7 in Figure 2) responded with a spontaneous sharp increase of macro-cell current, which in some of those specimens decreased afterwards. In others it remained constant over the observed period.

This sharp rise could be due to time dependent diffusion effects, which have led to a repeated accumulation of chlorides on the steel surface. These provoked further depassivation of the reinforcement which occurred coincidently at the same time when the coating was applied.

In order to clarify these questions as well as to completely investigate the corrosion behavior in the coated specimens further measurements have to be carried out.

REFERENCES

DBV German Society for Concrete and Construction Technology 2010. DBV-Bulletin: Multi-storey and Underground Car Parks.

Šavija, B. & Schlangen, E. 2012. Chloride ingress in cracked concrete- a literature review. In C. Andrade and J. Gulikers (eds.), *Advances in Modeling Concrete Service life: Proceedings of 4th International RILEM PhD Workshop held in Madrid, Spain, 19 November 2010.*

Warkus, J. & Raupach, M. 2008. Numerical modelling of macrocells occurring during corrosion of steel in concrete. *Materials and corrosion, 59, No. 2, pp. 122–130.*

Yu, L., François, R., Hiep Dang, V., L'Hostis, V. & Gagné, R. 2015. Development of chloride-induced corrosion in pre-cracked RC beams under sustained loading: Effect of load-induced cracks, concrete cover and exposure conditions. *Cement and Concrete Research, Vol. 67, pp. 246–258.*

Concrete Repair, Rehabilitation and Retrofitting IV – Dehn et al. (Eds)
© 2016 Taylor & Francis Group, London, ISBN 978-1-138-02843-2

Physical model for structural evaluation of R.C. beams in presence of corrosion

A. Cesetti, G. Mancini & F. Tondolo
Politecnico di Torino, Torino, Italy

A. Recupero & N. Spinella
Università di Messina, Messina, Italy

ABSTRACT

Corrosion of steel in concrete is a very common degradation phenomenon observed in reinforced concrete structures. It impairs the resisting section of the bars, the surrounding concrete around the bars and finally modifies the bond between steel and concrete. Some experimental tests are present in literature in which the structural behavior of reinforced concrete beams in presence of corrosion of longitudinal bars and stirrups was analyzed: a progressive reduction of the load bearing capacity was demonstrated with an increase level of corrosion. Furthermore a not-obvious rupture mechanism was observed due to the effect of rust formation; in particular with the increase level of corrosion a change in the type of collapse was observed.

In the present paper a plastic model for the analysis of corroded structure is employed. The model is able to take into account the interaction between shear and bending moment acting in the same section of the r.c. element.

The sectional model here used belonging to the theoretical group of formulations, originally proposed by Recupero et al. (2005) and in the following widely validated (Colajanni et al. 2014) for sound structures.

The effects of the reduction of bearing capacity of concrete in compression and the reduction of steel section together with the modification of the constitutive law for steel due to the corrosion are implemented.

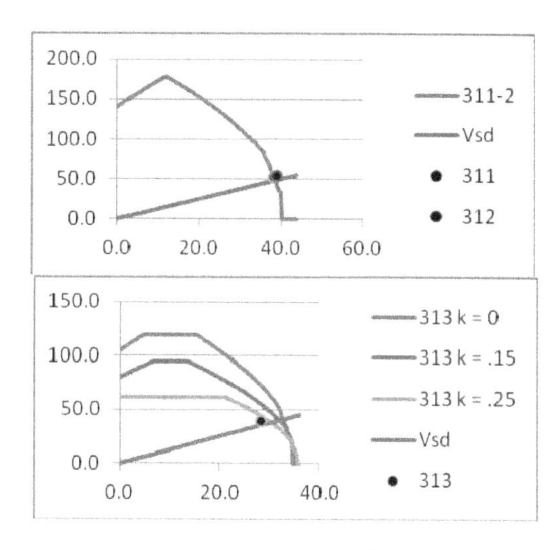

Figure 1. Interaction diagrams M-V for all specimens tested by Rodriguez et al. (1996).

Results show a good agreement between the M-V curve domains (Fig. 1) and the experimental data taken from reliable experiences available in literature (Rodriguez et al. 1996, El Maaddawy et al. 2005, Zhu et al. 2014). Some differences between tests and model are registered; they could be addressed to the scatter of the phenomenon of corrosion. The model could be furtherly improved in order to take into account other effects of corrosion as the bond loss between steel and concrete.

Concrete Repair, Rehabilitation and Retrofitting IV – Dehn et al. (Eds)
© 2016 Taylor & Francis Group, London, ISBN 978-1-138-02843-2

Bond-slip model for corroded steel in concrete

A. Cesetti, G. Mancini, F. Tondolo & C. Vesco
Politecnico di Torino, Torino, Italy

ABSTRACT

Bond in concrete is of outmost importance for the definition of resisting mechanisms in reinforced concrete (r.c.) structures. Bond performance is directly related to both serviceability and ultimate behaviour of r.c. members. In presence of corrosion between steel and concrete bond is modified. Previous models as Lundgren et al. (2012) and Ozbolt et al. (2014) give an important contribution to the simulation bond in presence of corrosion.

In the present work a new analytical model able to simulate the structural behaviour of embedded steel in concrete even in presence of corrosion with and without transversal confinement is introduced. It is able to take into account the effect of radial pressure due to: the mechanical action that arises between concrete and steel ribs during bond test, the corrosion of the longitudinal bar and the presence of transverse confinement. Local bond-slip relationships are related to a criterion in which the maximum bond strength is a function of the radial pressure. Pull-out failures and splitting failures can be reproduced. An interaction between the effects of corrosion and bond action is highlighted in terms of radial pressure. This model is used to simulate the experimental results of pull-out specimens characterized by the presence of transversal confinement (Mancini et al. 2014). The embedded bars of the specimens were subjected to a corrosion values ranging between 0% and 20% in terms of mass loss. In Figures 1 and 2 some results al plotted: the model simulation with the dashed line whereas a black and grey curves for the tests. The model is able to reproduce the increase of bond strength for low values of corrosion if they are unable to already crack the concrete before the bond test. A good agreement can be evidenced for all the corrosion values both for confined and unconfined specimens. Splitting failures are properly associated to unconfined samples with a minimum values of corrosion starting from 5% whereas for all the other tests a pull-out mechanism is evidenced.

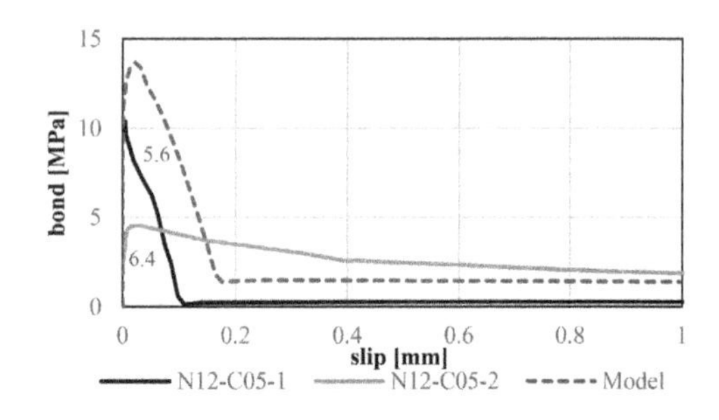

Figure 1. Bond slip curves for unconfined specimen, simulated mass loss of 5% for the model.

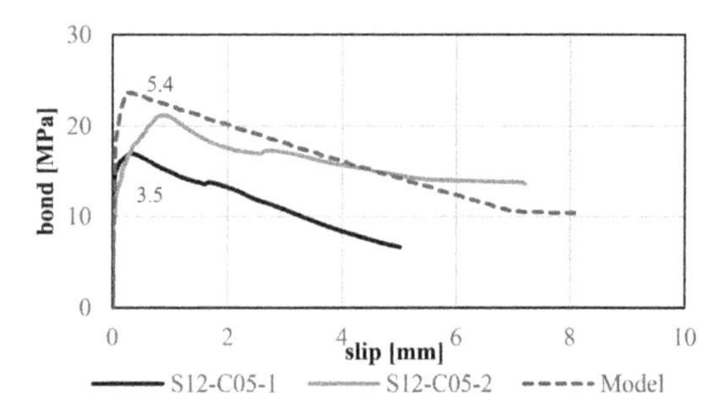

Figure 2. Bond slip curves for confined specimen, simulated mass loss of 5% for the model.

REFERENCES

Lundgren, K., Kettil, P., Hanjari, K. Z., Shulne, H. & Soto San Roman, A. 2012. Analytical model for the bond-slip behaviour of corroded ribbed reinforcement, *Structural and infrastructure Engineering*, 8(2): 159–169.

Ozbolt, J., Orsanic, F. & Balabanic, G. 2014. Modeling pull-out resistance of corroded reinforcement in concrete: Coupled three-dimensional finite element model. *Cement and Concrete Composites* 46: 41–55.

Mancini, G., Sabia, D., Pettinato, E. & Tondolo, F. 2014. Pull-out tests in presence of high level of corrosion. In IALCCE Fourth International Symposium on Life-Cycle Civil Engineering, Tokyo.

Concrete Repair, Rehabilitation and Retrofitting IV – Dehn et al. (Eds)
ISBN 978-1-138-02843-2

Pull-out tests on R.C. corroded specimens

A. Cesetti, G. Mancini & F. Tondolo
Department of Structural, Geotechnical and Building Engineering, Politecnico di Torino, Torino, Italy

ABSTRACT

The major cause of deterioration of reinforced concrete (r.c.) structures must be sought in corrosion of reinforcing steel. Corrosion level is considered a significant parameter for predicting the useful service life of r.c. structures. It influences not only the flexural and shear ultimate resistance, ductility and bond strength but also service behaviour. Reinforcement corrosion modifies steel transversal section with a reduction of effective area, concrete integrity by cracking of portions around corroded bar eventually with spalling and delamination of concrete cover and finally the interaction between concrete and reinforcement (bond reduction). Hence, performance of r.c. structures is deeply dependent on bond mechanism.

Analysing the present literature, some lack on bond study in presence of corrosion can be found. In fact, it is noticeable a limited number of reliable experimental tests, because of the use of too high level of current densities employed for the accelerated corrosion mechanism. Moreover, low amount of tests conducted on specimens with transverse reinforcement and an even lower number of corroded transverse reinforcement is present. Eventually, these tests result in an incorrect evaluation of transverse confining action and assumption of the actual conditions. Furthermore, the reference code, fib Model Code 2010 (MC2010), does not provide information for corrosion levels over than 5%.

In order to extend the knowledge about bond in presence of corrosion, tests with a specific setup are performed: low current density, confined test specimen with corroded transverse reinforcement, test performed up to a corrosion level of 20%.

Pull-out tests of RILEM type (1994) (see Figure 1) were carried out on 29 specimens in order to evaluate bond slip performance for samples under corrosion levels equal to 0/2/5/10/20% of mass loss and bond slip performance corresponding to different bar diameter as well. Corroded specimens were subjected to the procedure of electrochemical

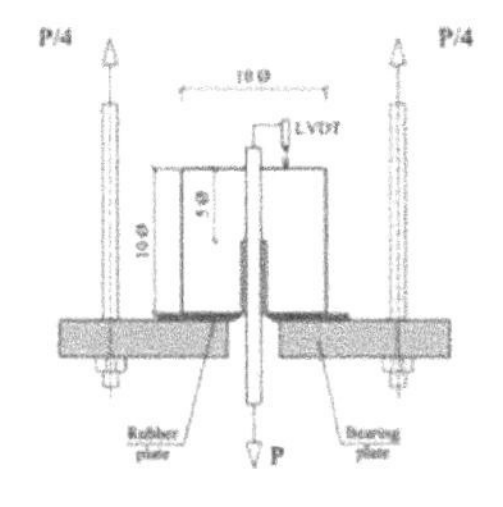

Figure 1. Overview of the specimen and testing machine.

corrosion: the amount of current density used (200 $\mu A/cm^2$) can be considered as an upper limit for electrochemical accelerated corrosion methods (Mancini & Tondolo, 2014). At the end of the pull-out test the embedded bars were removed and cleaned in order to determine the real corrosion rate in terms of mass loss in accordance with ASTM G1-90 procedure.

The bond test results show the effect of oxidation of the steel at the interface with concrete and the change of performance in terms of bond strength. The role played by confinement that ensure a residual bond strength in presence of corrosion is highlighted for confined specimens whereas a progressive reduction of bond performance for unconfined specimens is observed.

REFERENCES

ASTM G1-03. 2003. Standard practice for preparing, cleaning, and evaluating corrosion test specimens.
fib, 2012. MC2010 fib model code, Bulletin no. 65–66. Lausanne (CH), Switzerland.
Mancini, G. & Tondolo F. 2014. Effect of bond degradation in existing structures-literature survey. In Structural Concrete, Accepted for publication, DOI: 10.1002/suco.201300009.
RILEM 1994. Recommendations for the Testing and Use of Constructions Materials.

Concrete Repair, Rehabilitation and Retrofitting IV – Dehn et al. (Eds)
ISBN 978-1-138-02843-2

Corrosion resistance of BS 8500 compliant concretes

C. Christodoulou & D. Dunne
AECOM Ltd., Birmingham, UK

C.I. Goodier
School of Civil and Building Engineering, Loughborough University, Loughborough, UK

R. Yea
Kier Construction, UK

ABSTRACT

Corrosion of steel reinforcement in concrete, due to the ingress of chloride ions, is the most significant form of concrete deterioration. Fresh concrete however, provides a highly alkaline environment which facilitates a protective passive oxide layer around the steel reinforcement. Modern engineering standards provide guidance to designers on the specification of concrete mixes to meet a prescribed design life. More than one cement blend may however be available to meet the specific design life for a particular exposure classification, which can occasionally lead to confusion. This work investigated the corrosion resistance of BS 8500 compliant cement combinations for a XS3 environmental exposure in order to identify relative durability performance differences. The concretes investigated contained Portland Cement (PC), Fly Ash (FA) and Ground Granulated Blastfurnace Slag (GGBS). FA and GGBS were blended in binary combinations with the PC, at levels of 28% and 51% respectively. Two water/cement (w/c) ratios of 0.35 and 0.40 were investigated, together with total cementitious contents of 380 kg/m³, representing typical structural reinforced concrete. Specimens were cyclically exposed to a saline solution and tested for compressive strength, electrochemical potential, resistivity, and chloride ion content with depth.

INTRODUCTION

Corrosion of reinforcement is the most common cause of deterioration. Modern design standards aim to overcome this by means of prescribed controls such as minimum cementitious content, minimum cover, maximum water/cement ratio etc. This is generally known as a "deemed-to-satisfy approach" and assumes that the concrete composition selected will result in the desired service life. However, such an approach is not taking adequately into consideration the significantly different performance of cementitious binders, which will ultimately control the time to onset of corrosion.

However, the ultimate choice of concrete mix design lies with the designer, who may be unfamiliar with the effect of the numerous variables which will influence the overall durability performance of the final product. Furthermore, there is a common misconception that a greater cementitious content will result in significant improvements in concrete durability as a result of a potentially greater cementitious binder capability. However, this is not rigorously proven. In fact, significant increases in cementitious content may result in thermal cracking and early age shrinkage, which will therefore reduce durability.

Therefore, it is often the case—in particular for ordinary structures such as typical highway crossings – that the structural designer, who is also responsible for the selection of the concrete mix design, is presented with the option of a number of concrete mix designs conforming to the design standard requirements and are all "deemed to satisfy" the intended design life.

The aim of this research was to undertake preliminary research on the performance of resultant concrete mix designs from the above three main design categories for an XS3 environmental exposure class, in order to establish whether there were any differences in their performance.

METHODOLOGY

Based on the permissible design combinations, CEM I (Portland cement), CEM IIB-V (Portland cement with 28% fly ash) and CEM IIIA (Portland cement with 51% GGBS) were investigated. As the minimum concrete cover of CEM I is higher than that of the binary blends, an additional CEM I mix design with a similar minimum cover depth to CEM IIB-V and CEM IIIA was also investigated.

For each concrete mix design, two slabs with dimensions of 300 mm wide by 500 mm long and 125 mm deep were cast, together with 6 Nr 100 mm³ cubes. Each slab contained 2 Nr 16 mm diameter,

480 mm long ribbed reinforcing main bars, which were interconnected by 2 Nr 6 mm diameter, 295 mm long smooth transversely placed reinforcing bars.

The slabs were air cured for a period of 28 days. Following this initial conditioning, they all underwent a cyclic dry and wet conditioning which aimed to replicate XS3 environmental exposure conditions. The samples were wetted with an aqueous solution of 3.5% salinity using laboratory grade sodium chloride for a period of 3 days and then allowed to air-dry within the laboratory for 4 days. This procedure was followed over a total period of 125 days. Testing was undertaken by means of Compressive strength, Electrochemical potential, Surface resistivity, Chloride profiles

RESULTS

Compressive strength testing was undertaken on three sets of cubes for each mix design at 7, 28 and 125 days. The generic trends exhibited by each concrete were as would be expected, with an initially rapid increase in compressive strength in all cases up to 7 days, especially for the CEM 1 mix design.

More specifically though, it was be observed that the evolution of compressive strength for CEM IIIA (45) was the lowest both at 7 and 28 days with CEM I being the highest, as the former have the lowest amount of Portland cement. However, at 140 days this had reversed, suggesting that the rate of hydration curing of CEM III A has continued, after initial strength gain, at a faster rate than the other mix designs. This may be attributed to a prolonged period of moist curing.

Electrochemical potentials of the steel were measured against a silver/silver chloride electrode (Ag/AgCl) over a period of 125 days. It was observed that in all cases there was a decline in potentials after the first exposure cycle. Most notably, the non-compliant CEM I (45) concrete exhibited a significant drop in potential at 125 days. This suggests that corrosion of the steel reinforcement may have initiated.

On initial inspection of the evolution of resistivity over time it appeared that the resistivity of each concrete developed at a different rate. However, on closer review it appeared that the development of resistivity of the CEM I slabs was as expected very similar. The development of resistivity was also very similar for the CEM IIB-V and CEM IIIA concretes, which contained pozzolanic material.

Chloride profiles were based on drillings on 25 mm increments. It was observed that generally, all concretes exhibited relatively high chloride concentrations at the depth of reinforcement after 125 days. It was noted that there was significant variation in the chloride content of the two CEM I concretes, highlighting the inherent variability of this method.

DISCUSSION

The electrochemical potentials for all the mix designs have been depressed with time. This is likely to be the result of a large increase in moisture content, as strong capillary suction has drawn the additional moisture into the pore matrices of the samples. This would increase the conductivity of the electrolyte within the cover zone, allowing for an increase in corrosion intensity and a reduction in IR drop.

Another observation was that the evolution of resistivity of CEM IIBV (45) behaved in a similar manner to CEM I (65) up until approximately 50 days. From there it diverged, and an increased rate of resistivity was observed which may be synonymous with the 'incubation period'. This infers that it has taken approximately 50 days for the pH level within CEM IIBV (45) to develop to a level where dissolution of the PFA particles may commence, and the pozzolanic reaction of these may take effect.

The chloride profiles for all concretes gave mixed results, without very distinctive trends. It was observed though, that CEM I (45) had a significantly different performance than the CEM I (65) mix design at the 25–50 mm and 50–75 mm, suggesting that there may have been issues either with the production and casting of the specimens, the concrete dust sampling, or the chemical analysis.

Overall, the results of the monitoring have not yet given strong evidence which of the CEM I (65), CEM IIB-V (45) or CEM IIIA (45) offered greater durability performance characteristics. It is postulated that this is due to the fact that the specimens have thus far undergone only 9 wetting and drying cycles (i.e. 125 days) and therefore are considered to still be of a relatively young age.

CONCLUSIONS

To date, the research has not yielded tangible evidence to determine which of the complaint CEM I (65), CEM IIB-V (45), CEM IIIA (45) concretes offers the best durability performance. This may be expected as design standards consider that these mix designs provide a similar level of durability. However, it is noted that the potential mapping for the non-compliant CEM I (45) mix design has given early evidence that corrosion activity may have initiated.

The resistivity of all mix designs has increased slowly over time; however, it remained low. Of particular interest was the fact that the resistivity development of CEM IIB-V and CEM IIIA was similar to CEM I for the first 50 days. However, the former exhibited a greater rate of resistivity increase following this initial period, which is attributed to the "incubation period" of the pozzolanic material, as CEM IIIA is latent hydraulic.

Concrete Repair, Rehabilitation and Retrofitting IV – Dehn et al. (Eds)
© 2016 Taylor & Francis Group, London, ISBN 978-1-138-02843-2

Variations of humidity within a relatively large ASR-affected concrete cylinder exposed to a natural environment

Hiroyuki Kagimoto
Electric Power Development Co. Ltd., Kitakyushu, Japan

Yukihiro Yasuda
JPower Design, Kanagawa, Japan

Shigeru Kinoshita
CTEC, Kanagawa, Japan

Mitsunori Kawamura
Kanazawa University, Kanazawa, Japan

ABSTRACT

In order to elucidate changes of the value of Relative Humidity (R.H.) related to understanding of the behavior of ASR expansion within massive concrete bodies, humidity within relatively large ASR-affected concrete cylinders (φ450 mm × 900 mm) placed in a natural environment as well as in atmospheres whose temperature and humidity had been precisely controlled, was pursued by measuring R.H. values for long periods. When changes in R.H. value with time in the concrete with ASR cracks which had been placed in dry atmospheres were followed by the re-saturation in an atmosphere of >95% R.H., the humidity up to the depth of 100 mm from surfaces increased in early stages after the beginning of re-saturating process, but a sudden rise of environmental R.H. to >95% was considerably delayed at the depth of 200 mm. It is found that permeability of near-surface regions in the concrete cylinder with reactive aggregate was increased by the formation of ASR cracks. When the concrete was exposed to natural environments whose humidity changed from 20 to 80% throughout a year, humidity within the concrete changed very slowly. R.H. values in surface near portions up to 50 mm were smaller than 80%, but humidity in the portions deeper than 50 mm was

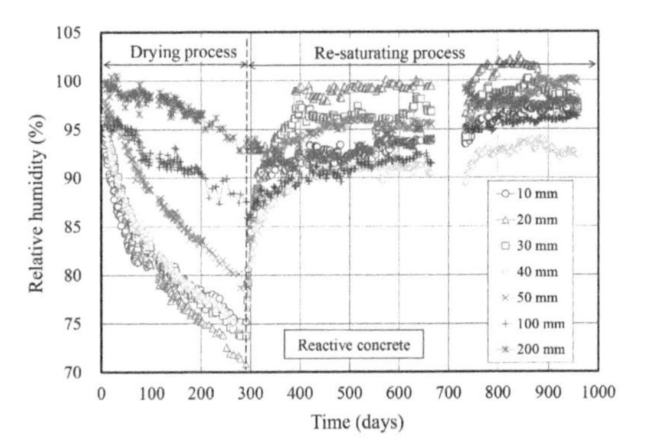

Figure 1. R.H. value vs. time curves at various depths in a ASR-affected concrete cylinders.

maintained at greater than 90% R.H. From a drying-re-saturating repetition tests in the laboratory, the permeability of the portions up to the depth of 50 mm was found to have decreased during the first drying-re-saturating process. The reduction in the permeability of near-surface layers thinner than 50 mm may be due to the impregnation of micro-cracks with ASR gels in the first drying- re-saturating process.

Concrete Repair, Rehabilitation and Retrofitting IV – Dehn et al. (Eds)
© 2016 Taylor & Francis Group, London, ISBN 978-1-138-02843-2

Influence of electrochemical lithium penetration from various kinds of lithium solution on ASR expansion of concrete

T. Ueda
Tokushima University, Tokushima, Japan

A. Nanasawa
Denki Kagaku Kogyo Kabushiki Kaisha, Niigata, Japan

M. Tsukagoshi
Tokushima University, Tokushima, Japan

ABSTRACT

It has been confirmed by many researchers that lithium salts have the effect of suppressing ASR-induced expansion of concrete. Supposing the application of the lithium salts as a repair additive for concrete structures deteriorated by ASR, sufficient amounts of Li^+ must be driven into the ASR-affected concrete. Then, an electrochemical technique to accelerate the penetration of the lithium ions (Li^+) in a lithium-based electrolyte solution into concrete has been developed for the purpose of suppressing ASR-induced expansion due to Li^+. However, from the results of past research works, the penetration area of Li^+ is limited around the concrete surface and it is difficult to carry Li^+ into the deeper part of concrete. In this study, experimental investigations were carried out aiming to grasp the influence of the kinds of lithium salts and the temperature of the electrolyte solution on the migration properties of ions in concrete and ASR-induced expansion of concrete. Moreover, chemical compositions of white products around the steel bar in reactive concrete were analyzed after the electrochemical treatment.

Distributions of Li^+ content in the concrete immediately after the treatment are shown in Figure 1. In this figure, Li content around the concrete surface increases with the electrochemical treatment. However, when the temperature of electrolyte is 30°C, the penetration depth of lithium is about 25 mm regardless of the kinds of lithium salt. On the other hand, this figure shows that 40°C electrolyte of $LiNO_3$ increase the amount and the depth of lithium penetration compared with the case of 30°C. The authors reported similar results from the experiment using Li_2CO_3 solution as the electrolyte.

Variation curves of concrete expansion rate after the treatment are shown in Figure 2. From this figure, regardless of the kinds of the electrolyte solution, the concrete expansion rate after the treatment was suppressed when compared with the non-treated case "N". Furthermore, it can be said that the 40°C electrolyte of $LiNO_3$ was more effective than other electrolytes used in this study to suppress the ASR expansion of concrete. Such tendency was agree with the results of lithium penetration due to the treatment shown in Figure 1.

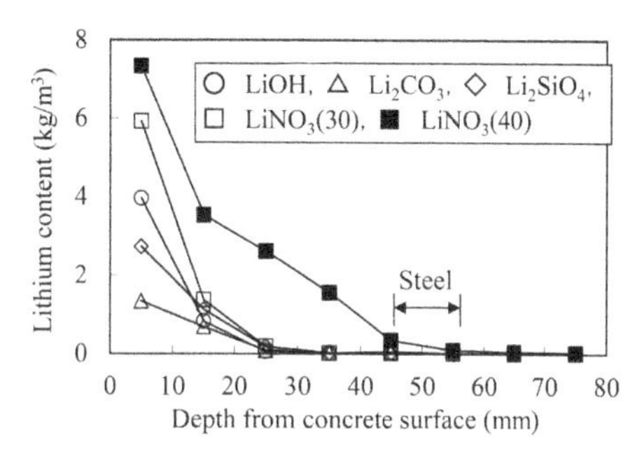

Figure 1. Distributions of lithium content in concrete after treatment.

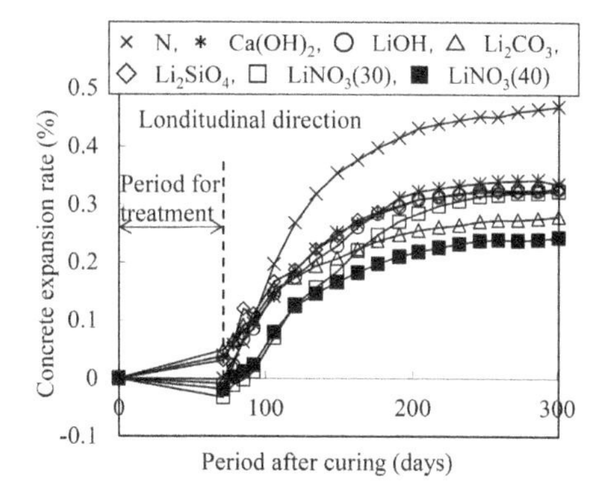

Figure 2. Variations of concrete expansion rate after treatment.

Concrete Repair, Rehabilitation and Retrofitting IV – Dehn et al. (Eds)
© 2016 Taylor & Francis Group, London, ISBN 978-1-138-02843-2

Effect of surface-applied inhibitors on anticorrosion performance of steel bars in sea sand concrete

Zixiao Wang, Shi Di & Zhiyong Liu
School of Civil Engineering, Yantai University, Yantai, China

ABSTRACT

The meaning of this paper is proving surface-applying corrosion inhibitors to sea sand concrete can effectively delay steel bar corrosion. By comparing the compressive strength of surface-applying corrosion inhibitors to sea sand concrete with that of river sand concrete, proving use sea sand in concrete has little adverse effect on concrete. By comparing the anticorrosion effect of several corrosion inhibitors in sea sand concrete with the anticorrosion effect of sodium nitrite, proving surface-applying corrosion inhibitors to sea sand concrete is an effect way to inhibiting corrosion of steel bar.

Surface-applied inhibitors include an Alcohol Amine Inhibitor (AMA), an amino carboxylic acid (PCI-2010) and the water solution of Monofluorophosphate (MFP). Sodium nitrite is used as Darex Corrosion Inhibitor (DCI) mixed in fresh concrete as a contrast of anticorrosion effect.

Using two kinds of sea sand with different contents of Chloride ion make three kinds of water binder ratios of concrete specimens. The contents of Chloride ion in two kinds of sea sand are 0.004% (Offshore Sand (OS)) and 0.12% (Artificial Sea Sand (ASS)), the water binder ratios of concrete are 0.35, 0.45, 0.55, the size of specimen is 100 mm × 100 mm × 100 mm. Slag and fly ash are used as admixtures in concrete; the specimens are grouping in 15 groups by mix designs, the specific mix designs are given in the full paper. Specimens in group 1# to 3# are River Sand (RS) concrete of three water binder ratios, these groups are used as control groups for 3 days and 28 days compressive strength. Specimens in group 4# to 15# are surface-applying or adding corrosion inhibitors.

At the age of 3 days, bring specimens in group 4#, 6#, 8#, 10#, 12#, 14# out of curing room, 3 specimens per group. Leave these specimens in the room temperature 1 day for drying, then, painting inhibitors on the face of specimens as shown in figure 1, one inhibitor per specimen per group. The total amount of inhibitor for one specimen is 2 g.The R_p and electrochemical impedance spectroscopies of steel bars are tested by electrochemical

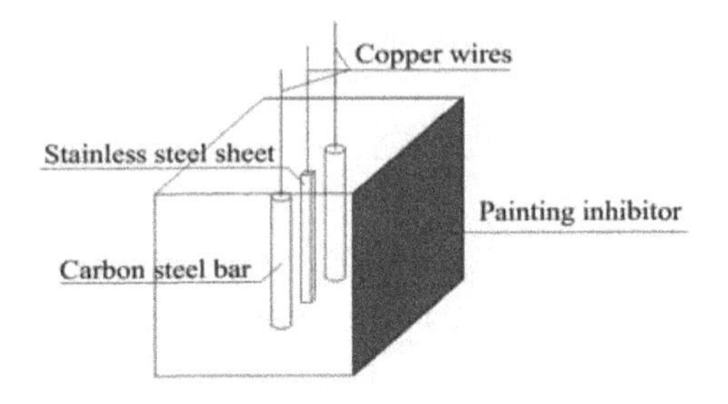

Figure 1. Sketch diagram of concrete specimen.

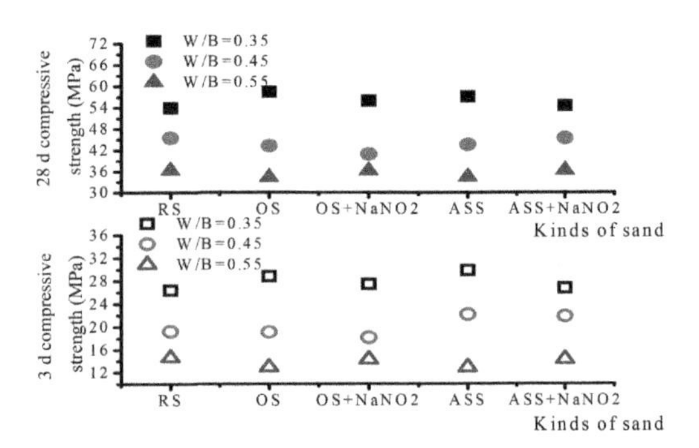

Figure 2. Compressive strength of concrete.

workstation at the concrete age of 30 days, 60 days, 120 days and 240 days. As to the steel bar in sea sand concrete, an I_{corr} value of 0.3μ A/cm^2 is considered as the threshold criterion for corrosion initiation.

Figure 2 shows the 3 days and 28 days compressive strength of rive sand concrete and two kinds of sea sand concrete with or without sodium nitrite. It can be seen that sea sand did not adversely affect the compressive strength of concrete; on the contrary, sea sand can improve the compressive strength at 3 days and 28 days. But using sodium nitrite in sea sand concrete can reduce the days and 28 days compressive strength of concrete.

Figure 3 and figure 4 are representing the I_{corr} of steel bar in two kinds of sea sand concrete in different water binder ratios and at different ages.

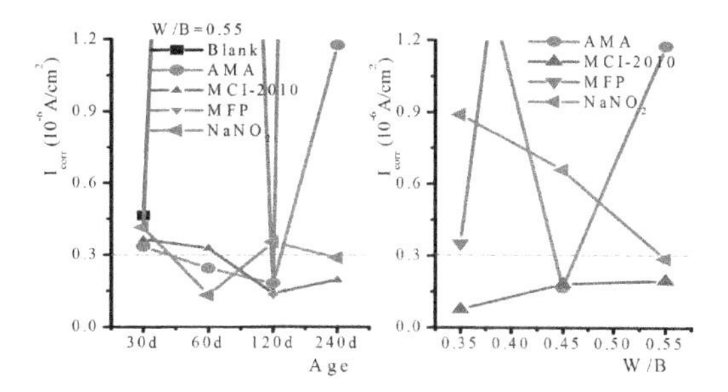

Figure 3. Icorr of steel bars in OS concrete.

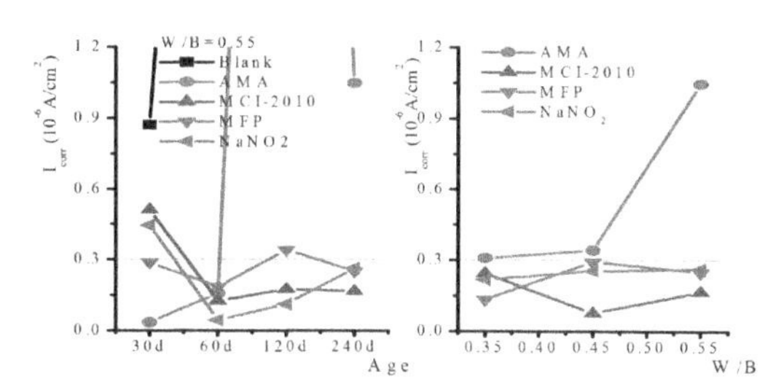

Figure 4. Icorr of steel bars in ASS concrete.

From these figures, it can be concluded that the anticorrosion effect of PCI-2010 for steel bar is similar with sodium nitrites; moreover, the anticorrosion effect of PCI-2010 is less sensitive to the change of water binder ratio or the content of Chloride ion in sea sand. The anticorrosion effect of AMA is decreasing with the increase of concrete ages, and the effect is greatly influenced by concrete water binder ratios. And the anticorrosion effect of MFP is affected by the content of

Table 1. Inhibition efficiency (%) values of steel bar in offshore sand concrete.

Time (d)	$NaNO_2$	PCI-2010	AMA	MFP
30	44.8	21.6	27.8	10.7
60	98.0	95.0	96.3	–
120	95.4	98.2	97.6	98.3
240	96.7	97.8	86.6	–

Table 2. Inhibition efficiency values of steel bar in artificial sea sand concrete.

Time (d)	$NaNO_2$	PCI-2010	AMA	MFP
30	49.0	41.5	96.0	66.9
60	99.7	97.3	96.6	96.0
120	98.7	98.0	21.6	96.0
240	96.7	97.6	86.8	96.9

Chloride ion in sea sand. Table 1 and table 2 are showing the Inhibiting Efficiencies (IE) of four inhibitors in sea sand concrete at different concrete age. The order of IE of four inhibitors is as follows: PCI-2010 > $NaNO_2$ > MFP > AMA.

The results of EIS test are accordance with the above results of LPR test. After crushing the specimens (W/B = 0.55) at the age of 240 days, the corrosion on the surface of steel bars is viewed directly by naked eyes; and the results agree with the results of LPR and EIS.

In summary, surface-applying migrating corrosion inhibitors PCI-2010 to sea sand concrete adding slag and fly ash is a very effective method of delaying corrosion of steel bars in the age of 240 days.

Concrete Repair, Rehabilitation and Retrofitting IV – Dehn et al. (Eds)
© 2016 Taylor & Francis Group, London, ISBN 978-1-138-02843-2

Durability design of concrete mixtures for sewer pipe applications: A review of the Life Factor Method

M.W. Kiliswa, M.G. Alexander & H.-D. Beushausen
Concrete Materials and Structural Integrity Research Unit, Department of Civil Engineering, University of Cape Town, South Africa

ABSTRACT

This paper reviews certain equations that are used for calculating key parameters influencing microbially-induced concrete corrosion in sewers. These equations form the basis of the so-called Life Factor Method (LFM) that is widely used by practising engineers for corrosion prediction in Portland Cement (PC) based concrete outfall sewers that flow partly full. The LFM quantifies sulphide (H_2S) build-up in stale sewage (US EPA 1985). Under certain conditions H_2S is readily released into the sewer headspace where it is biologically converted to sulphuric acid (H_2SO_4) on the moist pipe walls. This acid attacks the concrete matrix. Thus, the key parameters in this method are the rate of H_2S oxidation on the sewer walls and the alkalinity of the concrete matrix. In effect, this alkalinity amounts to the neutralisation capacity of the concrete matrix (Alexander & Fourie 2011). It therefore follows that the microbially-induced corrosion of the concrete sewer pipe is dependent on the rate at which the attacking acid forms and the effectiveness of the pipe binder system in neutralising the aggressiveness of this acid.

In PC based concrete with siliceous aggregates, the LFM (Equation 1) can be used accurately to predict the rates of microbially-induced corrosion.

$$c = 11.4k\frac{\phi_{sw}}{A} \tag{1}$$

where c = annual corrosion rate (mm/year); k = acid efficiency factor < 1; ϕ_{sw} = rate of H_2S absorption in the moisture at the concrete surface (g H_2S-S/m^2/h); A = alkalinity of the concrete material in units of g $CaCO_3$ per g concrete material (US EPA 1985).

Studies undertaken in South Africa for over 20 years clearly show a distinction in the performance of different binder systems used to combat H_2SO_4 attack in sewers. The performance of Portland Cement (PC) based systems with siliceous aggregates (whose alkalinity can readily be determined) closely relate to predictions based on the LFM. However, the performance of alternative binders such as PC based systems with calcareous aggregates and Calcium Aluminate Cement (CAC) based systems varied from the LFM predictions. CAC based systems performed up to 4 times better than PC based systems due to additional rate-controlling parameters such as control of bacterial metabolism by toxicity (bacteriostatic effect) of the CAC (Goyns 2014).

Thus, if the LFM is to be used to predict corrosion of concrete sewers with different binder systems, additional factors, such as material factors must be introduced to cater for additional rate-controlling parameters.

Concrete Repair, Rehabilitation and Retrofitting IV – Dehn et al. (Eds)
© 2016 Taylor & Francis Group, London, ISBN 978-1-138-02843-2

Moisture exchange in concrete repair system captured by X-ray absorption

M. Lukovic, E. Schlangen, G. Ye & K. van Breugel
Section of Materials and Environment, Faculty of Civil Engineering and Geosciences, Delft University of Technology, Delft, The Netherlands

ABSTRACT

One of the main reasons for premature failure of concrete repair systems are stresses due to differential shrinkage between the old (concrete or mortar substrate) and new (repair) material. These are primarily caused by moisture exchange between the repair system and environment or moisture transport within the repair system itself. However, during the setting and hardening of the repair material, moisture exchange will also determine the development of material properties in the repair material and interface between repair material and substrate. In this paper, moisture profiles are quantified using X-ray absorption technique. The output of X-ray attenuation measurement is a spatial distribution of linear attenuation coefficients expressed as grey scale values. First, preliminary studies are performed to determine the absorption rate of the mortar substrate. Cumulative water absorption and penetration depth of the water front are monitored as a function of time. Furthermore, absorption of the substrate when water is absorbed from the top is investigated. Main advantages and drawbacks of the method are presented and compared with results from the gravimetric test. Following, moisture exchange in the repair system is investigated. In one specimen, a primer was used. When curing is stopped, drying profiles in the repair system are quantified. Knowledge about the water exchange between the repair material and the concrete substrate, which takes place at the early stage, is only the first step towards more-justified decisions about substrate preconditioning and possible use of primers and curing compounds in experiments and field practice.

INTRODUCTION

In this paper, a possibility of using X-ray absorption method for studying the moisture dynamics in a repair system was explored. First, the rate of moisture absorption of the mortar substrate was quantified. Sample preparation and post-processing procedure for X-ray measurement data are explained. Furthermore, absorption of the substrate when a fresh repair material is cast on top of it is investigated. For improving repair performance, primers are commonly used in the field. In one test, commercially available primer was applied in order to investigate its role on reducing and stopping absorption of the substrate.

MATERIALS AND METHODS

Materials and sample preparation

The substrate used in the study was a two year old mortar. A standard mortar mixture (OPC CEM I 42.5 N, water-to-cement ratio 1:2, cement-to-sand ration 1:3) was used. Small prism specimens ($18 \times 19 \times 40$ mm^3) were slowly cut with a diamond saw from bigger mortar samples.

Before testing, mortar substrate was dried in the oven at 105°C until constant weight is achieved. This was done in order to remove all evaporable water and have a zero initial moisture content at the start of the experiment. As a repair material, cement paste with OPC CEM 42.5 N and a water-to-cement ratio of 0.5 was used.

X-ray absorption measurement

After sample preparation, mortar substrate or repair system were placed in the Phoenix Nanotom X-ray system (CT scanner) for measuring the water exchange. CT images are maps of X-ray absorption in the material. Here, only a single X-ray image is taken at certain time step. The sample was kept in the same position during the whole experiment. The resulting CT images are spatial distributions of linear attenuation coefficients, expressed by GSV (grey scale value). In this paper, an attempt is made to correlate the GSV change with change in moisture content as a function of time. This is done by taking advantage of a simple physical principle. In a dry sample, X-rays are attenuated *only* by the dry material. If there is some moisture intake in a porous material, the attenuating material consists of the dry material and a thickness of water layer

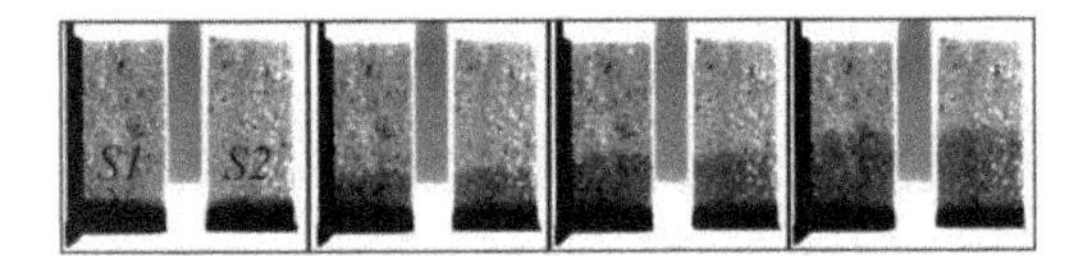

Figure 1. Qualitative representation of water change as a function of time (wetting from the bottom).

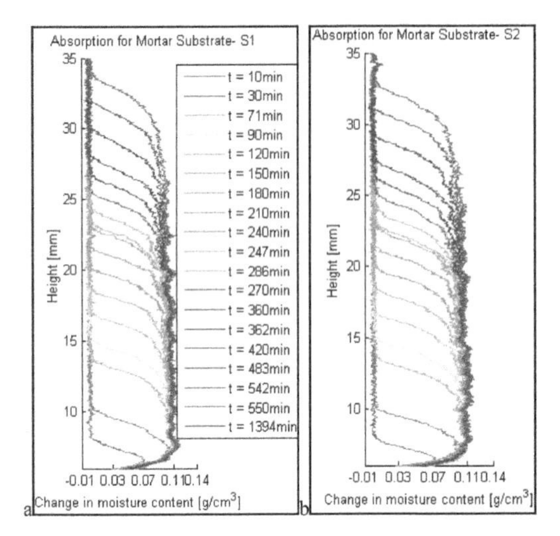

Figure 2. Moisture profiles in the left and right specimen (from figure 1) as a function of time.

equivalent to the (volume) moisture content of the material. This means that, by knowing the attenuation coefficient of water, and knowing the change in GSV, the (additional) moisture content inside the material can be determined.

RESULTS AND DISCUSSION

Absorption of the mortar substrate when water is absorbed from the bottom

First study is done on the absorption rate of mortar. Moisture change as a function of time is given in Figure 1. Moisture profile is averaged through the thickness of the sample. Moisture profiles are given in Figure 2.

Obtained curves resemble the curves obtained by neutron radiography or nuclear magnetic resonance method. This means that water profiles into mortar samples can be monitored very accurately by X-ray absorption.

Water exchange in a repair system

Cement paste with a w/c ratio of 0.5 and around 15 mm thickness was cast on the top of the substrate mortar (Figure 3). In one case, the substrate was treated with a commercial primer prior to application of the repair material. Immediately after casting, repair systems are sealed with aluminum self-adhesive tape and placed in the CT-scanner in order to investigate the moisture exchange between repair material and concrete substrate.

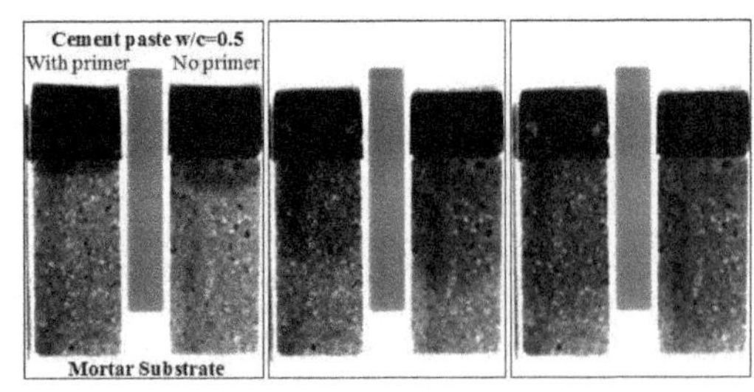

Figure 3. a) Obtained images for testing repair system (from left-immediately after putting the specimens in the CT scanner-26 min after casting, 421 min and 1283 min after casting).

Primer did not reduce the rate of water absorption. However, in the beginning, it did slightly postpone the onset of water loss from the repair material (Figure 3). In addition, drying profiles of repair material with and without primer were quantified.

CONCLUSIONS

Based on the presented study, the following conclusions can be drawn:

- X-ray absorption can be used to predict moisture transport in porous materials very accurately and with high resolution. Both absorption profiles and drying profiles can be quantified.
- With this technique, water absorption as a linear function of square root of time could be captured.
- Applied primer did not prevent or reduce the rate of water loss from the repair material, but only delayed the onset of water loss.

Saturation level of the concrete substrate is the critical parameter for development of properties inside the repair system and its influence needs to be further investigated.

ACKNOWLEDGMENT

Financial support by the Dutch Technology Foundation (STW) for the project 10981-"Durable Repair and Radical Protection of Concrete Structures in View of Sustainable Construction" is gratefully acknowledged.

Concrete Repair, Rehabilitation and Retrofitting IV – Dehn et al. (Eds)
© 2016 Taylor & Francis Group, London, ISBN 978-1-138-02843-2

Forecasting chloride-induced reinforcement corrosion in concrete—effect of realistic reinforcement steel surface conditions

U.M. Angst
ETH Zürich, Institute for Building Materials (IfB), Zürich, Switzerland

B. Elsener
ETH Zürich, Institute for Building Materials (IfB), Zürich, Switzerland
Department of Chemical and Geological Science, University of Cagliari, Monserrato, Italy

ABSTRACT

Over the last decades, considerable efforts have been made in science and engineering to predict initiation of chloride-induced reinforcement corrosion in concrete. After it was recognized in the second half of the last century that i) initiation of pitting corrosion in alkaline concrete requires the presence of chloride and that ii) some sort of correlation exists between the risk for corrosion initiation and the chloride concentration in the concrete (Hausmann, 1967, Richartz, 1969, Vassie, 1984), the concept of the *critical chloride content* or *chloride threshold value* was born. This concept is now well-established and reflects the wide-spread belief that initiation of chloride-induced reinforcement corrosion can be predicted directly as a function of the chloride concentration.

A large number of attempts to determine chloride threshold values were made, as e.g. reviewed in (Angst et al., 2009). It is very well known that these attempts generally got bogged down in many details such as specimen design, accelerating chloride ingress, measurement methodologies, etc. As a result, literature chloride threshold values scatter over several orders of magnitude and there exists still no accepted test setup or method to determine this parameter of uttermost practical importance. One of the major challenges lies in simultaneously satisfying the general desire for a fast and reproducible method and the requirement of practice-related conditions to ensure that the results are relevant for practice.

In the present work, ten series of "as-received" reinforcement steel originating from different countries and having been subject to different exposure histories (storage, transport, handling, etc.) were studied. There were clear differences in visual appearance. While some series were essentially free from red or brown visible corrosion products and exhibited a blank and shiny metal surface, the rebars of other series were partially coated with red-brown corrosion products. Finally, some series had rebars that were severely pre-corroded and exhibited relatively thick and flaky layers of red-brown corrosion products. These differences in visual appearance were reflected in electrochemical characterization by means of cyclic voltammetry after 40 d exposure in saturated calcium hydroxide solution (chloride-free).

Metallographic analyses indicated differences in terms of steel microstructure. Some rebar series exhibited an almost homogenous ferrite-pearlite microstructure. The rebars belonging to the other series, on the other hand, had a composite microstructure with typically three zones: a ferrite-pearlite core, an intermediate layer, and a surface layer of martensite. These bars obviously underwent a thermo-mechanical strengthening process (e.g. as known under trade-names such as "Tempcore") that lead to the tempered martensite surface layer.

Chloride threshold values were subsequently determined by continuously increasing the chloride concentration in the saturated calcium hydroxide solution. The results are shown in Fig. 1. Note that each series consisted of 2 rebars, with the exception of series A, where 3 bars were tested. Clearly, the measured threshold values scatter over a wide range. In most cases, the scatter within one series was, however, smaller than the overall variability.

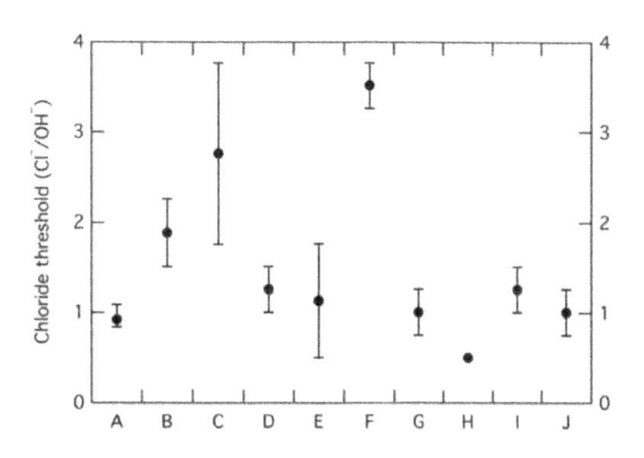

Figure 1. Chloride threshold values (min/mean/max) for each series of the as-received rebars.

Comparing these results with the observations made with microscopy and electrochemical surface characterization by means of cyclic voltammetry, one can conclude that there exists no systematic relationship between the chloride threshold value and the degree of initial pre-existing rust. Neither the amount of rust initially present on the surface nor the visual appearance of the rebars permits thus any prognosis of the corrosion behavior when exposed to chlorides.

From this it is inferred that another parameter rather than the presence of absence of initial rust plays an important role in corrosion initiation. There was indication that the microstructure of the steel may be such an influencing factor. In fact, the rebars with tempered martensitic surface layers appeared to yield lower chloride threshold values than ferritic-pearlitic microstructures. This is in agreement with literature results that indicated that martensitic steels are more prone to pitting corrosion than the ferrite-pearlite microstructure (Al-rubaiey et al., 2013). In this regard, however, more research work is clearly needed.

If all rebars tested in the present work are considered (circles in Fig. 2), the resulting distribution of chloride threshold values scatters considerably, in fact from Cl^-/OH^- = ca. 0.2 to almost a value of 4, thus by over one order of magnitude. Nevertheless, it appears that more than 80% of the values are within the range Cl^-/OH^- = 0.2 to 2, and that only a small number of rebars shows a significantly better corrosion resistance.

For laboratory testing, the rebar surface is only one of the numerous degrees of freedom in the design of experiments or test methods for chloride threshold values (Angst et al., 2009). It is clearly evident that the so-called "as-received condition" cannot guarantee repeatability and reproducibility: When *n* rebars from the same batch are tested in alkaline solution, *n* different chloride threshold values are obtained. The results from Hausmann (Fig. 2) indicate that this is similarly true also if the as-received surface state is eliminated by cleaning and burnishing. The observed variability in threshold values may simply be explained be the stochastic nature of pitting corrosion initiation.

The present results were determined with a setup in solution that can be considered as a homogeneous medium around the rebar. In concrete, on the other hand, further irregularities and defects arise at the steel surface such as macroscopic voids (either filled with air or with solution), cracks, aggregates, etc. These are well known to have local effects on corrosion initiation (Reddy, 2001, Soylev and Francois, 2003) and are thus expected to broaden the scatter even more.

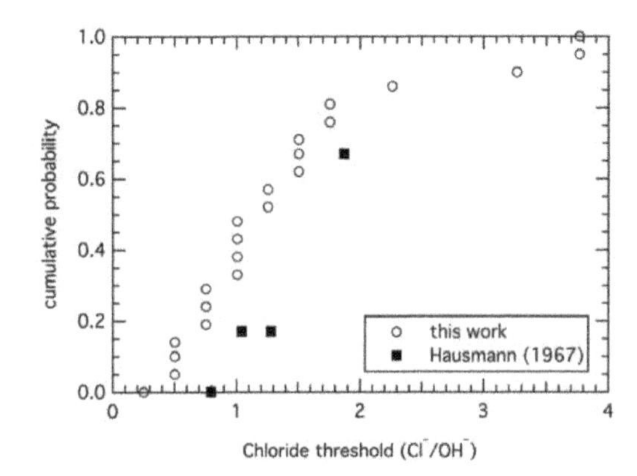

Figure 2. Cumulative probability distribution of chloride threshold values of all tested rebars, in comparison to the well-known results obtained by Hausmann (1967).

It was concluded that even the steel surface condition alone is a parameter that causes an enormous uncertainty in current conceptual service life modeling. The common approach of reducing the issue of predicting corrosion initiation to a mere matter of chlorides must thus be questioned. At present, the relevant parameters are obviously overlooked.

There is thus an urgent need for more refined methods to characterize in detail the fundamental properties and mechanisms that dictate why corrosion initiates on one rebar and not on another, given the same chloride concentration. Only with such an approach will it be possible to reduce the large uncertainties inherent to current concepts and models.

REFERENCES

Al-Rubaiey, S. I., Anoon, E. A. & Hanoon, M. M. 2013. The influence of microstructure on the corrosion rate of carbon steels. *Eng. & Tech. Journal*, 31, 1825–1836.

Angst, U., Elsener, B., Larsen, C. K. & Vennesland, Ø. 2009. Critical chloride content in reinforced concrete—A review. *Cement and Concrete Research*, 39, 1122–1138.

Hausmann, D. A. 1967. Steel corrosion in concrete. How does it occur? *Materials Protection*, 6, 19–23.

Reddy, B. 2001. Influence of the steel-concrete interface on the chloride threshold level, PhD thesis, Imperial College, London.

Richartz, W. 1969. Die Bindung von Chlorid bei der Zementerhärtung. *Zement-Kalk-Gips*, 10, 447–456.

Soylev, T. A. & Francois, R. 2003. Quality of steel-concrete interface and corrosion of reinforcing steel. *Cement and Concrete Research*, 33, 1407–1415.

Vassie, P. 1984. Reinforcement corrosion and the durability of concrete bridges. *Proc. Instn Civ. Engrs.*, Part 1, 76, 713–723.

Concrete Repair, Rehabilitation and Retrofitting IV – Dehn et al. (Eds)
© 2016 Taylor & Francis Group, London, ISBN 978-1-138-02843-2

Effect of reinforcement corrosion on serviceability behavior of RC beams and analytical model

L. Hariche
LDMM Research Laboratory, Djelfa University, Djelfa, Algeria

M. Bouhicha
Civil Engineering Research Laboratory, Laghouat University, Algeria

S. Kenai
Civil Engineering Research Laboratory, Blida University, Algeria

Y. Ballim
School of Civil and Environmental Engineering, University of the Witwatersrand, South Africa

ABSTRACT

The research work involved the testing of four series of RC beams subjected to reinforcement corrosion while maintaining a sustained load. The first three series of beams were subjected to the same test conditions except that the arrangement of the bottom tension reinforcement was varied. Series 1 contained two Y12 bars (226 mm²), Series 2 had a single Y16 bar (201 mm²) and Series 3 contained three Y10 bars (235 mm²). The beams in these three test series were subjected to a 4-point load of 20 kN. In the fourth test series, the tension reinforcement consisted of two Y12 bars and the beams were subjected to an increased 4-point load of 30 kN.

Twenty-four beams were cast with the dimensions and reinforcing layout shown in Figure. The beams were divided equally into four series of six beams. In each series, the main tension reinforcing steel of three beams was subjected to accelerated corrosion while the three beams served as control, un-corroded beams.

CONCLUSIONS

The following conclusions can be drawn from this experimental and analytical investigation:

1. In the process of acceleration of corrosion the current efficiency is more important for 1Y16 then 2Y12 and 3Y10. That's mean; to accelerated corrosion of small diameter steel we need more current than the big diameter steel to have the same level of corrosion. Also when there is vagabond current in RC structure there is less risk of corrosion in small diameter then the big one.

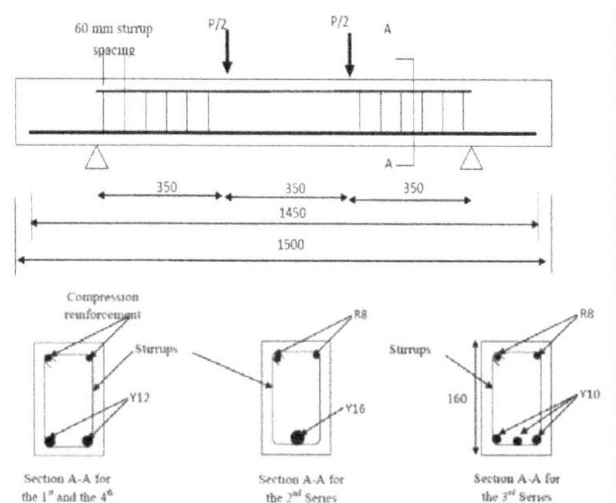

Figure 1. Dimensions and reinforcement details of the concrete test beams.

2. Conformed to the literature of corrosion [1, 14–16], the reinforces steel are corroded by pitting corrosion, because it was caused by chloride attack, and there is no carbonation of concrete.
3. From the crack on the surface of RC beams we can predict that the reinforced steel in this side is corroded. Than for perspective, it will be very interesting if we can find a relationship between the rat of corrosion and the crack width, from that we can estimate the degree of corrosion in the RC structure just from the crack width on the surface.
4. Under simultaneous load and accelerated corrosion, the deflections of the beams increase with progressive corrosion of the reinforcement. A large increase in deflection was noted during the early stages of corrosion as a result of propagation of transverse cracks on the beams caused by the flexural tension and the expansive stresses induced by the corrosion products.
5. From this investigation it was observed clearly the role of the arrangement of the steel in the section of concrete. In spite of air section of steel was nearly the same, the beams of 2Y12 and 3Y10 were more performed in deflection, curvature and rigidity then the beams of 1Y16.
6. The analytical analyses propose some correlation related to the rat of corrosion.

RESULTS

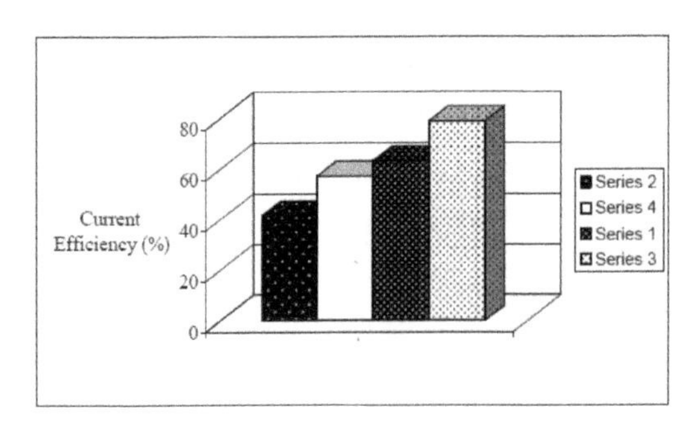

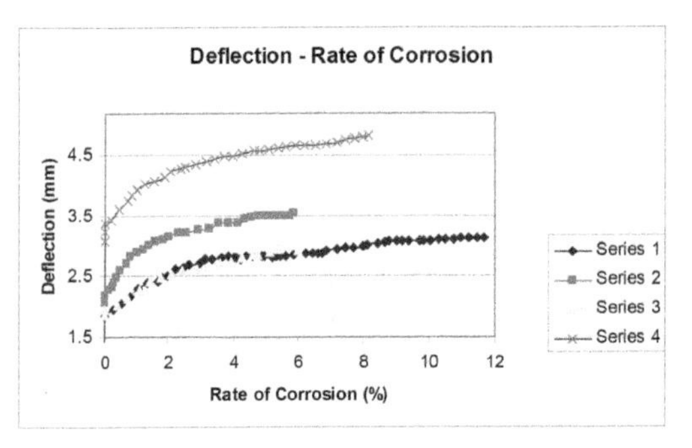

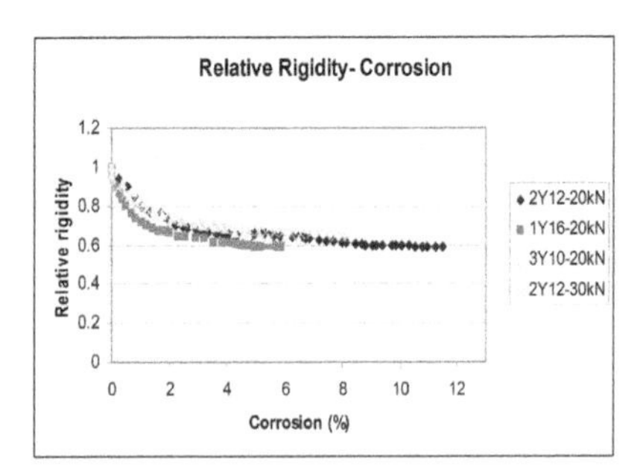

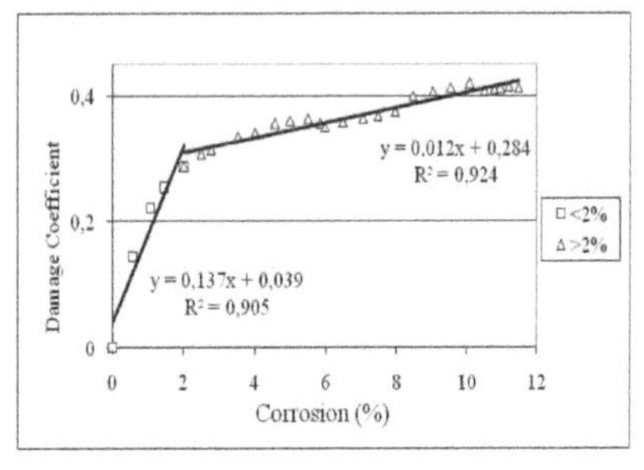

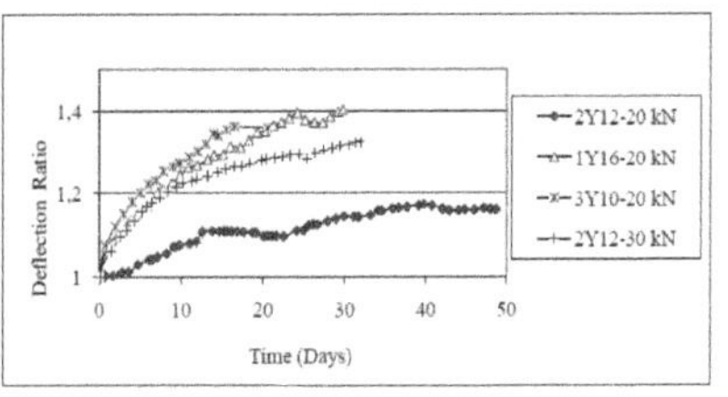

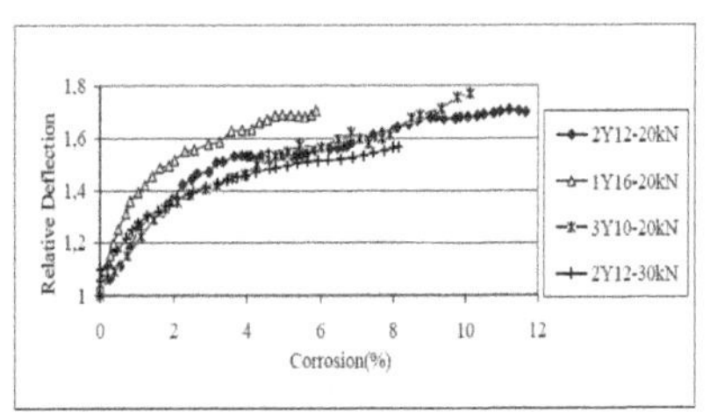

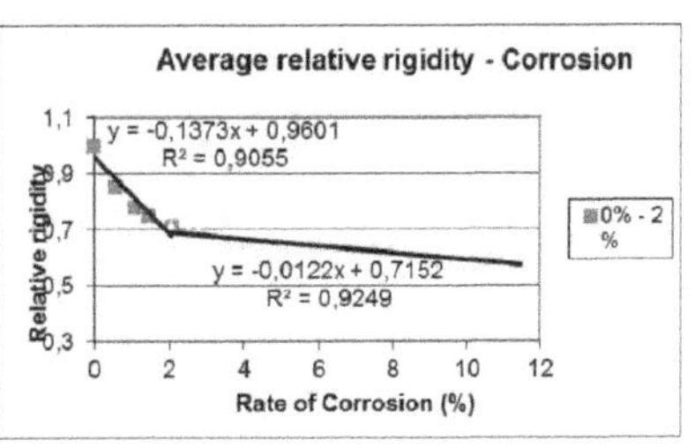

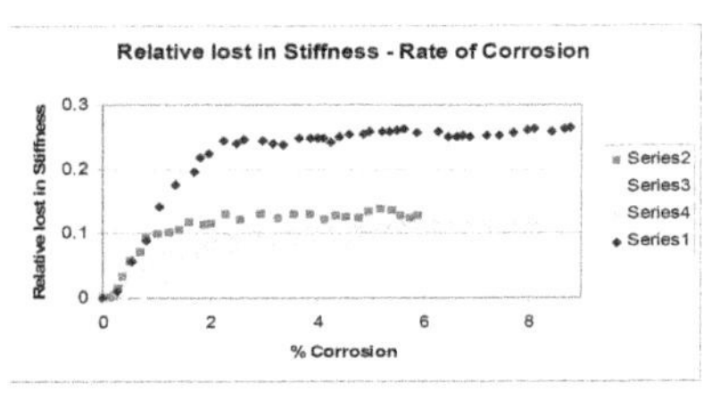

Concrete Repair, Rehabilitation and Retrofitting IV – Dehn et al. (Eds)
© 2016 Taylor & Francis Group, London, ISBN 978-1-138-02843-2

Damage evaluation for freezing-thawing affected concrete by automated panoramic fluorescent microscope

Shuguang Li, Gaixin Chen & Guojin Ji
State Key Laboratory of Simulation and Regulation of Water Cycle in River Basin, China Institute of Water Resources and Hydropower Research (IWHR), Beijing, P.R. China

Wanqiu Xia & Dong Zhang
Jiangxi Hongping Pumped-Storage Power Plant, State Grid Xinyuan Holding Co. Ltd., Nanchang, China

ABSTRACT

The deteriorating process of concrete suffered Freezing-Thawing (FT) is a process of initiation and development of microcracks. It's very important to establish the correlation between the macroscopic properties and the microcrack characteristics to evaluate the damage extent in concrete. The first step is to quantify the microcrack characteristics such as length, width and density. Though CT or MRI is a promising tool of analyzing microcracks nondestructively, the spatial resolution of either of them is far from satisfactory to detect FT-induced microcracks. Traditional microcrack-analyzing techniques such as Scanning Electron Microscope (SEM), Optical Microscope (OM), Fluorescent Microscope (FM) are all based on a limited number of small microscopic samples 1 or 2 mm in dimension cut from concrete slices. However, the limited samples are not representative and the microscopic analysis results are not reliable because concrete is a highly heterogeneous materials composed of mortar, Interfacial Transitional Zones (ITZ) and aggregates. Accordingly, this paper presented a method of acquiring panoramic microscopic images of concrete slices (10 cm * 10 cm in size) impregnated with fluorescent epoxy by Automated Panoramic Fluorescent Microscope (APFM) which is designed and developed by our team. The APFM is characterized by four automatic modules, i.e., auto-scanning module, auto-focusing module, auto-mosaicing module and auto-analyzing module. Panoramic images of the concrete slices can be obtained by APFM in about two hours, which is efficient and accurate. In addition the area, length, density of microcracks in the slices can also be extracted and analyzed in the auto-analyzing module by digital-image-processing technique once the panoramic image is obtained. The evolution of microcrack characteristics in concrete during the FT damaging process is obtained and the results show that the length density and area density of the microcracks increase with the increase of the FT damage degrees. Relationships between the mechanical properties and the microcrack density are also established. Quantitative microcrack analysis based on APFM is a promising tool in evaluating FT damage and revealing the essence of damage in concrete.

Concrete Repair, Rehabilitation and Retrofitting IV – Dehn et al. (Eds)
© 2016 Taylor & Francis Group, London, ISBN 978-1-138-02843-2

Modelling the service life of concrete until cover cracking due to reinforcement corrosion

E. Bohner, M. Ferreira & O. Saarela
VTT Technical Research Centre of Finland Ltd., Espoo, Finland

ABSTRACT

The service life design of reinforced concrete structures requires material models capable of reliably describing both mechanisms of damage and the general progression of damage over time. However, most models that are currently being used only capture the process of carbonation and chloride penetration into the uncracked concrete that is at the initial phase of degradation. Typically, these models disregard the actual damage, i.e. the corrosion of the reinforcing steel. As a result, the service life design established to date only considers the end of the initiation phase of the degradation process, i.e. the onset of damage (time of depassivation or onset of corrosion) as a critical limit state. The corrosion of the reinforcement and its consequences, i.e. the crack formation and spalling of concrete, are not considered, which may lead to a substantially shorter estimated service life of the structures. Comprehensive investigations were recently undertaken on the depassivation of steel reinforcement and on crack formation in concrete which have resulted in an analytical model for corrosion-induced cracking occurring in the surface zones of structural components. This paper presents a holistic approach in which two models used for determining the time to depassivation (initiation phase) and the time to cover cracking as a result of reinforcement corrosion (propagation phase) are combined. An example is provided of a semi-infinite reinforced concrete wall which has been designed for the serviceability limit state of concrete cover cracking.

A methodology is presented which contributes to the development of reliability-based approaches for the durability design and service life prediction of RCS that combines the probabilistic determination of two limit states (corrosion initiation and corrosion induced cracking) in a single analysis, thus removing the need for one of the limit states.

By combining initiation with propagation until cracking, the combined service life covers the period from $t = 0$ until the time for the first crack with a width of a minimum of 0.05 mm to appear on the surface of the concrete.

Assuming that corrosion is initiated at some time τ, for cracking to take place at time t the corrosion process has to have duration of $t - \tau$. Consequently, the combined probability distribution of the time of cracking is

$$f_{i,p}(t) = \int_{-\infty}^{\infty} f_p(t - \tau | T_i = \tau) f_i(\tau) d\tau \quad (1)$$

where $f_{i,p}$ = probability of failure; $f_p(t)$ = probability Density Function (PDF) of the propagation model; and $f_i(t)$ = PDF of the initiation model.

The start of corrosion propagation process depends on the corrosion initiation process. However, the mechanisms that describe each of these processes are independent of each other. Therefore the modelling and computation of this distribution can be simplified by replacing the conditional PDF with corresponding marginal probability distribution:

$$f_{i,p}(t) \approx \int_0^t f_p(t - \tau) f_i(\tau) d\tau \quad (2)$$

A consequence of this approach is that there is no need to establish a limit state requirement for corrosion initiation, and another for cracking. By combining the calculations, only one requirement for the outcome of both needs to be defined. Furthermore, the uncertainties related to both processes, as modelled in the PDF's, are properly combined.

In this paper a hypothetical semi-infinite reinforced concrete wall located in a XS3 environment (tidal/splash zone) is considered. This scenario enables the use of a 1-D analysis and diffusion being the main form of chloride transport in the concrete (convection zone not considered in this study).

Service life calculations have been performed considering two distinct concrete qualities: a CEM I 42.5 N with w/b ratio 0.50, and a CEM III/A 42.5 R with w/b ratio 0.55. Both concretes have similar mechanical performance, but differ significantly from a chloride ingress perspective, where CEM III out per-forms CEM I. For the corrosion propagation model, two distinct corrosion rates were chosen to simulate a "fast" (3.0 μA/cm²) and a "slow" corrosion process (0.3 μA/cm²). Corrosion rate is defined by a lognormal distribution (CoV = 0.3) with an average value of 0.0035 mm/year and 0.035 mm/year for "slow" and "fast" corrosion, respectively. More details on the parameters can be found in the full paper.

The calculation of the limit states for both individual models was performed with a direct Monte Carlo simulation. As an example, the results for the probability of limit state failure for corrosion initiation and for corrosion initiation combined with corrosion propagation are presented in Figure 1, for CEM I and CEM III concretes subjected to "slow" corrosion.

The service life obtained by combining the two individual service life calculations depends on both the quality of the concrete and the corrosion rate. However, the quality of the concrete, mostly determining the time to corrosion initiation, dominates as it is the precursor for the corrosion propagation phase.

Assuming a 10% probability for the limit state failure for corrosion initiation, it is normal to expect that the probability for the limit state failure for corrosion propagation should be lower since the consequences are more severe, and already damage inducing. The combination of both limit states should then be judged by the requirement for the second limit state since it is the phenomenon being observed, i.e. first cracking. A value of 7% is assumed for the limit state for corrosion propagation, for the sake of the analysis in this paper.

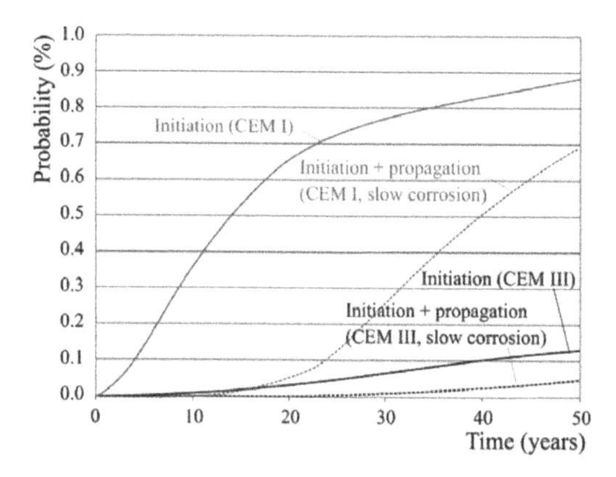

Figure 1. Probability of limit state failure for corrosion initiation and for corrosion initiation combined with corrosion propagation with a "slow" corrosion rate of the propagation model.

If considering a "slow" corrosion (see Figure 1), the combination of the two limit states leads to an extension of the overall service life of the RCS of approx. 18 years, when compared with the limit state for corrosion initiation. If considering a "fast" corrosion rate, an extension of 2 to 3 years in the service life is observed for a CEM I concrete. However, for the CEM III concrete, the combination of the two limit states results in a reduction of the overall service life of the RCS.

This result seems contradictory because the overall service life is shortened when compared with the service life of corrosion initiation. The calculation of the combined service life takes into account the probability of each limit state occurring simultaneously, which normal service life calculations cannot account for. For a traditional service life calculation, the limit state of propagation is always an extension of the limit state of corrosion initiation.

These results show that the choice of evaluation criteria for the limit state for corrosion initiation or for propagation cannot be chosen arbitrarily. It would be expected that a limit state for propagation results in an extension of the limit state of corrosion initiation. This raises questions regarding what are the values that should be used. Could the criteria differ based on the type of structural part, type and quality of concrete and environmental conditions?

It has been suggested that the limit state for propagation could be identical to that of the initiation because the appearance of the first crack does not affect the mechanical performance of the RCS. In this case, a combination of limit states would always result in an extension of the service life of the RCS.

There is currently not enough background information to understand the consequences of certain values, and while many values are being put forward based on those used in structural design they do not necessarily represent adequately the complexity of the deterioration mechanism and the economic con-sequences of limit state failure. However, the choice of limit state requirement is still relatively subjective. Could a lower reliability (higher probability of limit state failure) for cracking be considered?

While the benefits for the service life design of new concrete structures, and for the remaining service life determination of existing structures, are apparent, the criteria by which the limit states are de-fined are not yet mature.

Despite the necessary discussion regarding the requirements for service life determination, this approach improves modelling for the entire service life of a structure, enabling cost savings possibilities through the avoidance of unnecessary repair costs by an optimised timing of the repair measures.

Concrete Repair, Rehabilitation and Retrofitting IV – Dehn et al. (Eds)
© 2016 Taylor & Francis Group, London, ISBN 978-1-138-02843-2

A whole of life approach to concrete durability—the CIA concrete durability series

F. Papworth
BCRC, Perth, Australia

ABSTRACT

fib Model Code 2010 has introduced various methodologies for treating durability design as a whole of life requirement. The Concrete Institute of Australia (CIA) committee on durability has used the Model Code as a background for producing seven recommended practice notes that detail many of the requirements for ensuring durability is achieved by appropriate action from the planning stage through design, construction, maintenance, restoration to decommissioning. This paper shows how these various documents build to provide a whole of life durability approach. It covers deemed to satisfy requirements applicable to all concrete structure types based on standard input parameters for design life, reliability and exposure. The series includes details on project planning and good practice which if followed will increase the likelihood that the specification, design detailing and construction will be optimal to achieving the developer and community expectations regarding the long term performance of concrete structures. Also included are methods for modelling, degradation over time and crack control design. Thus the series provides what is described as a unified durability design process.

The durability series provides the required guidelines in seven documents.

Z7/01 DURABILITY PLANNING

Whist there has been various papers published on durability planning Z7/01 is the first industry guideline on the topic. Z7/01 sets out the process of planning to achieve the required level of durability. The durability planning outcomes will be delivered in a Durability Assessment Report specific for the project.

The Durability Assessment Report issued will explain the durability requirements and provide details to be included in the project design reports, specifications, design drawings, asset maintenance plans and/or operation and maintenance manuals. This report may be a page for simple structures or detailed for complex or critical or structures in severe exposures.

Designer and/or contractor provided durability requirements without adequate asset owner input may mean optimum functionality and whole of life cost may not be achieved. Involvement of owners in the durability design process is discussed.

Z7/02 EXPOSURE CLASSES

Australian Standard durability classes have three issues in how they deal with exposure classifications:

a) Exposure classes for different structure types, e.g. piling, marine structure, water retaining structures and general concrete structures, can have different exposures classifications according to current Australian codes. In reality the exposures are common to all concrete and use of different class types for the same exposure is confusing.
b) The same exposure class is used for different exposure types, e.g. a B2 exposure could be due to chloride exposure or sulfate exposure. This makes it difficult to provide a range of solution that will always be applicable to each exposure class.
c) The exposure classes are not in line with those used in ISO or European standards. This makes it difficult to compare Australian durability requirements with international requirements.
d) Hence Z7/02 provides different exposure classes for each type of exposure and applies them to concrete in all structure types. The ISO 16204 [10] exposure classes are adopted where possible but the ISO format for exposure classes is adopted in all cases.

The exposure classes discussed in Z7/02 are the criteria against which deemed to comply provisions in Z7/03 are defined. Hence Z7/02 can be considered a commentary on the exposures used in Z7/03. Exposure classes comprise:

XC – atmospheric carbonation inducing,
XS – seawater exposure,
XD – chlorides other than seawater,
XA – aggressive chemicals in ground exposure

XF – freeze thaw,
XG – exposure to liquids, vapours and gases,
XM – water migration
XX – metals in the cover zone
XR – abrasion
XI – moisture and ASR
XH – temperature and delayed ettringite formation

Z7/03 DEEMED TO COMPLY REQUIREMENTS

The following approach was followed:

- Provide requirements for each exposure class (e.g. chloride, sulfate, acid, atmospheric gases)
- Requirements linked to cement systems including GP cement, fly ash, slag and silica fume)
- Provide guidance for galvanised and stainless steel reinforcement, prestressing and steel fibres
- Give recommendations for effect of coatings on other durability requirements
- Define significance of curing methods on other durability requirements
- Provide advice based on minimum cover.
- Incorporate reliability as a factor in determining durability requirements
- Design lives of 25, 50, 100 and 200 years to be considered.

Z7/04 GOOD PRACTICE

Z7/04 has applicability to more general concrete design and construction as well as concrete requiring specifically higher levels of durability. Z7/04 details areas such as:

- the impact of specifications on the contract process,
- impacts of design on construction,
- details of materials requirements used in construction,
- material quality control processes,
- construction process and supervision,
- detailing issues in common structural elements that may present potential durability issues to the designer & constructor.
- reinforcement spacers and chairs. This is an area that has been demonstrated to cause weakness in durable construction and is rarely adequately specified.

Z7/05 MODELLING

The objectives of Z7/05 was to review commonly used models and their input parameters for chloride diffusion, carbonation and corrosion propagation process so as to be able to recommend the most suitable models and input parameters with their statistical distributions. However, considering the complex nature of concrete deterioration process and still insufficient researches this current guide is considered to be a working document which will be updated with more understandings and new findings in future.

Z7/06 CRACKING AND CRACK CONTROL

At present, there is no Australia Standard method for calculating crack widths. The methods outlined in CIRIA C660 are often used for early-age thermal crack control and a variety of methods have been proposed for estimating the width of load-induced cracks.

Z7/06 will provide a consistent and rational approach for predicting crack widths in concrete structures. It will provide guidance on how this approach can be used for the determination of final crack widths in a variety of common design situations, including example calculations for early-age cracks, restrained shrinkage cracks and load-induced cracks.

Z7/07 TESTING

The various test methods available are discussed by reference to when they might be used, i.e.:

- Mix Acceptance Tests
- Tests For Quality Assurance
- Tests Where Placed Concrete is Suspect
- Tests For Condition Monitoring

Often several test methods supply similar information. The limitations and advantages of these methods are reviewed, and recommendations provided on which test is the most suitable for project specifications.

CONCLUSIONS

The CIA durability series is an attempt to provide recommendations influencing durability into a consistent, conservative and comprehensive set of documents.

Concrete Repair, Rehabilitation and Retrofitting IV – Dehn et al. (Eds)
© 2016 Taylor & Francis Group, London, ISBN 978-1-138-02843-2

On the relationship between the formation factor and diffusion coefficients of Portland cement mortars

Z. Bajja, W. Dridi, B. Larbi & P. Lebescop
CEA, DEN, DPC, SECR, Laboratoire d'Etude du Comportement des Bétons et des Argiles, Gif-sur-Yvette, France

ABSTRACT

Despite the significant efforts deployed to improve the design, production and placement of concrete mixtures, the premature decay of infrastructures remains one of the main challenges facing the construction industry at the beginning of this 21st century [1]. Therefore, the durability of concrete and cement-based materials in general continues to receive significant international attention from both scientists and engineers.

Since these materials are porous, their durability is basically determined by their ability to resist the penetration and the transport of aggressive agents. The diffusion coefficients of ions and radio nuclides in cementitious materials are the most important parameters to evaluate the state of degradation of structures.

In this work, two different tracers (an ion, and a radionuclide) were tested on the same formulations of mortars (w / c = 0, 4 and sand volume fractions from 0 to 60%) by the trough-out diffusion, in order to determine the effective diffusion coefficients of each tracer and each formulation. The obtained results have proven the validity of the equation relating the formation Factor (F) to the effective diffusivity (D_e^{ion}, D_e^{rad}) of different ions and radionuclide in the cementitious material (D_0^i is the diffusion coefficient of the specie 'i' in an infinitely diluted solution -usually in pure water at 25 ° C):

$$F = \frac{D_0^{rad}}{D_0^{rad}} = \frac{D_0^{ion}}{D_0^{ion}}$$

This result is extremely interesting because once the geometric formation factor of a material is known; it will be possible to determine the values of the effective diffusion coefficients of any other diffusing species in this material.

In particular, in the framework of the storage of nuclear waste, (HTO) the liquid form of tritium, is considered as an ideal tracer for the characterization of the effective diffusivity in cementitious materials, because of its negligible interaction with the cement matrix [2] and the facility to apply Fick's law to a molecule such the HTO. However, these tests are expensive and very dangerous, thus, thanks to this formula; it would be possible to determine tritium diffusion coefficients from achieving diffusion test of an ion such as lithium Li^+.

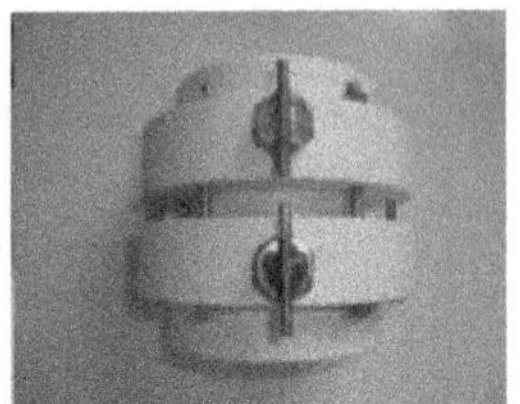

Figure 1. Picture of a diffusion cell used.

REFERENCES

[1] Young J.F., Bridging concrete into the 21st century Proceedings of the ECI Conference on Advances in Cement and Concrete, August 21–24, Copper Mountain, CO (2003), pp. 1–8.
[2] C. Richet, Ph.D. Thesis; University Paris XI Orsay (1992) (in French).

Concrete Repair, Rehabilitation and Retrofitting IV – Dehn et al. (Eds)
 ISBN 978-1-138-02843-2

Resistivity and water absorption of concrete

C.I. Goodier & C. Xueting
School of Civil and Building Engineering, Loughborough University, Loughborough, UK

C. Christodoulou & D. Dunne
AECOM Europe, Birmingham, UK

R. Yea
Kier Construction, UK

ABSTRACT

Concrete is a naturally porous material, and the size and distribution of pores depends upon the constituent materials, quality of compaction, materials used in the mix design, w/c ratio, degree of hydration, and curing (Christodoulou et al, 2013; Concrete Society, 2008).

Corrosion is a significant cause of the deterioration of reinforced concrete structures. The main cause of corrosion is the ingress of aggressive chemicals such as chloride ions from salts. However, there is a lack of knowledge and understanding regarding the relationship between compressive strength, resistivity and water absorption of different concrete types—all of which are critical parameters influencing the chloride ingress rate and development of corrosion in reinforced concrete. Concrete cubes and cylinders with varying proportions of water-cement (w/c) ratios, PFA (10–40% replacement), GGBS (20–70% replacement) and Silica Fume (SF) (5–15%) contents were cast, and tested for compressive strength, hardened density, bulk and surface resistivity, and water absorption. The results (Table 1) showed that increasing the PFA, GGBS or SF replacement contents significantly increased both the surface and bulk resistivity of the concrete (e.g. with the 70% GGBS replacement, up to 9 times greater when compared to the control concrete). The addition of SF or GGBS had a considerable positive effect on the water absorption (even at low dosages), lowering it by up to a factor of 10. The PFA however, had little, or even an adverse, effect on the water absorption.

Table 2 shows, where surface resistivity of the PC concretes are <12 kΩ·cm, they are all classed as 'high' chloride ion penetrability. The surface resistivity of the 10–40 PFA concrete is 11–22 kΩ cm, hence classed as 'moderate'. The GGBS, penetrability is classed as 'low-very low', whilst lastly, the SF concretes possessed an even greater resistance to the penetrability of chloride ions.

Table 1. Summary results for compressive strength, resistivity and water absorption.

Mix	Comp. strength (MPa) 7d	Comp. strength (MPa) 28d	Surf. res. kΩcm	Bulk res. kΩcm	Abs. kg/m^2/hr$^{0.5}$
W/C 0.5	36.9	45.4	9.58	31.60	38.84
W/C 0.55	32.0	39.5	9.41	29.23	47.50
W/C 0.6	25.8	34.1	9.36	30.73	59.13
W/C 0.65	20.0	25.5	7.90	29.57	62.38
PFA 10	22.3	31.4	10.69	36.50	85.57
PFA 20	19.8	25.9	13.76	43.67	57.01
PFA 30	13.8	22.8	17.49	56.23	54.70
PFA 40	10.3	20.4	21.88	68.6	85.57
GGBS 40	21.3	35.0	37.37	68.57	10.11
GGBS 50	21.6	38.9	46.52	155.9	6.56
GGBS 60	19.7	29.5	56.90	197.3	12.16
GGBS 70	18.9	31.4	76.63	269.6	9.24
SF 5	26.0	41.8	21.82	71.63	7.32
SF 10	25.6	44.5	40.30	144.2	5.35
SF 15	26.5	47.9	67.90	229.1	4.23

Table 2. Chloride ion penetrability based (AASHTO, 2011).

Chloride ion penetration	Resistivity (kΩ·cm)
High	<12
Moderate	12–21
Low	21–37
Very low	37–254
Negligible	>254

REFERENCES

Concrete Society 2008. Technical Report 31, Permeability testing of site concrete. Surrey, UK.

Christodoulou, C., Goodier, C., Austin, S.A., Webb, J., Glass, G.K. 2013, Long-term performance of surface impregnation of reinforced concrete structures with silane, Construction and Building Materials, Vol. 48, pp.708–716, http://dx.doi.org/10.1016/j.conbuildmat.2013.07.038, https://dspace.lboro.ac.uk/2134/13023.

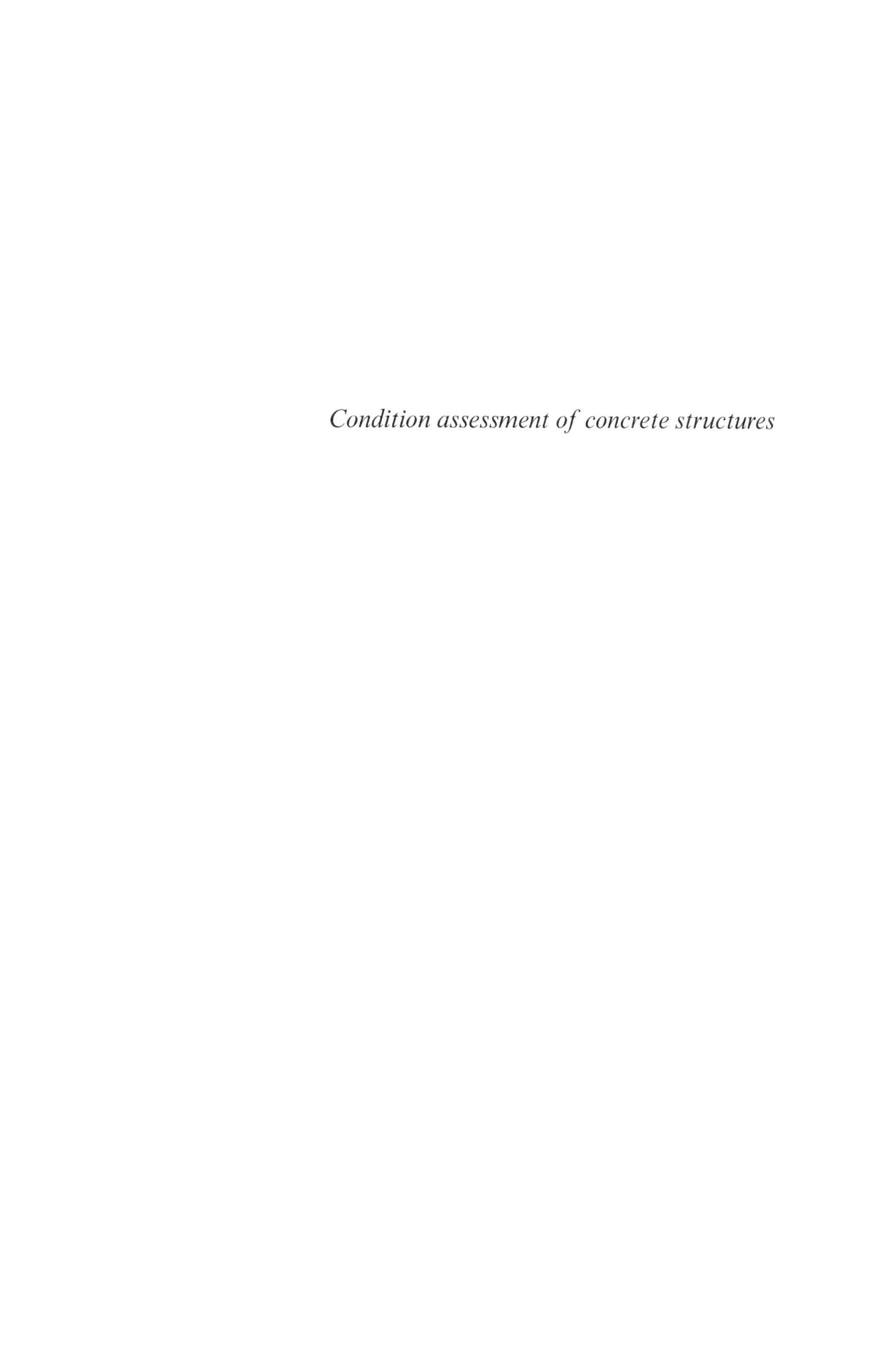

Condition assessment of concrete structures

Concrete Repair, Rehabilitation and Retrofitting IV – Dehn et al. (Eds)
© 2016 Taylor & Francis Group, London, ISBN 978-1-138-02843-2

An investigative study into the application of Non-Destructive Testing techniques for integrity assessment of RC piles

E. Okwori, P. Moyo & K. Matongo
Department of Civil Engineering, University of Cape Town, South Africa

ABSTRACT

Reinforced Concrete (RC) pile foundations may generally be rendered ineffective in performing their design function of transferring loads from a superstructure to bedrock, due to inconsistency in the pile material or discontinuity in the piles physical dimensions. The Integrity of a pile is defined by the consistency of its material and continuity of its physical dimensions. The presence of defects in the piles can severely reduce the structural load carrying capacity of the pile. The integrity of a pile may, therefore, be compromised. These defects occur during the construction process and life cycle of the pile.

Construction, quality assurance, troubleshooting problems and integrity testing of piles have always been problematic because of their underground placement, allowing only limited access and observation of piles. Existing methods of pile integrity assessments are based largely on the sampling of concrete to perform standard tests on properties of concrete. This approach does not provide information regarding the in-situ properties of concrete piles. Non-Destructive Testing (NDT) methods offer the advantage of providing information regarding the in-situ integrity of reinforced concrete piles. The low strain integrity method, also known as the sonic echo method, is popular in applications of NDT for integrity evaluation of reinforced concrete piles. However, research towards improving the effective application and feasibility of the technique is still ongoing.

This study investigates the effectiveness of the sonic echo method in defect detection of reinforced concrete piles, pattern characterization of defects in piles, and an identification of the influences of defect size and defect location. This study attempts to establish that the ability to identify defects using the sonic echo method will improve the effectiveness of the method in pile integrity evaluations. This is performed experimentally through the construction of laboratory mock-up piles specimens, consisting of one non-defective pile and eight piles containing of defects associated with changes in cross-section. The experiment consisted of three necking piles where the necking defect occupied 63%, 42% and 23%, respectively. Two bulbing piles with bulbing defects increasing the cross-section by 12% and 50%. Three honeycombing defect piles had defects occupied 63%, 42% and 23%, respectively. Figure 1 below shows a schematic of the sonic echo system for integrity assessment of RC piles.

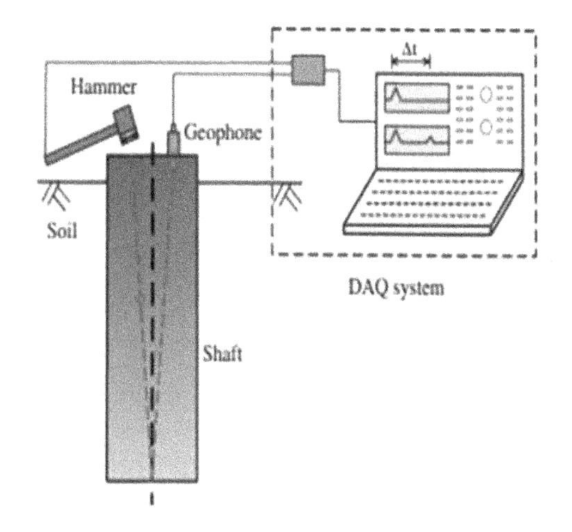

Figure 1. Schematic of sonic echo system.

Findings reveal that the defects investigated can be identified in terms of their appearance in the wave signals from testing with the sonic echo method. The findings also show that the larger the magnitude of a defect the higher the chance of detection. Furthermore, necking defects appeared to be the most easily detectable defect, followed by bulbing, and honeycombing. Defect size showed a more significant influence on defect detectability compared to defect location.

Generally, all defect locations and pile lengths investigated were identified with reasonable accuracy.

Concrete Repair, Rehabilitation and Retrofitting IV – Dehn et al. (Eds)
© 2016 Taylor & Francis Group, London, ISBN 978-1-138-02843-2

Studies on the key technical problems of asphalt concrete facing slabs in the upper reservoir of Huhhot Pumped-Storage Plant in North China

Shifa Xia, Yihui Lu, Zhengxing Wang & Fucheng Zhang
China Institute of Water Resources and Hydropower Research, Beijing, China
Beijing IWHR-KHL Co. Ltd., Beijing, China

ABSTRACT

Asphalt concrete facing slabs have been widely used as anti-seepage structures worldwide in Pumped-Storage Plants (PSP). In China a great many PSPs are going to be built in the next 5–10 years according to the national PSP developing plan to meet the increasing need for power and energy. Huhhot PSP which was built in 2013 in Inner Mongolia in North China where the weather is very cold, the anti-fracture property of the asphalt concrete in low-temperature is the critical issue to be solved to use asphalt concrete as seepage-proof facing slabs. In this paper, firstly the developing process of construction technology of asphalt concrete facing slab in China as well as its application in home and abroad are briefly introduced. Then three main issues in paving asphalt concrete facing slabs in Huhhot PSP, namely, the anti-fracture design for asphalt concrete slab in low temperature, the compaction technology of impermeable layer with modified bitumen and the construction technology in low temperature are addressed.

In China about 44.1% and 14.7% of the PSPs built and being built use reinforced concrete facing slab and asphalt concrete facing slab as impervious body, respectively. Though it seems that the reinforced concrete facing slab is preferred in China, asphalt concrete facing slab has a very broad application prospect in cold region, because of its excellent anti-seepage performance, excellent low-temperature crack resistance, deformability adapt to bad foundation and water level change, faster construction speed and similar unit price compared with reinforced concrete facing slab.

Modified bitumen and mixing ratio design is the key factor of the low-temperature crack resistance of asphalt concrete. The modified bitumen applied in Hohhot PSP upper reservoir represents the highest level of material properties in low-temperature crack resistance of bitumen in China. Compared with unmodified bitumen, properties of aged bitumen showed reverse rule and needs further study. The mixing design technologies of asphalt concrete have been mature domestic, yet the properties of modified asphalt should be further studied.

In cold regions, modified asphalt is usually adopted at impermeable layer of facing slab. However, its high viscosity and the high content could result in difficulties in compaction during the construction process. Based on the experiences of the Huhehot project, it is difficult to control the porosity of joints between strips of impermeable layer. To meet the design requirement, perfect construction technology and elaborate treatment are necessary. The construction technologies of rolling right after the asphalt paver should be further studied to improve the quality of the joints.

In cold regions, the construction period is short if asphalt concrete is paved strictly according to the requirements in related specifications. Therefore, construction technology of leveling layer under low temperature was explored in the Huhehot project. Such measures as low-temperature reformation of mixer, thermal insulation with quilts and quick rolling to ensure the initial rolling temperature of asphalt concrete were adopted during the construction process, which ensured the quality of mixtures and asphalt concrete after rolling compaction. The test results of the Huhehot project indicate that the qualification rate of onstie paving and rolling quality is higher than 95%, which proves the practicability of above measures. However, the construction technologies of impermeable layers under low temperature should be further studied.

Concrete Repair, Rehabilitation and Retrofitting IV – Dehn et al. (Eds)
© 2016 Taylor & Francis Group, London, ISBN 978-1-138-02843-2

Evaluation of sulfate damages in a tunnel concrete segments

F. Moodi & A.A. Ramezanianpour
Concrete Technology and Durability Research Center, Amirkabir University of Technology, Tehran, Iran

Q. Bagheri Chenar & M. Zaker Esteghamati
Department of Civil and Environmental Engineering, Amirkabir University of Technology, Tehran, Iran

ABSTRACT

Interaction of sulfate ions with cement paste matrix causes sulfate attack which threatens integrity and durability of concrete by expansion of cement matrix and progressive loss of strength. This paper discusses the devastating effects of sulfate attack in a tunnel which is located in the South west of Iran. During tunnel excavation, a vast volume of groundwater and H_2S gas entered into the tunnel that leaded to the failure in the concrete segments. At severely damaged parts of the tunnel, surface layers of concrete segment were liquefied and accumulated (Figure 1).

In order to assess the damages in concrete wall segments of the tunnel, 69 core specimens were extracted and tested. Compressive strength, electrical resistance and Water absorption capacity tests have been performed on 64 specimens and 85 sulfate ion tests have been implemented.

Compressive Strength test result (Figure 2) shows that 56.3% of specimens had lower compressive strength than the expected value. While all the specimens from concrete segments stored in the site had compressive strength above 35 MPa. Therefore, it can be deducted that the lack of sufficient strength is mainly due to an external factor like sulfate attack.

The water absorption test measures concrete porosity and permeability by indicating its ability to absorb water (and possibly hazardous water-soluble). In current study, Water absorption test result shows that 92% of specimens had a water absorption value over 4% (Figure 3). Persian Gulf concrete durability standard recommends water absorption value under 2%, which concludes a high potential for concrete permeability.

Iranian concrete standard (ABA, 2000) states that water-soluble sulfate content (in the term of So_3) should not exceed than 4% of cement weight and total existing sulfate content should not exceed than 5% of cement weight. Therefore, based on the documentations of the tunnel concrete mixture, it is deducted that soluble sulfate content and total sulfate content should be less that 0.72% and 0.9% respectively. Result shows that, except for 3 specimens, soluble sulfate content in tunnel inner sections specimens is less than 72% in all depths (Figure 4). Also, soluble sulfate content deviation from the value prescribed by the standard in that 3 specimens is less than 0.1%. However, the majority of specimens had total sulfate content more than 0.9% (Figure 5). The difference between two values is mainly due to calcium sulfate, which is the main product of sulfuric acid attack. Therefore, it can be concluded that the

Figure 1. Liquefaction of the concrete segments surface layers.

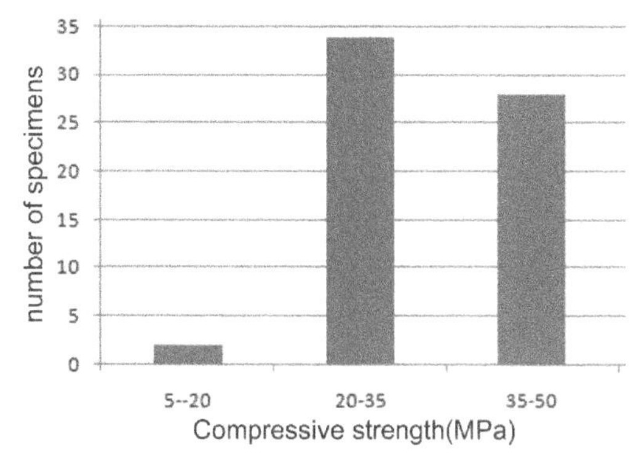

Figure 2. Compressive strength values of core specimens.

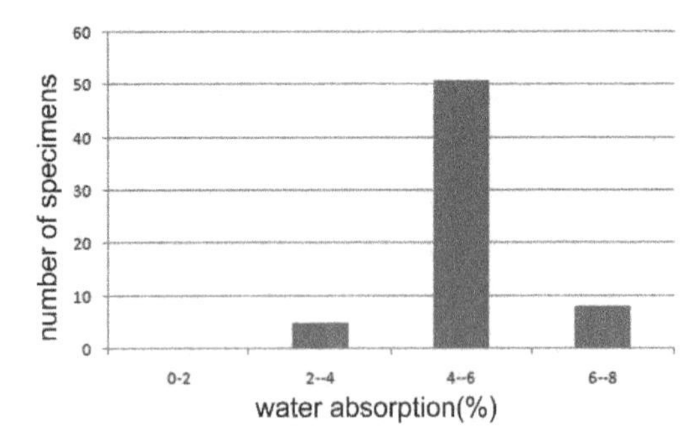

Figure 3. Water absorption values of core specimens.

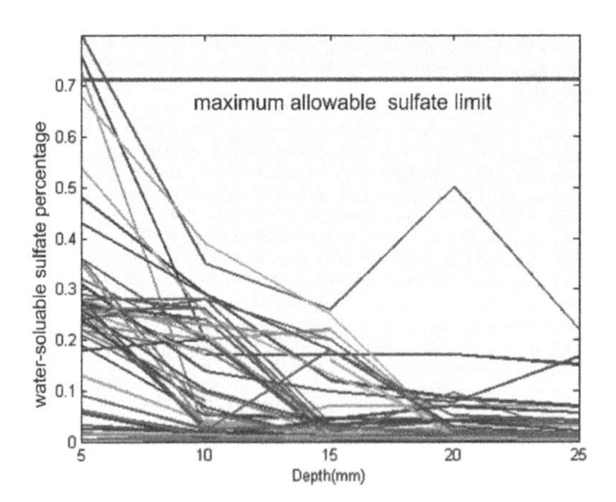

Figure 4. Water-soluble sulfate content of core specimens from inner sections of the tunnel.

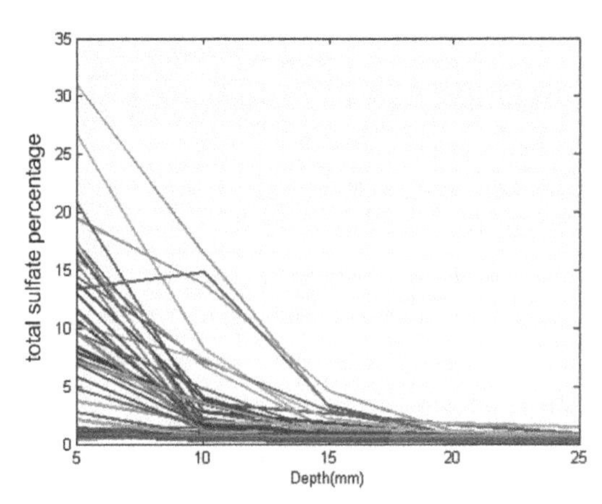

Figure 5. Total sulfate content of core specimens from inner sections of the tunnel.

sulfuric acid attack which is produced by bacterias from H_2S could be the cause of tunnel segment damages. A comparison of total sulfate content between specimens from tunnel segments and segments stored in the site, supports this hypothesis. Stored segments have much less total sulfate content.

Also, because of sulfate ions, lime aggregates used in the tunnel concrete segments, low temperature in some parts of the tunnel and humid environment, it is possible that TSA has happened. The tunnel which is discussed in this paper is located below ground water level which helps seepage. Seepage provides the necessary water to carry chemical substances and accelerates chemical damage mechanism. Field observation of concrete Liquefaction supports TSA occurrence.

Concrete Repair, Rehabilitation and Retrofitting IV – Dehn et al. (Eds)
 ISBN 978-1-138-02843-2

In-situ concrete strength assessment based on Ultrasonic Pulse Velocity (UPV), rebound, cores and the SonReb method

F. Papworth
BCRC, Perth, Australia

David Corbett
Proceq, Switzerland

Reuben Barnes
PCTE, Perth, Australia

ABSTRACT

This paper gives a brief outline of the key aspects of strength assessment. CIA Z11, the Australian guide to assessment of insitu concrete strength, was based on CSTR 11. CSTR 11 has been replaced by BS EN 13791 and its commentary BS 6089. The Concrete Society issued Advice Note 43 in 2013 which summarises some of the approaches in the new European standards. In this paper some of the differences between European Standards and the methods used in Australia based on CSTR 11 are given.

BS EN 13791 now provides clear guidance on the use of ultrasonic pulse velocity and rebound hammer testing. These methods enable rapid scanning of the concrete to detect variations in strength and BS EN 13791 provides guidance on how their use can be incorporated into reducing the number of cores required.

EN 13791 requires at least 15 cores to be taken to establish the insitu concrete strength. This may be reduced to 9 when used in combination with NDT tests such as rebound hammer or ultrasonic pulse velocity.

In the German national annex to EN 13791 there is also the possibility to assign a compressive strength class based on rebound hammer testing alone, as in many cases, it is not allowed to take cores.

In 2007 an electronic version of the original rebound hammer was introduced. This instrument measures rebound value (Q) as the quotient between the velocity of the hammer mass just before and after impacting the contact rod. The manufacturer provides a strength relationship between Q values and compressive strength based on a lower 10th percentile curve.

The direct method is the most reliable method of measuring ultrasonic pulse velocity. The pulse transit time is measured across the entire element and gives an average velocity for a large thickness of concrete. Direct UPV can also be measured on cores to give a direct correlation between UPV and core strength. However in cases where access to only one face is possible a direct velocity measurement is not possible. Yaman developed a method of measuring the indirect velocity over 4 head spacings (200, 250, 300 and 350 mm). The slope of a straight line plot of time vs head spacing gives an indirect velocity that is very close to the direct velocity for homogeneous specimens.

One issue with combined use of NDT and cores for strength assessment is the number of core correlations required according to EN BS 13791. By reducing uncertainty by combining two NDT measurements the number of correlation points decreases. Breysse describes the SonReb method of combined UPV and Rebound measurement. The preferred approach is to base assessment on compressive strength = aV^bQ^c where a,b and c are constants, V is UPV and Q is rebound number. Although there have been a number of other propositions on combined UPV and rebound over the last 30 years.

An outline of the strength assessment of seven projects is given in the paper to show the variety of approaches that may be appropriate:

1. Structural assessment of new concrete structures at a mine in Ghana was called for when, shortly after construction, failure of the 25 MPa primary crusher approach slab led to a preliminary investigation of the strength of concrete. A detailed investigation identified the potential issues of using uncalibrated rebound hammer results and use of rebound hammer on trowelled surfaces. Core strengths enabled a reasonable assessment of insitu strength. A structural assessment identified that for the 10 year design life the estimated 15 MPa actual cylinder strength was adequate except for one structure and after further testing this was strengthened.
2. On a mine plant in Botswana detailed NDT was requested but on checking the original cube results

and limited coring and rebound results a consistent and coherent picture of actual insitu strength of 25 MPa concrete was established. High water demand of the mix used indicated that the cause of low strength was consistent. The project moved directly to structural assessment based on the original cube results. This showed the value in detailed assessment of existing results before launching into a further test programme. The structural assessment showed that the estimated 15 MPa cylinder strength was generally adequate but on two structures strengthening was required. One due to inadequate development length and one due to inadequate moment capacity.

3. The concrete strength on an Australian mine plant was called into question when the supervisor recorded that after taking cylinders for strength assessment the concrete water was added to the mix for placement. Cores and NDT measurements were used to establish the actual characteristic strength of 18 MPa for most elements, well below the minimum specified strength of 32 MPa. This serves as an important reminder about ensuring no water is added outside of that called for in the mix design and that an independent supervisor will checks all parts of the concrete supply chain to verify specification compliance. Structural assessment identified structural reliability was reduced to less than would have been expected in code based designs and risk assessment showed that the high consequence of failure meant the structures should be replaced.
4. Cylinder results had not been provided for parts of the 'top down' slab construction of an Australian shopping centre floor slab. Only the top surface of a slab could be tested. Widespread testing using indirect ultrasonic pulse velocity to show potential variations in strength was undertaken. Cores were taken for calibration. The calibrated UPV results showed that the strengths across the slab achieved the required 32 MPa. Rebound hammer results taken over a wide area showed two distinct areas where the 10 percentile strengths were 25 and 38 MPa. These variations were not seen in the UPV results. The rebound variations were probably due to variations in finishing and curing of the two pours and hence limited to surface effects, important where abrasion is a concern but not for the structural assessment of this slab.
5. Questionable cylinder results on S65 concrete in an Australian office tower led to a detailed investigation using the SonReb method. The testing gave a clear indication of the consistency of results and value of the assessment method and validated that there were no low strengths insitu. Based on limited cores it was possible to test concrete over a wide area and give certainty about strength.
6. Addition of floors as part of a refurbishment of a 50 year old structure led to the strength assessment of existing piles, columns and walls. The strength assessment was in two parts. Cores were taken from some walls, columns and slabs to confirm current strengths were at least equivalent to that designated on drawings. For the more critical columns an assessment of all columns was required. This was achieved by correlating direct UPV results with the core strengths and then using UPV results to give an indication of in-situ strengths. This verified that all columns had adequate strength. Piles were tested using a force vibration test to give structural capacity and geometry of the pile including confirmation that a bulb was present at the end of the pile.
7. Testing was undertaken on ground floor columns of a shopping centre in Queensland to assess the capacity for planned extensions. No coring was allowed in the structure and so a completely non-destructive NDT method was proposed. NDT results depend on the materials used and so cylinders made using local materials were tested to create a calibration between NDT (UPV and rebound) and cylinder strengths. The calibration co-efficient were assessed for an equation developed in Australia prior to the SonReb equation development. Based on the cylinder strength, UPV and rebound relationships ten columns were tested by NDT and shown to be of adequate strength.

CONCLUSIONS

All structures are different and the structural assessment requirements will depend on the circumstances of each. Testing required is unclear until historic information available is reviewed and possibly not fully understood until some preliminary tests are undertaken. An overall process to structural assessment might be:

a) Determine precisely what outcome is sought.
b) Review existing data to determine what is known about the concrete and what additional information is needed. Existing construction data may be adequate.
c) Prepare a test programme that will provide the information required. A preliminary test programme may identify if more detailed is required and if so what test methods should be used. On some structures it may be necessary to base the assessment on cores alone. On others rebound or UPV measurements, calibrated against cores or cylinders of a similar mix, may be particularly useful in providing a global assessment not practical with cores. On many structures SonReb will provide the most efficient assessment method.
d) From the strength assessment undertake a structural assessment to determine if the reliability is reduced below code requirements and if so undertake a risk assessment to determine if the risk is acceptable.

Concrete Repair, Rehabilitation and Retrofitting IV – Dehn et al. (Eds)
 ISBN 978-1-138-02843-2

Impact loads on concrete bridge caps—studying load distribution for recalculation of existing bridges in the ultimate limit state

M. Niederwald & M. Keuser
University of the German Armed Forces Munich, Germany

K. Goj
Building Authority, Bavarian Ministry of the Interior, Germany

S. Geuder
Central Office for Bridge and Tunnel Construction, Southern Bavarian Motorway Office, Germany

ABSTRACT

The superstructures of road bridges have to be designed for different impact situations according to DIN EN 1991-2 (2010). There are two accidental load cases, which have to be considered for vehicle leaving the road and either hitting the curb of the sidewalk or the crash barrier system, which is usually mounted onto the concrete bridge cap. Especially the load assumptions for the impact on the vehicle restraint systems have changed within the last two decades and the applied impact loads have been increased significantly. This is because of the increasing demand towards higher containment capacity according to the structural codes. Therefore, new crash barrier systems were developed and were brought to the market, which can transmit higher impact forces into the bridge superstructure. As a consequence in Germany the amount of connection reinforcement of ø12–40 between cap and superstructure required by the codes for the design of new bridges was nearly tripled to ø14–20 in the year 2009 (RiZ-ING 2009).

If existing bridges are considered to be retrofitted with vehicle restraint systems with a higher containment level, it is often not possible to perform the necessary verification successfully, that the existing connection reinforcement is able to safely carry the higher impact forces. The bending and shear capacity of the bridge cantilever is often unsufficient in that case. In order to find ways to minimize such costly remediation in the future a study has been initiated at the University of the German Armed Forces to investigate the potential of possible load distribution of impact forces in cap longitudinal direction. First results are presented in this paper.

According to DIN EN 1991-2 (2010) there are two different impact situations, which have to be investigated. The first one is the impact of a tire on the curb of the sidewalk and the second one the impact on the vehicle restraint system. The second one is divided into two different load assumptions for the structural design, like for example the design of the bridge cantilever, and the design of structural details such as the connection reinforcement between the concrete cap and the superstructure.

According to DIN EN 1991-2 (2010) there should be a horizontal load applied in transverserly direction in oder to cover the stresses due to vehicle impact on the curb of the sidewalk. Because of them usually being smaller than the ones for the impact on vehicle restraint system and because of the mechanism of load distribution being the same, in the following only the impact on vehicle restraint system is investigated in detail.

For the structural design there are equivalent horizontal loads provided which are split up into four classes A to D. The class has to be chosen in dependence to the vehicle restraint system used. Besides the horizontal force a vertical load also has to be applied, which results from an axle load of the load model LM1.

For the local design of the bridge cap and the connection reinforcement a load of 1.25 times the cross-sectional resistance (plastic shear force and bending moment) of the post has to be applied, engaging in the post axis at the top surface of the cap. Because there are no normative regulations to be found in DIN EN 1991-2 (2010) regarding the number of post forces which have to be applied, it is mandatory to analyze the load assumption first, before load distribution can be investigated.

In order to estimate the required number of post forces which have to be applied on the concrete cap, a practical solution was chosen by performing calculations based on the plastic hinge theory using a strut and tie model, which is intended to reflect one specific vehicle restraint system. The system, which is called "Super-Rail Eco Bw" was chosen according to Gütegemeinschaft Schutzplanken (2014). For further information regarding this approach see Niederwald et. al. (2014).

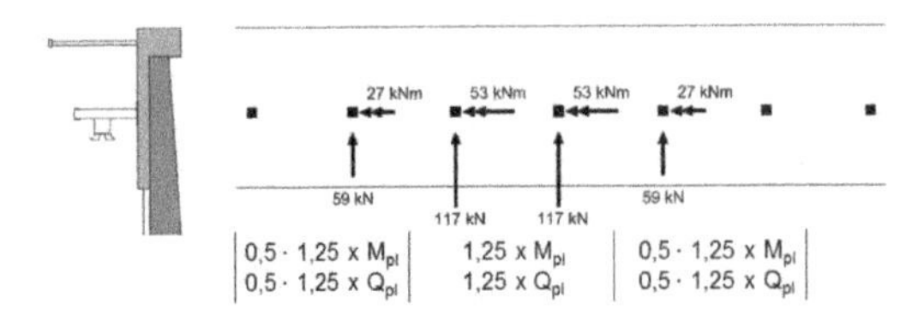

Figure 1. Proposed load arrangement for the vehicle restraint system "Super-Rail Eco Bw".

a)

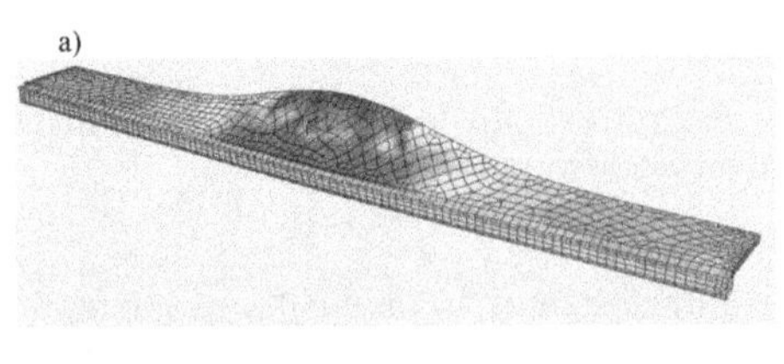

b)

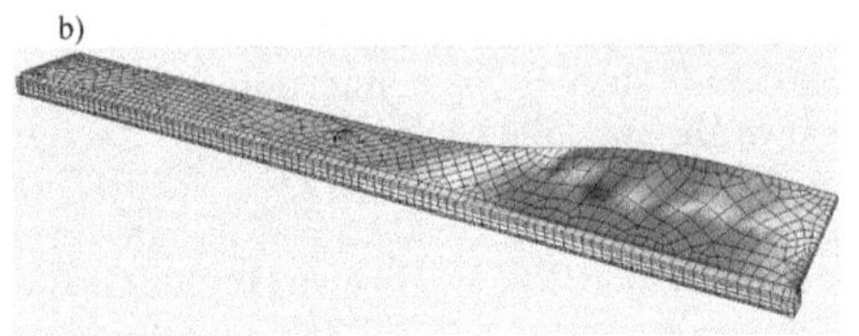

Figure 2. Deformation figures for a) impact in the middle of the cap and b) impact at the beginning of the cap.

The proposal so derived for a safe load arrangement for the investigated vehicle restraint system is shown in Figure 1.

In order to study the effect of the load distribution in longitudinal direction within the bridge cap finite element simulations are getting performed using a folded plate structure of the cap. The simulations were done for two different impact situations: impact in the middle of the cap and impact at the beginning of the cap. In Figure 2 the finite element system and also the deformation figures for both impact situations can be seen.

In order to study how the load distribution in the local component design is influenced by the geometry of the cap, a parameter study was performed, for which first results are presented in the following figures. In Figure 3 the tension forces acting in the connection reinforcement between the cap and the bridge cantilever are plotted in reference to the length of the cap for different heights of the cap of 15, 20 and 25 cm to consider the increasing in-plane stiffness of the cap.

In Figure 4 the maximum tension forces acting in the connection reinforcement are plotted in reference to different cap widths ranging from 2.0 m to 5.0 m for the two impact situations and two different heights of the cap of 15 cm and 25 cm.

a) Impact in the middle of the cap

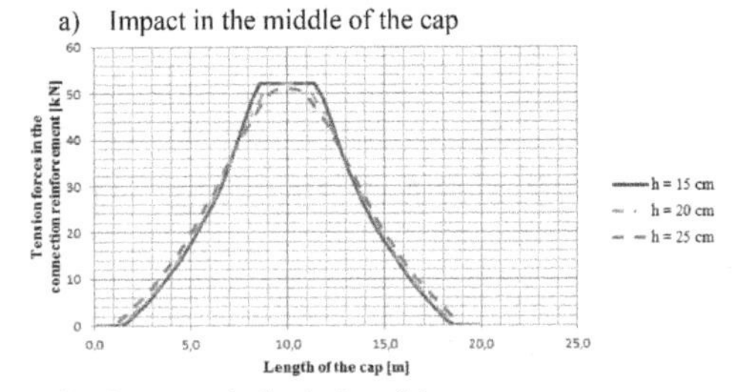

b) Impact at the beginning of the cap

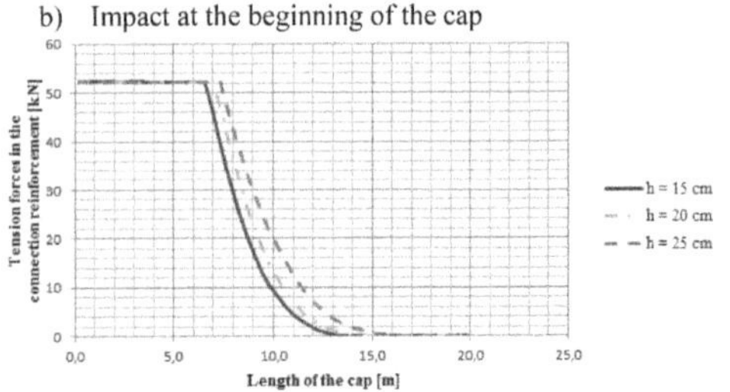

Figure 3. Tension forces in the connection reinforcement for a) impact in the middle of the cap and b) impact at the beginning of the cap.

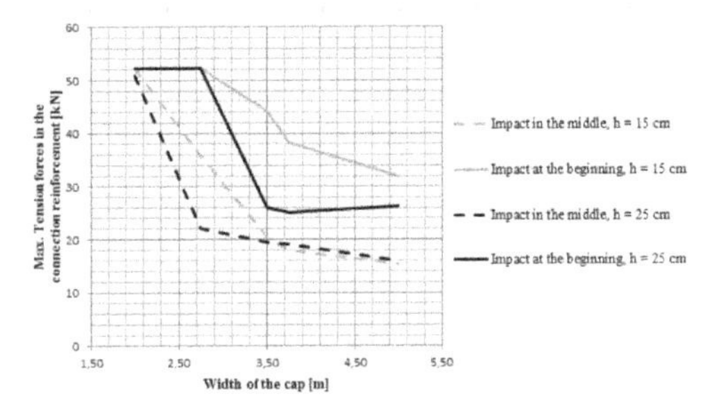

Figure 4. Maximum tension forces in the connection reinforcement for different cap widths.

REFERENCES

DIN EN 1991-2. 2010. Einwirkungen auf Tragwerke—Teil 2: Verkehrslasten auf Brücken; Deutsche Fassung EN 1991-2:2010. DIN, Deutsches Institut für Normung e.V.

Gütegemeinschaft Schutzplanken. 2014. Gütegemeinschaft Stahlschutzplanken e.V.: Stahlschutzplanken-Info 2/2014, Siegen.

Niederwald, M. & Keuser, M. & Goj, K. & Geuder, S. 2014. Investigation of the load bearing behaviour of existing bridge caps subjected to accidental loads due to impact on vehicle restraint systems. In Fischer, O. & Keuser, M. & Mangerig, I. & Mensinger, M. & Siebert, G. & Yamaguchi, T. (eds.), Proceedings of the 10th Japanese German Bridge Symposium, Munich, Germany.

RiZ-ING. 2009. Richtzeichnungen für Ingenieurbauten. Bun-desministerium für Verkehr, Bau und Stadtentwicklung, Verkehrsblatt-Verlag Borgmann GmbH & Co KG, Dort-mund.

Concrete Repair, Rehabilitation and Retrofitting IV – Dehn et al. (Eds)
© 2016 Taylor & Francis Group, London, ISBN 978-1-138-02843-2

Laser Induced Breakdown Spectroscopy (LIBS)—innovative method for on-site measurements on chloride contaminated building materials

S. Millar, T. Eichler, G. Wilsch & C. Gottlieb
BAM, Federal Institute for Material Research and Testing, Berlin, Germany

ABSTRACT

Concrete buildings are designed in consideration of a specific expected lifetime. Usually for reinforced concrete structures in civil engineering as e.g. multi-storey car parks a minimum life time of approximately 50 years is strived for. Environmental factors like weather, location or general exposition can force aging- and damage-processes due to deterioration of concrete and/or reinforcement. The ingress of harmful species like CO_2 or Cl^- causes corrosion of the embedded reinforcement in the concrete structure. In this case especially older constructions with insufficiently concrete cover are not able to attain the assessed life time. Laser Induced Breakdown Spectroscopy (LIBS) can be applied as a fast and reliable method in order to identify harmful species and their accompanying damage processes.

The LIBS method is a combination of material ablation, plasma formation and analysis of the emitted radiation by spectroscopic methods. To generate plasma a high energy pulsed NdYAG-laser is focused on a surface. The radiation of the plasma is analyzed by a spectrograph and a scientific CCD-camera. As a matter of principle all elements are detectable and with the use of reference samples a quantitative analysis is possible.

LIBS may be used on gases, liquids or solids. There are a lot of applications in different fields like environmental analysis, pharmaceutical investigation, biomedical investigation, forensic investigation or industrial applications like process control, recycling, sorting and quality control during manufacturing (Hahn & Omenetto 2012, Noll 2012).

At BAM LIBS has been successfully applied for the investigation of distributions and transportations of different ions in building materials (Wilsch et al. 2005 & 2011).

The challenge of separating aggregates and cement-matrix in studying element distributions of concrete based materials, which is e.g. wet chemical analyzes unfit to do, solves LIBS at BAM with a scanning system. Taking the heterogeneity of concrete into account is necessary and elementary, because the main transport-processes occur in the cement-matrix via pores. In damaged concrete the transport processes can be supported by cracks. With a spatial resolution of 50 μm and the aforementioned scanning technique LIBS is able to image the transport of harmful ions by cracks (figure 1).

To give the engineer a tool for on-site inspections of concrete buildings, a mobile LIBS-system is in development at BAM. This mobile setup is intended to help engineers making fast statements about the actual condition, the necessity of maintenance arrangements and the durability of structures. The set-up of the mobile system is shown in figure 2. The laser is located in the measuring head, which is fixed on a scanner system. The laser head is moved by the scanner to take the heterogeneity into account. The maximum area, which currently can be scanned, is 17,0 cm × 15,0 cm. With the help of compressed air the scanner can be fixed on vertical surfaces or even overhead. The head is

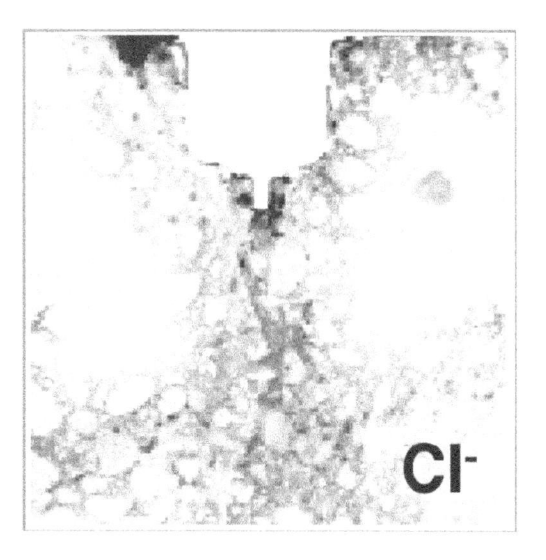

Figure 1. Chloride-ingress at a cracked concrete specimen, crack width = 350 μm; sample-size = 140 mm × 140 mm; applied with a NaCl-solution; darker grey represents a higher chlorine content.

Figure 2. Mobile scanner with fixed laser head (left) to scan surfaces an take the heterogeneity of concrete into account, connected with the control unit (right) via cable.

connected with the control unit via a 5,00 m cable. The laser energy is 3 mJ per pulse and can run with a frequency up to 100 Hz. The plasma radiation is delivered to an optical fibre and analysed by the detection unit. To maintain laser safety regulations, a cover in connection with a brush shields the laser beam path.

The performance of LIBS is demonstrated by investigation of a number of concrete cores drilled out of a chloride contaminated parking deck. A second example illustrates the detection of the carbonation depth. Both results are compared with the standard methods in civil engineering (wet chemical analysis and phenolphthalein-test). Due to the possibility to detect several elements at the same time allows the engineer to make statements about different damage processes or their probability with one measurement. Especially for the assessment of corrosion endangered reinforced concrete buildings LIBS may is an advantage with detecting the chloride ingress and the carbonation front simultaneously.

LIBS as a promising technique is currently on the step from pure laboratory applications to an on-site analyse technology. In cooperation with system developers and companies a mobile LIBS-system is in development at BAM. The mobile system is able to give quantitative results, comparable to the lab-system. First measurements on a bridge, located at the North Sea, are shown.

The setup of rules and standards is the next step to establish LIBS as a standard procedure for chemical investigations of building materials.

Concrete Repair, Rehabilitation and Retrofitting IV – Dehn et al. (Eds)
© 2016 Taylor & Francis Group, London, ISBN 978-1-138-02843-2

Enhancing the interpretation of Torrent air permeability method results

R. Neves
Barreiro Technology School, Polytechnic Institute of Setúbal, Barreiro, Portugal

ABSTRACT

The Torrent method is a non-destructive method applicable on site, which is suitable to assess near-surface concrete air permeability. Nevertheless, due to the usual scatter of gas permeability, a careful interpretation of results shall be ensued. The aim of this study was to investigate the influence of using different instruments on the air permeability coefficients.

Two samples of four concrete mixes were considered, totaling 120 specimens. The air permeability tests were carried out using two commercially available instruments: the Torrent Permeability Tester (TPT) and the Permea-Torr (PT). Each sample (sized 15) was divided: 7 specimens were tested with TPT and 8 specimens were tested with PT. Each specimen was tested by only one instrument.

An observation of the test data shown that, although higher pressure variation occurred with TPT instrument, the final rate of pressure variation was similar with both instruments and that for absolute pressures less than 35 mbar, the rate of pressure variation with the square root of time with the TPT instrument was not constant (Figure 1), which is in opposition to the theoretical background of the test.

Air permeability coefficients were calculated, considering as initial testing pressure the lowest testing pressure and 35 mbar. The symbols for these coefficients are kT and kT_{35}. These air permeability coefficients are presented in Figure 2.

The median was used as the representative value of a sample from an air permeability population and nonparametric statistical test were carried out.

It was found that calculating the air permeability coefficient considering an initial pressure of 35 mbar, does not affect the air permeability coefficients when PT is applied. Moreover it makes the air permeability coefficients obtained testing with TPT equivalent to those obtained testing with PT.

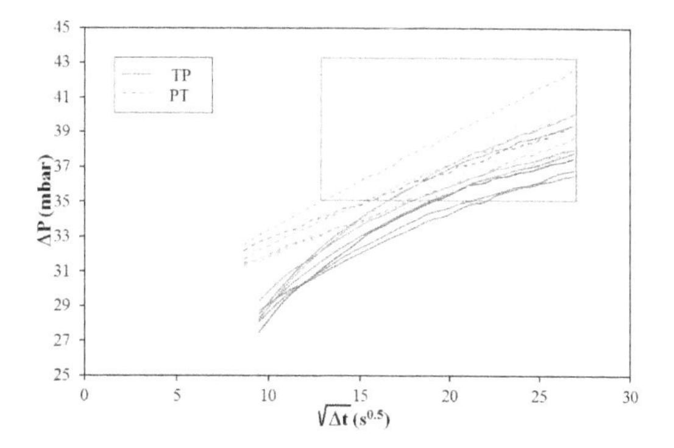

Figure 1. Example of pressure variation with time in a sample.

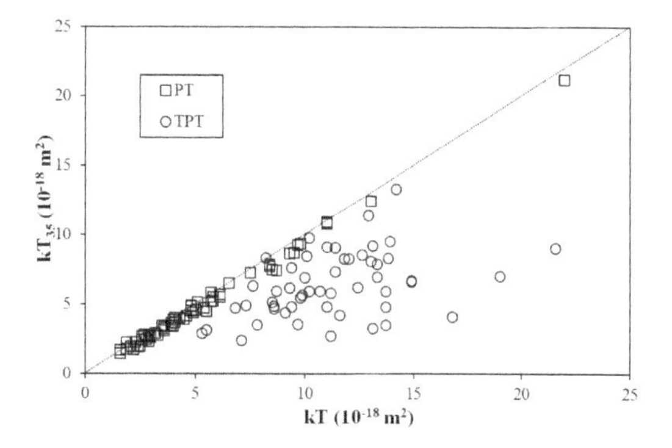

Figure 2. kT values versus kT_{35} values.

Thus, an adjustment of already calculated air permeability coefficients based on data obtained with TPT instrument, considering an initial pressure of 35 mbar, is advised, as it corresponds to an improvement of such air permeability assessments.

Concrete Repair, Rehabilitation and Retrofitting IV – Dehn et al. (Eds)
© 2016 Taylor & Francis Group, London, ISBN 978-1-138-02843-2

Combined Non Destructive Testing for concrete compressive strength prediction

G. Concu, B. De Nicolo, N. Trulli & M. Valdés
Department of Civil Engineering, Environmental and Architecture, University of Cagliari, Italy

ABSTRACT

Non Destructive Testing (NDT) techniques are widely used in civil engineering for evaluating materials and structural mechanical parameters such as in situ concrete compressive strength. A combination of two or more NDT techniques is generally recommended, in order to reduce errors dependent on materials, aggregate dimension and size, environmental parameters, so that the accuracy of in situ concrete compressive strength estimation is improved.

In this study the reliability of the combination of Ultrasonic Testing (UT) and Rebound Hammer Test (RHT) has been investigated. An experimental campaign has been carried out on 90 concrete cubic specimens having three different design concrete strength (C8/10, C25/30 and C45/55 according to EN 206) in order to determine robust correlations between compressive strength (f_c), rebound hammer Index (I_r) and ultrasonic pulse Velocity (V) and to evaluate the reliability of the proposed correlations for estimating in situ concrete compressive strength.

UT measurements have been carried out on the specimens at 7, 14 and 28 curing days in order to take the influence of the curing process into consideration. Then, each specimen has been placed in a press in order to avoid movements caused by the impact of the hammer and for each face of the specimen the rebound index I_r has been measured. After performing both UT and RHT, compressive tests have been carried out on the specimens. It has been noted that V increases during the curing process and both V and I_r increase with f_c. Regression analyses V– f_c and I_r – f_c have been performed and it has been noted that both V and I_r well correlate with f_c (Figures 1 and 2). Then, the results of UT and RHT have been combined by applying the SonReb method, which implements a relationship between f_c, V and I_r. A double power model has been chosen and Equation 1 has been achieved. The coefficients of Equation 1 minimize the sum

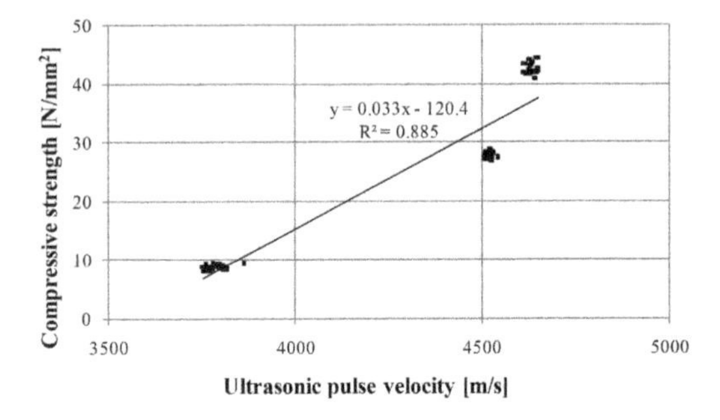

Figure 1. Correlation between V and f_c.

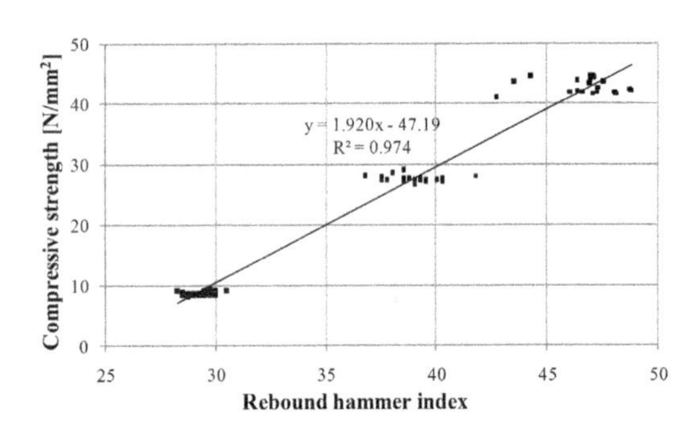

Figure 2. Correlation between I_r and f_c.

of squared differences between theoretical and experimental concrete compressive strength f_c. A coefficient $R^2 = 0.994$ has been achieved.

$$f_c = 1.10 \cdot 10^{-16} \cdot V^{4.05} \cdot I_r^{1.628} \tag{1}$$

Results show that a slight increase in R^2 coefficient can be achieved when the SonReb method is applied.

Concrete Repair, Rehabilitation and Retrofitting IV – Dehn et al. (Eds)
© 2016 Taylor & Francis Group, London, ISBN 978-1-138-02843-2

Estimation of concrete strength and stiffness by means of Ultrasonic Testing

G. Concu, B. De Nicolo, N. Trulli & M. Valdés
Department of Civil Engineering, Environmental and Architecture, University of Cagliari, Italy

ABSTRACT

The need to assess the structural condition of existing concrete structures strongly supports the development of Non Destructive Testing (NDT) techniques, in order to evaluate the degree of decay and to estimate mechanical parameters such as strength and stiffness (Concu, 2014, Puccinotti, 2015). Among NDT, Ultrasonic Testing (UT) is frequently applied, because the velocity of ultrasonic signals travelling through the material is directly related to mechanical and physical parameters, e.g. dynamic elastic modulus, Poisson's ratio and density.

In this study an experimental campaign has been carried out on 26 concrete cylindrical specimens having two different size: (i) small cores having a diameter of 65 mm and a length of 130 mm; (ii) big cores having a diameter of 100 mm and a length of 260 mm. Specimens have been cored from cubic specimens having a side length of 300 mm which have been casted by using different design concrete strength.

Ultrasonic velocities of both longitudinal (V_l) and shear waves (V_s) propagating through the specimens have been measured in order to evaluate the dynamic modulus of elasticity (E_d). The static modulus of elasticity (E_s) and the compressive strength (f_{ck}) have been experimentally determined and the theoretical modulus of elasticity (E_t) has been evaluated too.

It has been noted that as the compressive strength increases, an increase of the ultrasonic parameters (V_l, V_s, E_d) and of both the theoretical and the experimental modulus of elasticity (E_s and E_t) is achieved.

The regression analysis between non destructive parameters and mechanical parameters pointed out that the correlations $V_l - f_{ck}$ and $E_d - f_{ck}$ fit better for smaller cores than for bigger cores.

An overestimation of E_t compared to E_s can be observed for bigger cores, while smaller cores show an opposite behavior (Figure 1). Similarly, E_d is overestimated compared to E_s for bigger cores, while it is underestimated for smaller cores (Figure 2).

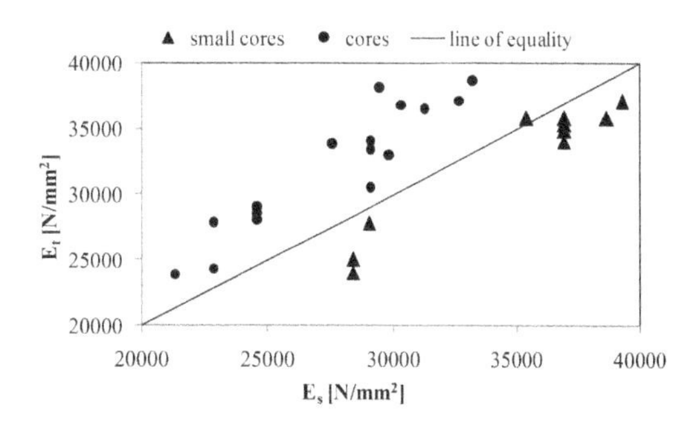

Figure 1. E_s vs E_t for each specimen.

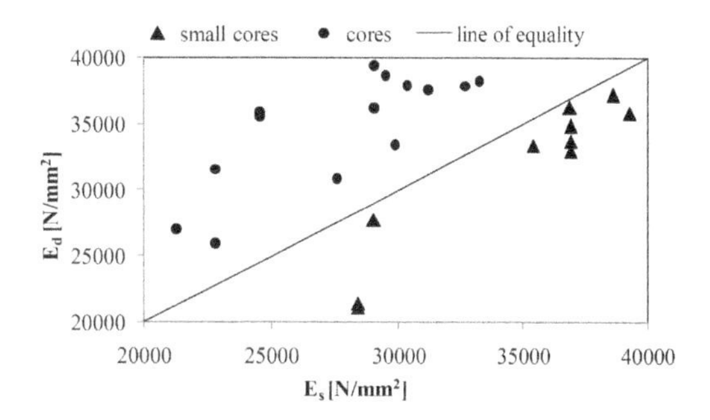

Figure 2. E_d vs E_s for each specimen.

REFERENCES

Concu, G., De Nicolo, B., Pani, L., Trulli, N., Valdés, M. 2014. Prediction of Concrete Compressive Strength by Means of Combined non-destructive Testing. *Advanced Materials Research* 894: 77–81.

Pucinotti, R. 2015. Reinforced concrete structure: Non destructive in situ strength assessment of concrete. *Construction and Building Materials* 75: 331–341.

Concrete Repair, Rehabilitation and Retrofitting IV – Dehn et al. (Eds)
© 2016 Taylor & Francis Group, London, ISBN 978-1-138-02843-2

A repair quality control with elastic waves based methods vs. concrete substrate quality

A. Garbacz & T. Piotrowski
Warsaw University of Technology, Warsaw, Poland

L. Courard
University of Liege, Liege, Belgium

B. Bissonnette
CRIB—Department of Civil Engineering, Laval University, Quebec (QC), Canada

INTRODUCTION

An adhesion in repair systems is one of the most important factors that affect the durability of repair. The adhesion depends on many phenomena taking place at interface zone. According to the standards and guidelines, e.g. European Standard EN 1504-10 and ACI Concrete Repair Manual, both bond strength and bond quality should be evaluated. The pull-off test is recommended for assessment of a bond strength. The use of pull-off test, due to its semi-destructive character, is restricted by owners and managers. Therefore, the elaboration of reliable non-destructive method for an adhesion mapping is one of the most important tasks. A majority of NDT methods mentioned in EN 1504-10 and ACI Concrete Repair Manual for assessment of concrete structures are based on propagation of stress waves. Particularly ultrasonic methods (UPV), Impact Echo (IE) and Impulse-Response (IR) methods are recommended for evaluation of repair quality.

To select the appropriate NDT method for repair quality control, the following factors should be taken into account: type and size of defects at the interface zone to be investigated, thickness of overlay, type of repair material, quality of concrete substrate. First two factors depend mainly on NDT method used. The next is not so important as it has been shown that in the case of many commercial polymer-cement (PCC) and polymer (PC) repair mortars their acoustic impedances are similar to the concrete substrate.

The aim of this paper is to analyze of the effect of a quality of concrete substrate on propagation of stress waves in repair system and their influence on a possibility of the bond strength estimation.

REPAIR SYSTEM AS AN OBJECT OF NDT ASSESSMENT

Repair system is difficult to test with NDT methods because of many factors influencing stress wave propagation. Two main types of defects can occur in this system:

- adhesion type: various types of disboands (e.g., voids, delaminations, weak adhesion areas due to presence of dust, oil, etc.);
- cohesion type (in repair material or/and concrete substrate).

Above defects are often result of the operations that have to be performed prior to repair as well as an application of repair material. It has been widely demonstrated that a surface preparation of concrete substrate can influence significantly on the microcracking level, surface roughness, the substrate saturation level and, as a consequence, it may affect the bond strength between repair material and concrete substrate.

Using stress waves based methods for evaluation of bond strength needs to find whether the interface quality affects the stress wave propagation and is it possible to extract from the signal any information related to the bond strength.

TESTED REPAIR SYSTEMS

In the framework of several project conducted at Warsaw University of Technology in cooperation with the University of Liege various repair systems different in concrete surface and interface quality, were tested. In the first stage, a commercial PCC repair mortar (thickness 10 mm) containing glass microfibers was applied on relatively weak concrete substrates (C20/25) subjected, prior to repair, to various surface preparations. As result concrete

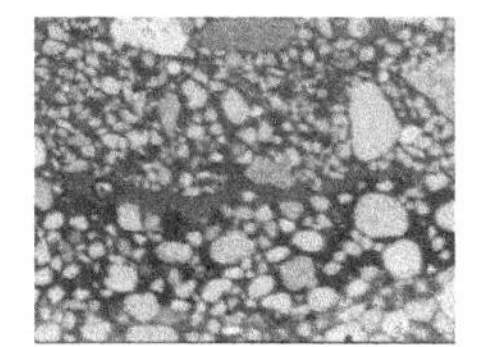

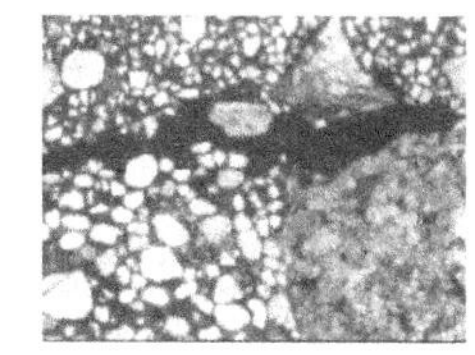

Figure 1. View of the interface between repair material and concrete substrate after milling without (left) and with (right) bond coat.

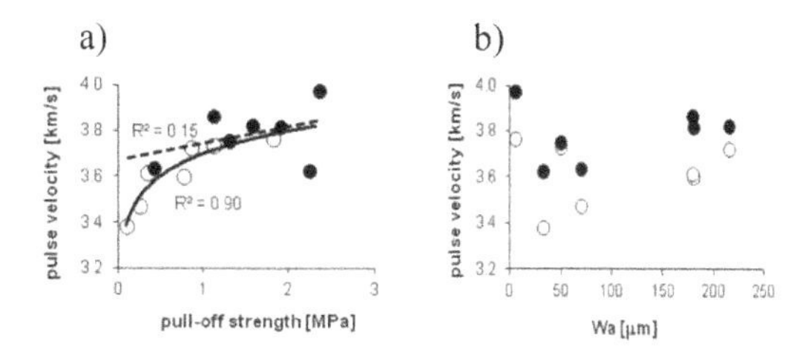

Figure 2. Relationships between pulse velocity IE and a) pull-off strength, b) mean waviness of profile, Wa; overlays with (●) and without (○) the bond coat.

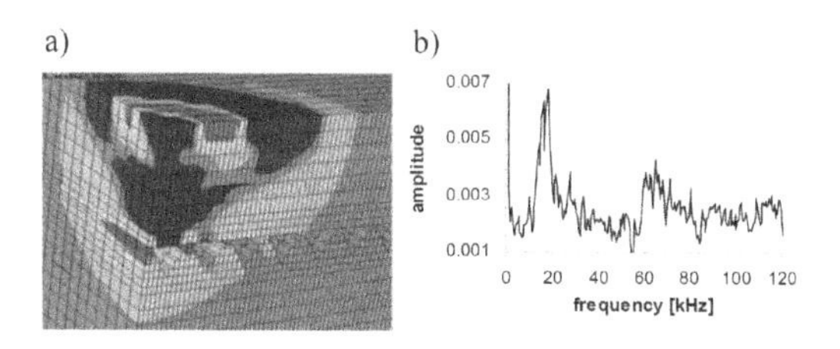

Figure 3. FEM simulations: a) examples of disturbance in wave propagation in the case of air voids presence at interface, b) frequency spectrum for substrate irregularities unfilled—presence of air voids at the interface.

substrates with different roughness and microcracking levels were obtained. The overlay was applied on the concrete substrate with and without a bond coat to obtain different interface (Fig. 1).

NDT MEASUREMENTS

After 28 days of hardening, the IE and ultrasonic measurements were carried out. The test results indicate that there is no correlation between the pull-off strength and the P wave velocity for the repair system with bond coat (Fig. 2a). The bond coat filled irregularities of concrete substrate properly and air voids at the interface were not observed. The statistically significant relationship was obtained for systems without the bond coat—the P wave velocity increased as the pull-off strength increases. In this case, the fraction of air voids at the interface increases when the roughness increases. In both types of systems the pulse velocity was not correlated with the substrate roughness (Fig. 2b).

The above relationship was investigated for stronger concrete, C40/50. Four types of surface preparation techniques were used: polishing, sandblasting, scabbling and very high pressure waterjetting. The concrete slabs have been covered by a self-compacting commercial PCC mortars (3-cm thick). For the repair systems, two specific ranges of the IE frequency spectrums were analyzed: around the bottom peak frequency and around frequencies corresponding to the interface. The relationships between amplitudes of either bottom or interface peaks and parameters describing quality of repair systems were not statistically significant for any of the tested repair systems.

The effect of substrate roughness and presence of air voids at the interface on stress wave propagation were also investigated using Finite Element Model (FEM) of repair system. The results of simulations indicate that the presence of relatively big air voids at the interface can significantly influence the stress wave propagation (Fig. 3).

SUMMARY

The multi-variants investigations showed that for both IE and ultrasonic methods, the roughness and microcracking of the concrete substrate does not affect significantly the P wave propagation through the repair system if the bond quality is sufficient—absence of large voids at the interface. However, a few authors confirmed that parameters describing roughness and microcracking of concrete substrate can be considered as important for improvement of reliability of the bond strength evaluation using stress wave based NDT methods.

Concrete Repair, Rehabilitation and Retrofitting IV – Dehn et al. (Eds)
 ISBN 978-1-138-02843-2

Condition assessment of a 100-year-old RC building in Hiroshima

T. Kinose & K. Imamoto
Department of Architecture, Graduate School of Engineering, Tokyo University of Science, Japan

T. Noguchi
Department of Architecture, Graduate School of Engineering, University of Tokyo, Japan

T. Ohkubo
Department of Social and Environmental Engineering, Graduate School of Engineering, Hiroshima University, Japan

ABSTRACT

Former Army Clothing Depot constructed in 1913 locates within about 2.7 km of radius from the blast of atomic bomb, and there are lots of trace of iron doors and windows at the blast center side. Walls of interior and exterior are built with masonry and the main structures such as column, beams and slabs are built with reinforced concretes. This building is one of the oldest reinforced concrete structures in Japan. Due to the above reasons, this building has high academic and historical values. Authors it was surveyed this building for the purpose of accumulation of the date which contribute preservation and repair as this building.

INTRODUCTION

It is well known all over the world that Hiroshima is the only country where American army dropped an atomic bomb in 1945. About 80 bombed buildings are left including Atomic Bomb Dome within 5 km of radius from the blast center. Former Army Clothing Depot constructed in 1913 is left within about 2.7 km of radius from the blast center, and there are lots of trace of iron doors and windows at the blast center side. Walls of the interior and exterior are built with masonry and the main structures such as columns, beams and slabs are built with reinforced concretes. This building is one of the oldest reinforced concrete structures in Japan. Due to the above reasons, this building has high academic and historical values. Authors surveyed this building for the purpose of accumulation of the date which contribute preservation and repair as this building. This paper deals with a condition assessment of this oldest RC building from the view point of durability.

EXPERIMENTAL

Outline of buildings

The building site includes 4 types of building. In this study, building No. 2 and No. 3 were surveyed.

Experiments

Following measurements were applied on the surveying.

Crack investigation
Clack width of concrete was measured with the crack scale. The clack widths were measured at 3rd floor of the building.

Re-bar corrosion
This survey aims at preservation and needs for repair of the building. Re-bar corrosion in concrete was surveyed by visual observation. Grades of corrosion rates were classified based on the standards proposed by guideline of Architectural Institute of Japan (AIJ).

Differential settlement
Differential settlement was measured with a laser level.

Carbonation depth and compressive strength
Carbonation depth was measured with concrete core samples according to Japanese Industrial Standards (JIS) A1152. Carbonation velocity is based on the square root theory shown in Equation (1). Compressive strength test was performed with core specimens after carbonation test according to JIS A1107.

$$C = A\sqrt{t} \quad (1)$$

C: Carbonation depth (mm)
A: Carbonation velocity (mm/years$^{1/2}$)
t: Age at survey (years)

RESULTS AND CONFERENCES

Crack of concrete

Cracks were remarkable at a column on the 3rd floor of building No. 3. Slices-like cracks with constant pattern were observed at all over the. The average width of concrete cracks was about 2.2 mm. This value was much larger than crack width induced by concrete shrinkage.

Differential settlement

Measured lines of differential settlement with laser level in each floor in building No. 2 are shown in figure 1. The results on 1992 and 2014 are shown in figures 2 to 4. The maximum differential settlement at center part of floor's ranged from 10 to 15 cm. The tendency of X axial direction was close to that of Y axial. The amount of the settlement increased by 5 to 10 cm since 1992. In this building, there were independent footing at central part of the building and connected footing by underground beams locates outer zone of the building as shown in figure 1. This might be a reason that the amount of settlement of the inner columns was larger than that of exterior walls. Hence, tensile stress occur at the column. In addition, it would be one of the causes of crack that diameter of column on the 3rd floor which diameter was smaller than those on the 1st and 2nd floors.

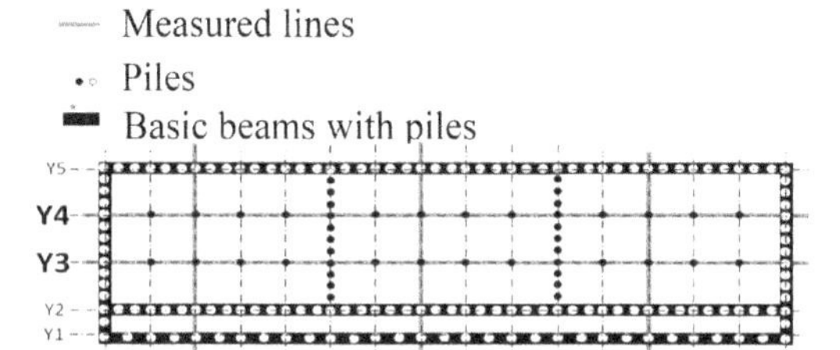

Figure 1. Measured lines of differential settlement and location of the piles and the basic beams.

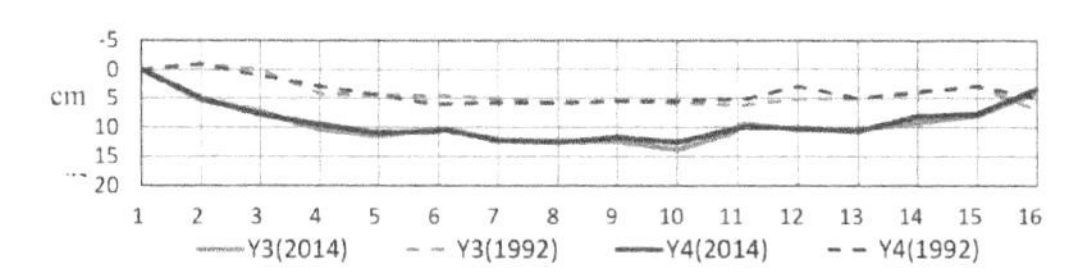

Figure 2. Amount of the settlement (No. 2, 3F).

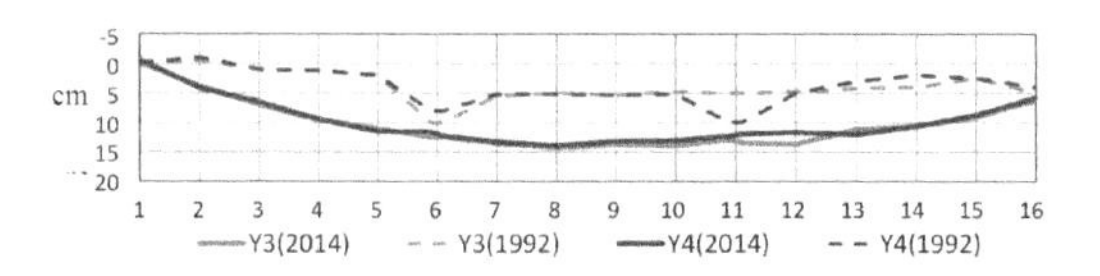

Figure 3. Amount of the settlement (No. 2, 2F).

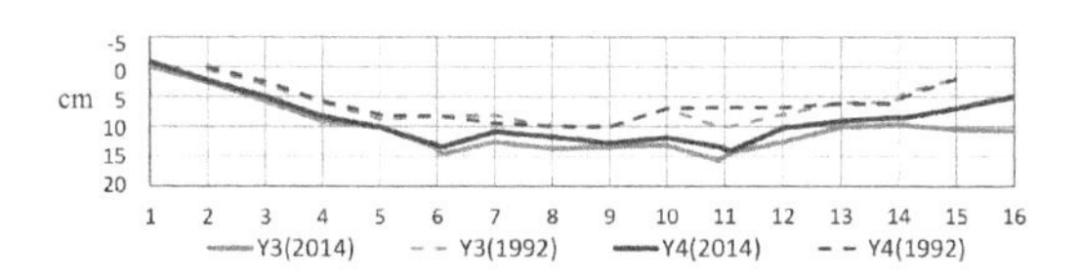

Figure 4. Amount of the settlement (No. 2, 1F).

Compressive strength and carbonation

Although the tested points are different in 2014 and in 1992, overall carbonation proceeded with time. Average compressive strength of concrete was 14.4 N/mm^2.

Re-bar corrosion

Re-bar corrosion rater were evaluated with exposed bars due to loss of cover concrete. Re-bar corrosion grade based on AIJ guideline. States of re-bar corrosion were classified into 5 grades. A re-bar corrosion at a roof of 1st floor of building No. 2 quite significant due to leak of water from the crack of roof concrete. Figures 5 to 6 show relationship between carbonation depth and cover thickness of concretes of existing structures which authors previously surveyed. The tendency that the re-bar corrosion progresses remarkably at the part where cover thickness minus carbonation depth was small. On the other hands, at the point where the cover depth of concrete was large enough, re-bar corrosion cannot be observed. Corrosions of columns were slight (grade 3). This might be due to relatively moderate R.H. at indoor. According to the above results, frames of this building would be sound and counter measurements for the differential settlement and water leakage from roof should be necessary to conserve this building.

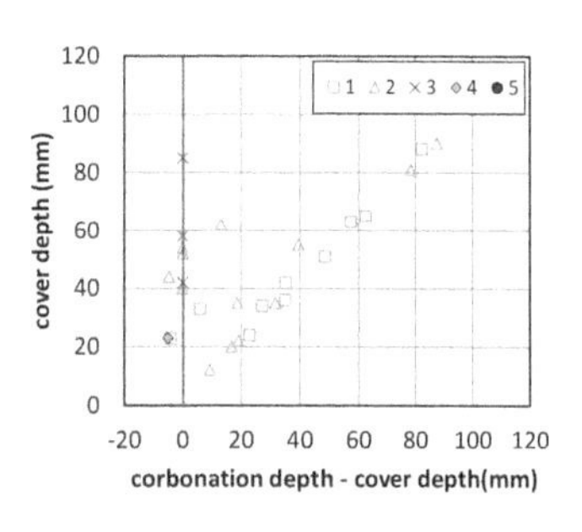

Figure 5. Relationship between a loss of carbonation and cover depth of A-D (indoor condition).

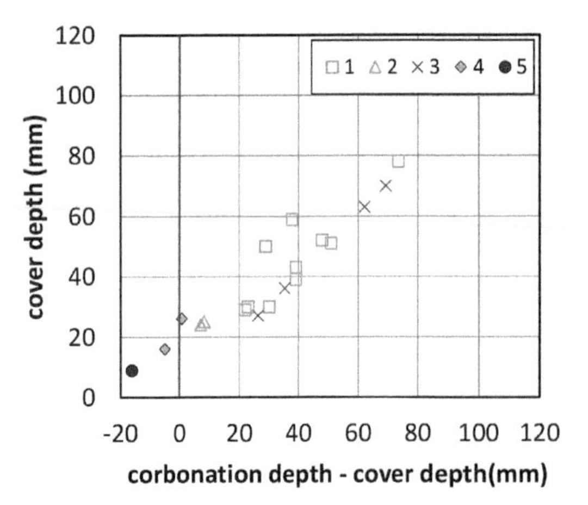

Figure 6. Relationship between a loss of carbonation and cover depth of A-D (outdoor condition).

CONCLUSIONS

1. Cracks of column were remarkable at 3rd floor of building. Sliced cracks with constant pattern were observed at all over the column.
2. The maximum differential settlement at central part of each floor ranged from 10 to 15 cm. The amount of the settlement increased by 5 to 10 cm since 1992.
3. This differential settlement might cause tensile stress induced cracks of columns.
4. Re-bar corrosion was quite significant due to water leakage. On the other hands, at the point where the cover depth of concrete was large enough, re-bar corrosion cannot be observed.
5. Frames of this building would be sound and counter measurements for differential settlement and water leakage from roof should be necessary.

Concrete Repair, Rehabilitation and Retrofitting IV – Dehn et al. (Eds)
© 2016 Taylor & Francis Group, London, ISBN 978-1-138-02843-2

A technical review of seven cathodic protection systems in Jersey

John Drewett
Concrete Repairs Ltd., London, UK

Kevin Davies
CorroCiv Limited, Manchester, UK

Kevin Armstrong
States of Jersey, UK

INTRODUCTION

The reinforced concrete infrastructure managed by the States of Jersey suffers from steel reinforcement corrosion initiated by chloride ions from the marine environment.

There have been seven impressed current cathodic protection systems installed to enhance durability. The structures are; three multi storey car parks, three marine berths and one airport landing light support facility. A range of anode systems have been used:

a) Discrete conductive ceramic anodes
b) Discrete MMO coated Titanium (MMO/Ti) anodes
c) Conductive cementitious overlay
d) MMO/Ti mesh with cementitious overlay
e) MMO/Ti ribbon mesh
f) Conductive paint

The installed ICCP systems are monitored using LD15 or LD25 silver/silver chloride reference electrodes and computer control systems.

THE PROJECTS

London Berth, Victoria Pier

This was the first ICCP system to be installed on the Island in 1998 after a two year trial using a range of anode systems. Fifteen anode zones were installed to protect the slab soffits, beams and columns using a combination of two anode types

Slab soffits, beams and high level pile caps—1520 m^2 proprietary CP60 conductive mortar spray applied.

Piles, columns and low level pile caps—950 m^2 MMO/Ti mesh with a sprayed overlay.

The system is monitored using 40 No. LD25 reference electrodes and powered by a multi-channel output, remote controlled, remote-operated and monitored, DC power supply and data acquisition system.

Maintenance
In 2011 the computer controlled power and control system was replaced, as it had become obsolete and the ICCP system was reconfigured from 15 zones to 8 zones to simplify operation. No works have been necessary to the original installed anode or monitoring system after 16 years in the tidal marine environment.

Green Street Multi Storey Car Park

In 1999 an ICCP system was installed to protect the reinforced concrete decks, columns, walls and internal faces of the parapets.

The anode system for the deck was the Zebra Deck Anode System supplied by Protector AS complete with a wearing surface. This system is a conductive paint system applied in strip coats on the deck surface and overlaid with the wearing surface. The columns and walls were protected using a conductive paint system again supplied by Protector AS with a decorative topcoat. In total 10,500 m^2 of deck surface and 1,600 m^2 of walls and columns were protected.

Maintenance
The cathodic protection system has received little maintenance since installation. One datalogger has been replaced after 14 years of service. The top deck wearing course was replaced 3 years after the system was installed after the original surface failed to sufficiently accommodate thermal movement in this location.

New North Quay

The system was installed in 2000 to protect the deck soffit and sub-structure. The original design was for 34 anode zones using 1,000 m MMO/Ti mesh ribbon anodes buried in chases cut into the concrete.

The system was monitored using 16 No. LD15 reference electrodes, a remote controlled central control unit and three networked substations.

Maintenance
In 2011 the obsolete computer controlled power and control system was replaced and the zoning reconfigured to 8 zones for operational reasons. The existing field wiring, AC power and telephone line were re-used.

Jersey Airport

The reinforced concrete Instrument Landing System (ILS 27) Localiser Plinth is located in a very harsh marine environment.

An ICCP system was installed in 2003 to provide long term protection to the steel reinforcement in the exposed parts of the piers, beams and deck slab soffits. The anode system comprises MMO/Ti ribbons and proprietary MMO/Ti tubular discrete rod anodes placed in horizontal rows at regular spacing. There are two anode zones, each monitored by 4 LD15 reference electrodes.

The ICCP system has been functioning correctly since the installation was completed.

Maintenance
In 2011some minor maintenance was undertaken to re-embed some of the discrete anodes and repair some localised acidification around some of the MMO/Ti ribbon.

Berth no. 2, Albert Pier

An ICCP system was installed in 2005 to protect the reinforced concrete soffit, beams and columns. The selected anode system comprises 300 m of MMO/Ti ribbons chased into horizontal rows at regular spacing on the deck slab soffits and transverse beams. The longitudinal beams and columns are protected using 550 No. conductive ceramic and MMO/Ti discrete anodes. There are 4 anode zones rated at 3.0 A/15 VDC with 2 LD15 reference electrodes monitoring each zone.

The power and control unit is fully automated but with a manual option for on-site interrogation if required.

Maintenance
By 2011 some of the discrete anodes had failed and so were replaced with MMO/Ti tubular anodes in acid-resistant CP15 grout.

Pier Road Multi Storey Car Park (MSCP)

Specific exposed areas of the car park deck slab and beams were protected using ICCP systems to enhance the durability.

The anodes are all MMO/Ti durAnode 3 discrete anodes, 12,060, in 18 zones with 108 No. LD15 reference electrodes used to monitor the system.

The power and control unit supplied by CPI has one central monitoring station and 3 sub stations.

The system was installed in 2005 and is functioning correctly in accordance with European Standard EN12696:2012.

Maintenance
No maintenance has been undertaken since installation.

Sand Street Multi Storey Car Park

In 2006 an ICCP system was installed to provide long term corrosion protection to the steel reinforcement of the longitudinal floor beams on the upper floors and the seaward facing edge columns on all floors. Previous concrete repairs had failed to achieve long term corrosion protection.

The anodes were durAnode Type 3 proprietary discrete anodes installed into the floor ribs and columns. A total of 12,080 anodes were installed in 5 floors decks and 864 anodes in 6 levels of columns. The system is divided into 18 zones and monitored using 126 LD15 reference electrodes.

The power and control unit has one central station and 5 substations.

The ICCP system has performed well with significant negative shifts in the steel as expected, indicative of adequate corrosion protection. The output currents are quite low and the system is operating on constant voltage output mode set at 4.0 VDC.

Maintenance
No maintenance has been undertaken since installation except one off-the-shelf low voltage power supply unit was replaced in one substation

SUMMARY

The ICCP systems have performed very well in harsh marine environments. The earlier control and monitoring units have become obsolete and were replaced but this is only to be expected with the rapid development of micro-processor based operating systems.

The variety of anode systems used have, in the majority of cases, worked well. A few anodes have needed to be replaced but there have been no whole system failures and there is every indication with the low voltage and currents required to achieve corrosion protection to the steel that these anodes will continue to operate for many years with minimal maintenance.

The reference electrodes have had some failures (<3%) but on the whole have performed well over the last 17 years.

The initial investment in ICCP systems, targeted at specific corrosion problem areas has reduced the long term costs of managing these structures and has extended the effective service life and residual value. With regular monitoring and maintenance the systems should continue to provide corrosion protection for many more years.

Concrete Repair, Rehabilitation and Retrofitting IV – Dehn et al. (Eds)

Corrosion survey of the bridge deck "Viadotto Colle Isarco"/"Autobahnbrücke Gossensaß" on the Motorway called "Autostrada del Brennero" in North-Italy

Roberto Giorgini
CorrPRE Engineering BV, Reeuwijk, The Netherlands

ABSTRACT

The scope of this corrosion survey was to detect and quantify zones with high corrosion probability of the normal reinforcement and if possible the prestressed tendons of a total of 8000 m^2 surface area of the upper bridge deck of the "Viadotto Colle Isarco". This viaduct is part of the A22 Motorway called "Autostrada del Brennero" and lays 10 km's south of the Austrian-Italian border in Italian territory approximately 1200 m high in the Alps. In case serious problems were found no recommendations were requested at this point of the survey due to the size and the structural complexity of the concrete structure.

Zones with high probability of corrosion were identified by the potentials measured and inspected through scarification and sandblasting. Based on this verification it was estimated an approximate 9% of the total surface area with high probability of corrosion in the prestressed tendons. This amount is based on adding up all measuring points with a potential being more negative than –450 mV divided by the total amount of measuring points.

Seeing very negative potentials at certain areas we feared that a particular situation could occur at these locations which is called macro corrosion cells, which could accelerate the corrosion rate of the prestressed tendons.

This was confirmed after scarification and sandblasting these areas. Clearly could be seen the highly corrosive tendons and very passive reinforcement bars laying on top of it. Several corrosion pits seen in the bars had a depth of 6 mm which indicates severe chloride attack

These "verification" zones where severe corrosion was found indicated no cracks in the tendons. This fundamentally indicates that the steel used is not susceptible for hydrogen embrittlement. The steel used is St 85/105 (32 mm tendons) and has a perlitic microstructure. Perlitic steel is known for its low susceptibility for hydrogen embrittlement.

During previous inspections fractured prestressed tendons were found. All the fractures were found in the threaded region of the coupler joints but no fractures were found in corroded tendons laying in the "active" zones away from the coupler joints. Samples taken from the fractured tendons were sent to the lab for metallurgical analysis and confirmed no susceptibility for hydrogen embrittlement of the steel used and also confirmed the perlitic microstructure of the tendons.

Fractures found in the tendons in previous inspections are most probably caused by major stress fatigue cycles from traffic loads in combination with corrosion due to the stress-sensitive couplings. Exactly similar cases (*Heerdter crossing in Düsseldorf in 1976*) were found in old reports, e.g. published in the technical report 26 "Influence of material and processing on Stress Corrosion Cracking of prestressing steel – case studies" by the International Federation for Structural Concrete (fib) in Lausanne.

Concrete Repair, Rehabilitation and Retrofitting IV – Dehn et al. (Eds)
© 2016 Taylor & Francis Group, London, ISBN 978-1-138-02843-2

Mechanical performance of deep beams damaged by corrosion in a chloride environment

Linwen Yu
Université de Toulouse, UPS, INSA, France
LMDC, Toulouse, France
Department of Civil Engineering, CRIB, Université de Sherbrooke, Québec, Canada

Raoul François
Université de Toulouse, UPS, INSA, France
LMDC, Toulouse, France

Richard Gagné
Department of Civil Engineering, CRIB, Université de Sherbrooke, Québec, Canada

Vu Hiep Dang
Faculty of Civil Engineering, Hanoi Architectural University, Hanoi, Vietnam

Valérie L'Hostis
CEA Saclay, CEA, DEN, DPC, SECR, Gif-sur-Yvette, France

ABSTRACT

Corrosion is the upmost threat for the durability of reinforced concrete structures exposed in a marine environment. Structural performance and serviceability of reinforced concrete structures is affected by corrosion from several aspects, cross-sectional loss of reinforcements, disbonding between concrete and reinforcements caused by the expansion of corrosion products, and brittle performance of corroded reinforcements. Some published research (Wang et al., 2011, 2012; Higgins et al., 2012) is focused on the shear performance of corroded reinforced concrete beams damaged only by corrosion on stirrups or by partial length corrosion on tensile reinforcements. However, normally not only tensile reinforcements or stirrups are damaged in the corrosion damaged RC structural elements in-service. And all the investigations mentioned above accelerated corrosion with impressed current, which lead to different corrosion distribution comparing with natural corrosion(Yuan et al., 2007).

In the current research, four short beams corroded with an artificial climate accelerated method under sustained load in a chloride environment were studied. The four beams have different span to effective depth ratios (1.55, 1.94 and 2.03) and different corrosion degrees. The net span of beam Bs04-A and beam Bs04-B was 1000 mm and the span/effective depth ratio was 1.94. The net span of beam Bs02-A and Bs02-B was 1050 and 800 mm respectively while the span/effective depth ratio was 2.03 and 1.55 respectively.

The cracking maps and corrosion maps of the four beams were depicted. Gravimetric method was used to calculate cross-sectional loss of corroded steel bars and a vernier caliper was used to measure the diameter loss of stirrups.

A three point load was applied on the beams to carry out mechanical test. It was found that Bs04-A and Bs04-B failed suddenly due to the failure of front tensile bar closed to the middle of the beam. However, beam Bs04-B experienced a much longer yield period before failure. It could be reflected on the load-deflection curves of the two beams, which are presented in Figure 1.

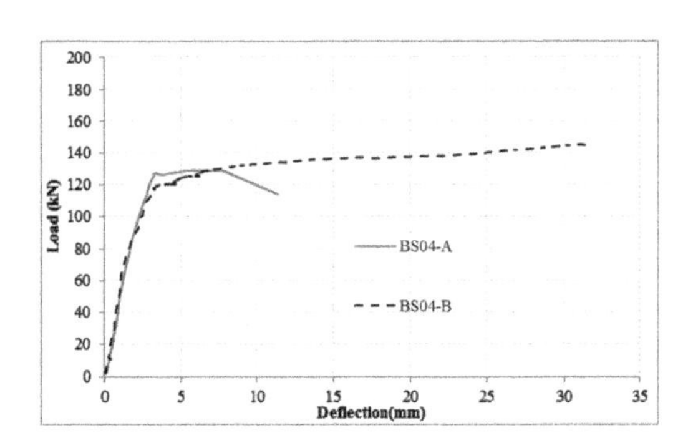

Figure 1. Load-deflection curves of beam Bs04-A and Bs04-B.

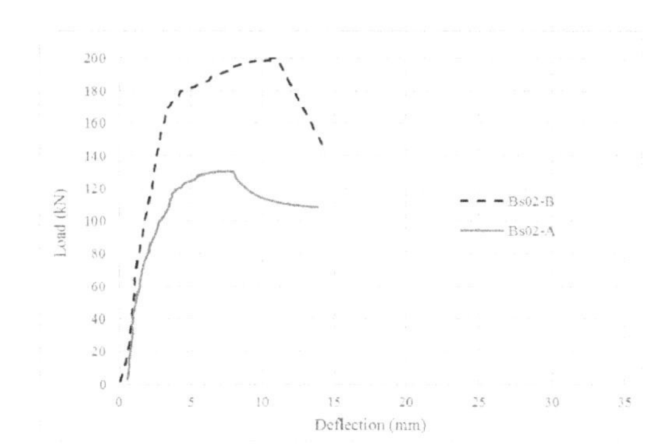

Figure 2. Load-deflection curves of beam Bs02-A and Bs02-B.

The beam Bs02-A failed due to the rupture of two stirrups and the disbonding between concrete and steel. There were obvious pitting corrosion at the failure points on the stirrups, the diameter loss of stirrups at the failure points was 42% and 17% respectively. The beam Bs02-B failed due to the disbonding between concrete cover and steel reinforcement. Both the two beams failed in a brittle way, the loading-deflection curves are presented in Figure 2.

Although beam Bs02-A had a greater span to effective depth ratio than beam Bs04-A and beam Bs04-B, it failed in a shear failure mode while beam Bs04-A and Bs04-B failed in a flexure mode. It could be contributed to different corrosion degree on the tensile bars. The corrosion degree on the tensile bars of beam Bs04-A and Bs04-B was more severe than that on beam Bs02-A. The maximum cross-sectional loss of tensile bars in beam Bs04-A and Bs04-B was 26.6% and 22.8% respectively, both of them located closed to the middle of the beams. The maximum cross-sectional loss of tensile bars in the middle of the beam Bs02-A was only 6.3%. Due to the serious corrosion on tensile bars of beam Bs04-A and Bs04-B, the tensile bars yielded before the formation of diagonal cracks. Finally the two beams failed due to the rupture of tensile bars in the flexural manner.

Although the failure mode and ultimate load of beam Bs04-A and Bs04-B were nearly the same, significantly different ductility was presented on the two beams. The possible reason is the different ductility of corroded tensile reinforcements. Ductility of steel bar depends on both corrosion degree and geometry of residual cross-section.

REFERENCES

Higgins, C., Farrow, W.C., Turan, O.T., 2012. Analysis of reinforced concrete beams with corrosion damaged stirrups for shear capacity. Struct. Infrastruct. Eng. 8, 1080–1092. doi:10.1080/15732479.2010.504213

Wang, X.-H., Gao, X.-H., Li, B., Deng, B.-R., 2011. Effect of bond and corrosion within partial length on shear behaviour and load capacity of RC beam. Constr. Build. Mater. 25, 1812–1823. doi:10.1016/j.conbuildmat.2010.11.081

Wang, X.-H., Li, B., Gao, X.-H., Liu, X.-L., 2012. Shear behaviour of RC beams with corrosion damaged partial length. Mater. Struct. 45, 351–379. doi:10.1617/s11527-011-9770-5

Yuan, Y., Ji, Y., Shah, S.P., 2007. Comparison of two accelerated corrosion techniques for concrete structures. Aci Struct. J. 104, 344–347.

Concrete Repair, Rehabilitation and Retrofitting IV – Dehn et al. (Eds)
 ISBN 978-1-138-02843-2

Literature overview on the application and limitations of stress wave propagation theory for conditional assessment of concrete structures and elements

E. Okwori
Department of Civil Engineering, University of Cape Town, South Africa

ABSTRACT

It becomes critical to in certain circumstances examine concrete structures for their continued suitability and sustainability of intended use. The examination however must not cause any damage to the structure in the process. In such circumstances non-destructive testing is utilized. There are various non-destructive tests available for concrete structures, but are selected based on its suitability and feasibility of application.

Owing to the high rate of material failure that is variable during initial stages of construction due to manufacturing defects and during the service life of structures resulting from induced damages. The deterioration and damage of concrete structures has become a prominent problem in recent decades preventing concrete structures from achieving their design specifications. Therefore the need to develop efficient methods to understand the nature of the deterioration and damage mechanisms effectively is essential to providing solutions to mitigate deterioration and damages in concrete structures. This paper presents an overview, which provides information regarding the application of stress wave propagation theory in Non-Destructive Conditional Evaluation (NDE) of concrete structures. Current trends, technologies and some limitations to the application of the theory, and illustrate the need for further research to improve the efficiency and applicability of the theory in NDE of concrete structures.

Concrete Repair, Rehabilitation and Retrofitting IV – Dehn et al. (Eds)
© 2016 Taylor & Francis Group, London, ISBN 978-1-138-02843-2

Concrete cultural heritage in France—inventory and state of conservation

E. Marie-Victoire & M. Bouichou
Laboratoire de Recherche des Monuments Historiques, Champs-sur-Marne, France

T. Congar & R. Blanchard
Cercle des Partenaires du Patrimoine, LRMH, Champs-sur-Marne, France

ABSTRACT

Concrete being a major construction material of the XXth century, more and more buildings and structures are listed as historical. This cultural heritage, even if it could appear "young" with respect to more ancient constructions such as medieval or antic monuments, is decaying, partly due to incipient construction technology knowledge and partly to environmental aggressive conditions.

To try to cope with this problematic, a European project named REDMONEST, gathering 3 countries (Belgium, France and Spain), was launched in 2014, through the JPI-CH platform. Within this project, 3 thematics will be addressed: inventory, diagnosis and conservation treatments of concrete cultural heritage.

One of the first steps of the project was to carry out a survey to make a precise inventory and to evaluate the state of decay of these monuments. In this paper the results obtained in France will be presented, showing a considerable amount of structures, with a great variety of construction periods and building uses (apartments, churches, industries...). Data on their main pathologies and current state of conservation will be discussed. Finally, potential diagnosis tools and conservation strategies will be put into perspective.

The first results of the survey performed in France, evidenced a strong policy of protection of buildings and structures partly or completely made of concrete, introduced in the 1990ies (Figure 1), leading to an amount of 816 monuments listed in 2015.

Twelve typologies of monuments were identified, with a predominance of domestic or sacred architecture. Most of these monuments were erected between the 2 world wars and are located in urban and suburban environments.

Based on 175 answers to a questionnaire submitted to architects and curators in charge of this cultural heritage in France, more than 60% of these monuments are in a good or in a satisfactory sate of conservation, but 20% are considerably damaged or endangered.

The main source of decay identified was carbonation induced corrosion, which lead to considerable loss of original materials.

This first series of results targeted the next steps of the project to diagnosis and conservation strategies focusing on rebars corrosion evaluation and treatment.

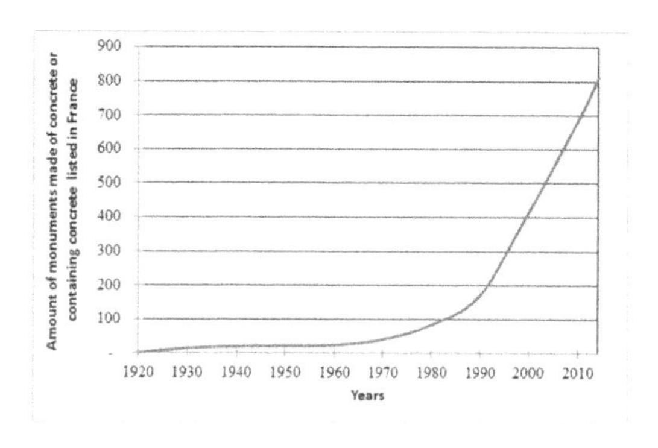

Figure 1. Increasing amount of monument completely or partially made of concrete listed in France.

Concrete Repair, Rehabilitation and Retrofitting IV – Dehn et al. (Eds)
© 2016 Taylor & Francis Group, London, ISBN 978-1-138-02843-2

The development of a new Dutch guideline for the conservation of historic concrete (URL 4005)

H.A. Heinemann
Department Building Solutions, Bartels Engineering, Apeldoorn, The Netherlands

ABSTRACT

In recent years, concrete conservation has been emerging as an own discipline from concrete repair. The relationship between the two fields is still strong, as they use similar approaches, techniques and repair materials. Yet what differs are the diverging aims of concrete conservation and repair, and the properties of historic concrete (Heinemann, 2013).

No generally accepted standards exist which clarify when to speak of concrete repair or of concrete conservation. Instead, higher aesthetical demands are often considered as the main aspects of concrete conservation. That such practice neither reflects the principles nor the complexity of historic concrete conservation and concrete repair has been addressed previously (Heinemann, 2013). Many of these demands have been discussed on an academic or governmental level, but have not yet been disseminated sufficiently to practitioners or owners.

Many historic concrete buildings have been repaired in the Netherlands. The success of the interventions varies. Some show premature failure of repairs, others involve a large loss of historic material, others received wide international recognition (Figures 1–3). Isolated case studies (e.g. Heinemann, 2013) showed several weak links in the conservation process, such as wrong diagnosis, lack of knowledge, or postponing interventions.

A cause is the little synergy between the fields of concrete repair and construction history. By better understanding the peculiarities of historic concrete, former state-of-the-art and the consequences for durability, investigations could be better targeted and the suitability of repair techniques better evaluated. Noteworthy is that in Dutch practice, technical surveys are not always carried out, and repair strategies then chosen by concrete repair companies.

As a reaction towards the increase of concrete conservation projects and to improve their quality, a guideline for the conservation of concrete is currently being developed: the *URL 4005 Conservation of historic concrete*. A *uitvoeringsrichtlijn* (URL)

Figure 1. Premature failure of a crack repair due to thermal movement of plain concrete (Fort Bezuiden Spaarndam, part of the Defense Line of Amsterdam, 1897-1901, UNESCO World heritage site, intervention 1998, situation 2006).

Figure 2. Large-scale removal of historic concrete during repair (Circumambulation passageway Pilgrimage church, Brielle, H. Goossens (reconstruction), E.J. Margy, 1912).

Figure 3. Prefab column made with coloured concrete and specially chosen aggregates, now showing cracks and leaching, indicating ASR (Giraffe compound, Zoo Blijdorp, Rotterdam, S. van Ravesteyn, 1940).

is a Dutch guideline which defines construction works. For restoration works separate URLs exist. In these guidelines the additional requirements for assessment, conservation aims, execution of repairs, properties of repair materials and intended outcome are described. Contrary to normal repair guidelines, the cultural heritage values and different types of historic material are explicitly addressed.

Although many of these aspects seem self-evident, in practice they are often neglected due to ignorance, or limited time and financial budgets. With the guideline, it is expected that the quality of concrete conservation will improve as the additional demands are outlined and stakeholders supported to define suitable conservation strategies.

The demand for more support for the execution of concrete conservation had been recognised by the parties involved in the conservation process. In 2013 it was decided to develop a URL for historic concrete. Aims were to distinguish the more challenging concrete conservation from normal concrete repair, and enhance the quality of concrete conservation. As starting point, the problems and wishes known from practice were outlined:

- Availability of technical surveys
- Material information in historical surveys
- Better legal support when using non-certified repair mortars
- Data-base with best-practice
- Execution of trial repairs
- Limited knowledge about historic concrete

The problems reflect that more structure and guidance is needed in the concrete conservation process.

The URL will address inspection and consultancy in one part, and the execution of the works in another part. Initially, this guideline was supposed to come into effect in the Netherlands by January 2015. Due to unexpected obstacles and the recentness of the field its release is delayed.

Obstacles were to embed the URL into existing (concrete repair) guidelines, avoid loop-holes and to focus on its main aim. One loop-hole was to allow construction companies to deviate from proposed repair strategies and materials. Another obstacle is the recentness of the field, the lack of guidelines related to concrete conservation, and involvement of stakeholders without knowledge of (historic) concrete during conservation. This tempted to incorporate too much information and demands in the URL, turning it into an unreadable document.

The development and introduction of a guideline for the conservation of concrete is a welcomed signal. It shows that historic concrete is being acknowledged as a historic material, requiring equal care and investigation as traditional materials. Yet, concrete conservation cannot rely yet on experience to identify best practice or loop-holes.

During the development of the URL, underlying problems within the conservation field frequently surfaced. Many of these problems originate from the aspect that no training exists for concrete conservation, explaining the varieties and properties of historic concrete, the consequences for durability and options for conservation. Most available experience is case-based and generic knowledge is limited.

The introduction of the URL can push the concrete conservation market, by requesting sound research prior to repair works, and by creating awareness that concrete conservation involves more than concrete repair.

REFERENCES

English Heritage, et al., 2012. *Practical Building Conservation: Concrete*. Ashgate Publishing, Ltd.

Heinemann, H. A., 2013. Historic Concrete From concrete repair to concrete conservation. Delft: Delft Digitalpress.

ICOMOS, 1964. International Charter for the Conservation and Restoration of Monuments and Sites (The Venice Charter 1964).

NEN-EN 1504:2004 Products and systems for the protection and repair of concrete structures—Delft: Netherlands Normalisatie-instituut.

Tilly, G. & Jacobs, J., 2007. Concrete repairs. Performance in service and current practice. Watford: IHS BRE Press.

Concrete Repair, Rehabilitation and Retrofitting IV – Dehn et al. (Eds)
 ISBN 978-1-138-02843-2

Evaluation of moisture and gas permeability of surface treated concrete and its application to historical reinforced buildings in Japan

Kei-ichi Imamoto & Chizuru Kiyohara
Tokyo University of Science, Japan

Kaori Nagai
Nihon University, Japan

Maiko Misono
Constec Engi. Co., Japan

INTRODUCTION

In this study, authors focus on surface-impregnation materials and coating agents that would not change the appearance of concrete. Authors investigated the applicability of these agents in the conservation of historical reinforced concrete buildings (Misono, 2013). In Japan, 'Keeping the original structure' is advertency when con-serving and repairing historical buildings. Therefore, authors verify the durability of surface-impregnation materials and coating agents in aggressive environments in order to conserve a historical RC building in Japan.

EXPERIMENTAL APPROACH

In this study, concrete samples with a water-cement ratio of 60% were prepared. Specimens were cured in water of 20°C for 28 days and then in air with 60% R.H. at 20°C and for 28 days. They were then stored in a climate controlled chamber with the following conditions: 20°C, 60% R.H., and 5% CO_2 concentration. Types of impregnation and surface coating agents are shown in Table 1. Fig. 1 shows an outline of the specimens.

Table 1. Types of impregnations and surface coatings.

Type	Main Component	Symbol	Construction Technique			
			Pretreatment	Frequency	Dilution	Application Quantity [kg/m^2]
Surface-Impregnation Materials	silane, siloxane	SS-1	dry	1	-	0.2
		SS-2	dry	1	-	0.2
	silane	S-1	dry	2	-	0.3~0.4
		S-2	dry	2	-	0.6
		S-3	dry	2	-	0.9
	silicate	K-1	wet	3	-	0.35
		K-2	wet	2	○	0.15~0.1
	acrylic, sodium, potassium	AKN	dry	1	-	0.28
Surface-Coating Agents	aqueous acrylic, silicon	E-AS	dry	3	-	0.1~0.15
	fluorine	E-F	dry	3	○	0.1~0.15
No Application	-	N	-	-	-	-

Outline of experiment

The specimen was exposed to 100 cyclic heat rays and water sprinkling (Photo 1). The protocols of heat rays and water sprinkling cycles are shown Fig. 2.

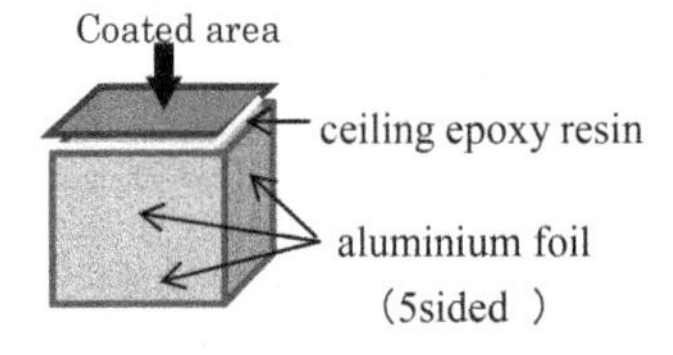

Figure 1. Schematic drawing of a structural specimen.

Photo 1. Left: heat rays; right: water sprinkling.

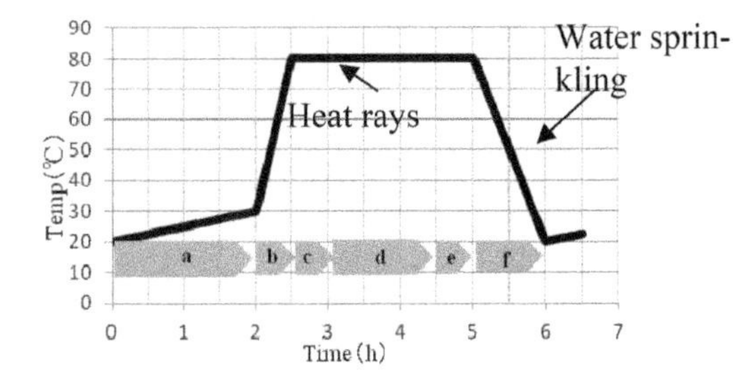

Figure 2. Protocol of heat deterioration.

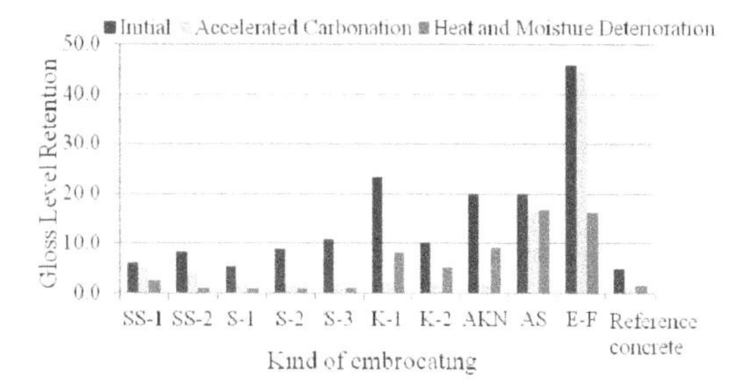

Figure 3. Gloss level retention.

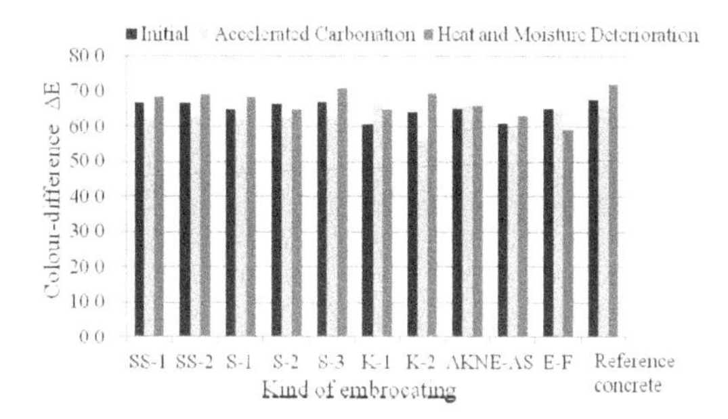

Figure 4. Color-difference (ΔE).

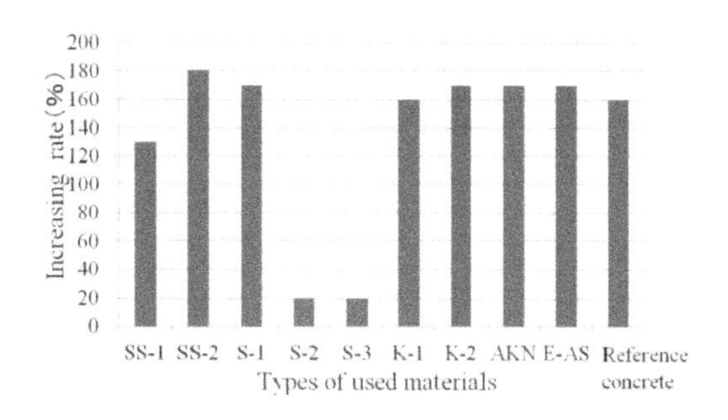

Figure 5. Increasing rate of moisture content.

RESULTS AND DISCUSSION

Figs. 3 and 4 show the results for gloss level retention and color difference. The surface-coating agents (E-AS, E-F) showed relatively high gloss levels. In contrast, the surface-impregnation materials except K-1, K-2, and AKN exhibited close values to reference concrete without treatment. Glossiness reduced overall after the accelerated carbonation test and heat and moisture deterioration test. In this study, gloss level retention was chosen as the color change index.

Photo 2. Appearance of the NMWA.

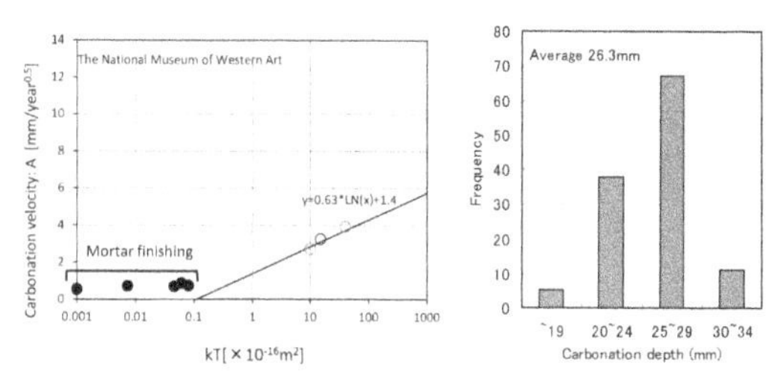

Figure 6. Carbonations of concrete evaluated with NDT.

Figure 5 shows changes in the moisture content of concrete measured with electrostatic capacity of ceramic sensors. It can be seen that the surface-impregnation materials S-2 and S-3 effectively prevented moisture ingress into concrete.

APPLICATION TO NATIONAL WESTERN MUSEUM (NMWA) IN JAPAN

The NMWA (Photo 2) was constructed in 1959 with concrete design strength of 18 MPa. The National Museum of Western Art (NMWA) is an important cultural landmark of Japan. In order to maintain the authenticity of the building, carbonations of concrete were evaluated with small core samples and Non-Destructive Test methods (NDT).

Figure 6 (Left) shows a good agreement between carbonation velocity and the air permeability coefficient with NDT. Based on this finding, it was clarified that concrete carbonation significantly proceeded as shown in Fig. 15 (Imamoto, 2012).

Based on the previous researches, It can be expected that the impregnation agent (S2) would be effective to prevent the ingress of moisture vapor into concrete. As shown in Fig. 7, there

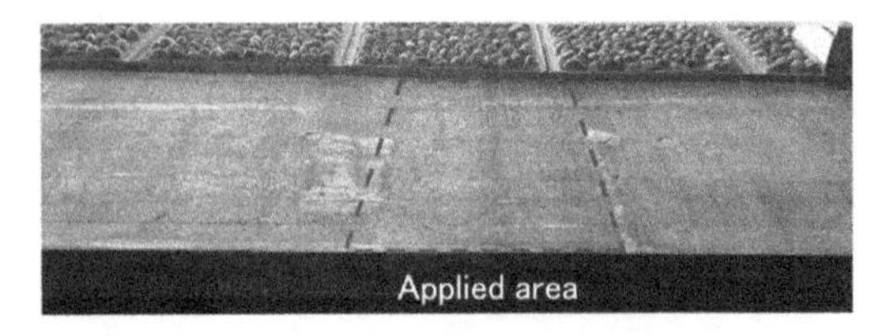

Figure 7. Applied area of impregnation agent.

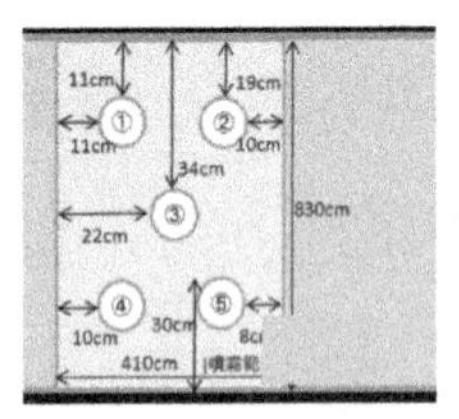

Figure 8. Surface electric resistivity test (Left: Test points).

was no color change after the applica tion of the agent "S2". The effectiveness of impregnation agent was confirmed with on-site with surface electric resistivity after water sprinkling as shown in Figs. 8 and 9. In this study, authors

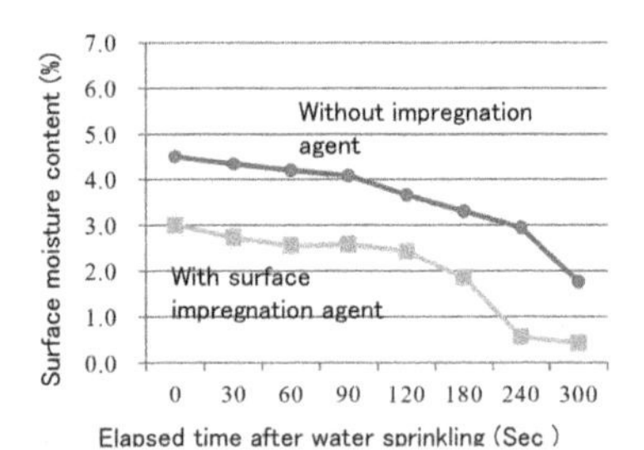

Figure 9. Change in electric resistivity after water sprinkling.

showed an example way to conserve the carbonated RC building.

CONCLUSIONS

An approach towards to conserve a historical RC building in Japan was introduced. Impregnation agent tested in this study well prevented the ingress of moisture which might cause steel corrosion of carbonated concrete of NMWA.

Modern materials technology

Concrete Repair, Rehabilitation and Retrofitting IV – Dehn et al. (Eds)

 ISBN 978-1-138-02843-2

A new restrained shrinkage test for HPC repair materials

A. Reggia, F. Macobatti, F. Minelli & G.A. Plizzari
DICATAM, University of Brescia, Brescia, Italy

S. Sgobba
CTG S.p.A., Bergamo, Italy

ABSTRACT

Repairing, maintaining and retrofitting existing Reinforced Concrete (RC) structures and infrastructures arises today not only as a possible perspective, but also as a necessity for the growth and development of our cities. In particular, the need to upgrade infrastructures to the increasing demand of traffic and to the performance requirements introduced by new technical regulations requires the development of innovative materials and technologies for the rehabilitation and retrofitting of existing structures. Among new materials, the use of High Performance Concrete (HPC) for structural reinforcement may be advantageous because, on one hand, it has a high chemical-physical compatibility with the substrate and, on the other, for the possibility to apply the reinforcing material in layers of a reduced thickness.

However, in the design of such interventions, the volumetric compatibility between the structure and the repair material must be considered. Restrained shrinkage, due to temperature and humidity conditions of the renewed structure, can cause shrinkage cracking when the free deformation of the new HPC layer is prevented by the presence of the old RC structure. The higher the degree of restraint provided by the substrate, the larger the effects of shrinkage of the overlay material. The reference standard methods for the evaluation of the relative likelihood of early-age cracking of cementitious mixtures are based on the so-called ring test, e.g. AASHTO Designation: PP 34–99 (1999) and ASTM Designation: C1581/C1581M-09a (2009). However, this methods do not provide a direct measure of the crack opening and do not allow to evaluate its development in time.

The goal of the work is to pursue research with an additional effort to make the ring test more suitable for its use as a reference standard for HPC and FRC. For this reason, the present work has two specific objectives: the first is the reduction of the time-to-cracking, through the increase of the degree of restraint of the test set-up; and the second is the need to measure the crack opening by means of the predetermination of its location. The introduction of a preformed notch in the concrete ring that, on the one hand, increases locally the degree of restraint and, on the other, allows to place a transducer for the measuring of the crack opening. The proposed change of the test method aims at developing a sufficiently robust testing technique able to emphasize the role of concrete as well as that of fibers.

The work presents the results of the new test performed on High Performance Concrete (HPC), Fiber Reinforced Concrete (FRC) with the additional use of Internal Curing (IC). The effects of the addition of an hybrid reinforcement of steel and plastic fibers and pre-wetted Light Weight Aggregates (LWA) are discussed with reference to the risk of cracking and to the development of the cracking process in the HPC repair materials. The HPC considered in the experimental study has consistency class S5 and strength class C80/95, with a slump of 240 mm and a cubic compressive strength of 101,5 MPa respectively. The HyFRC has consistency class S5, with a value of slump equal to 230 mm, and a strength class C80/95, with a cubic compressive strength of 104,1 MPa. The ICHy-FRC has lower consistency class S3 and strength class C80/95, with a slump of 150 mm and a cubic

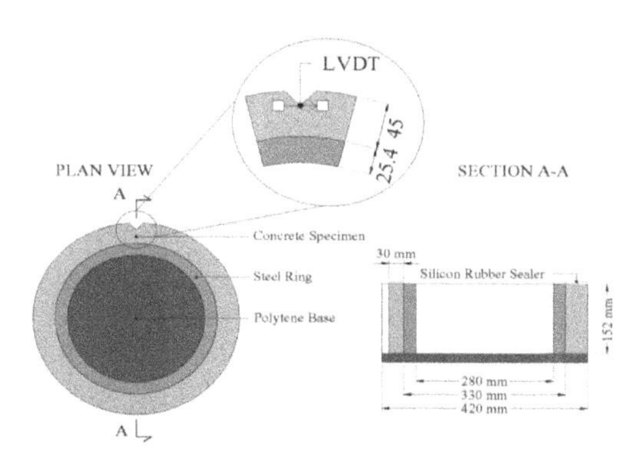

Figure 1. Notched ring test set-up: plan view and section.

Figure 2. Linear voltage displacement transducer on the tip of the notch (CTOD): HPC-2 specimen at 21 days.

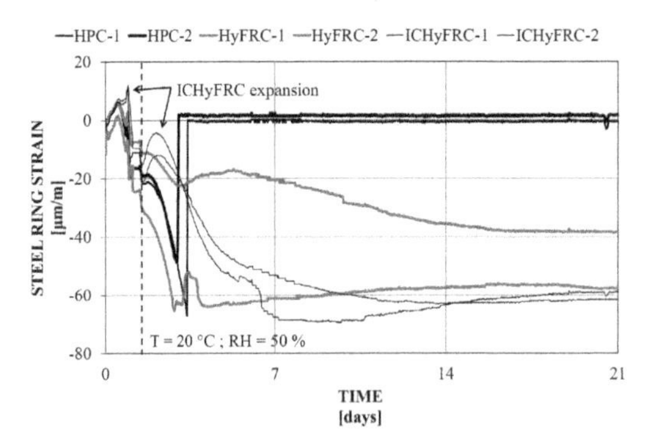

Figure 3. Notched ring test: steel rings strains.

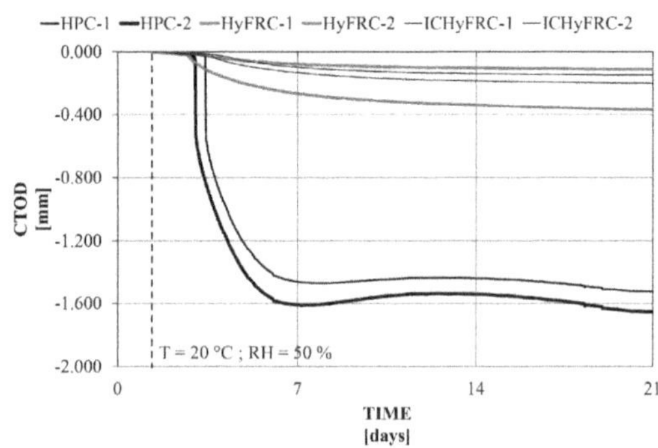

Figure 4. Notched ring test: crack development with time.

Table 1. Performance indexes: time-to-cracking (T_c) and crack opening rate (v_c).

Material	Tc days	$v_c(7)$ µm/day	$v_c(14)$ µm/day	$v_c(21)$ µm/day
HPC-1	3,4	399	133	85
HPC-2	3,0	393	137	90
Mean value	3,2	396	135	88
	(ref.)	(ref.)	(ref.)	(ref.)
HyFRC-1	3,3	16	8	5
HyFRC-2	2,8	59	29	19
Mean value	3,1	38	18	12
	(–3%)	(–90%)	(–87%)	(–86%)
ICHyFRC-1	4,0	25	12	7
ICHyFRC-2	3,5	29	15	10
Mean value	3,8	27	13	9
	(+19%)	(–93%)	(–90%)	(–90%)

compressive strength of 97,6 MPa. The average value of secant elastic modulus in compression is 36 GPa in HPC, 37 GPa in HyFRC and 31 GPa in ICHyFRC.

In conclusion, the resistance to restrained shrinkage cracking is defined on the basis of two experimental measurements: the cracking initiation time or time-to-cracking (T_c); and the average velocity of the crack opening or the crack opening rate (v_c). The time of initiation of the cracking process T_c is intended as the time interval between the exposure of the material to the environment which determines the shrinkage and the time in which a sudden variation of the crack opening (or a variation in the rate of crack opening) is measured.

T_c can be related to the likelihood of the cement matrix to crack due to restrained shrinkage: the more the material is prone to cracking, the more T_c will be short. The time-to-cracking is influenced by the time development of shrinkage and mechanical properties of concrete. In particular, concretes characterized by a rapid development of shrinkage and elastic modulus will be characterized by short T_c (e.g. HPC), while materials with a slow development of shrinkage or a gradual development of the elastic modulus will be characterized by longer T_c (e.g. ICHyFRC).

The rate of crack opening v_c defines, instead, the material's ability to counteract cracking: the lower v_c, the higher the material ability to control the cracking process. This parameter depends largely by the shrinkage and by the residual strength of the material after cracking. For instance, concretes characterized by a brittle post-cracking behavior will experience high v_c (e.g. HPC), while concretes characterized by high residual strength will be identified by a slow development of the cracking process (e.g. HyFRC and ICHyFRC).

Concrete Repair, Rehabilitation and Retrofitting IV – Dehn et al. (Eds)
© 2016 Taylor & Francis Group, London, ISBN 978-1-138-02843-2

Studies on creep deformation of ultra-rapid-hardening cement-type bonded anchor

S. Ando & T. Tamura
Sumitomo Osaka Cement, Chiba, Japan

K. Nakano
Chiba Institute of Technology, Chiba, Japan

T. Tanuma
Urban Renaissance Agency, Kanagawa, Japan

ABSTRACT

The creep deformation of ultra-rapid-hardening cement-type post-installed bonded anchor was studied, and the influences of the sustained load magnitude, deformed bar diameter, and anchor type were examined for an embedment length of seven times the nominal diameter. To investigate the influence of the sustained load magnitude, sustained loads were applied at magnitudes of 0.33 to 0.95 times the ultimate load magnitude. In each test, the load magnitude and the displacement of the anchor at the free end were measured. For the influence of deformed bar diameter, D13 and D19 reinforcing bars were used for the anchors. And the three types of anchors were prepared: the Cast-In-Place anchors (CIP), the inorganic-type Post-Installed Bonded anchors (InPIB), and the organic-type Post-Installed Bonded anchors (OrPIB).

For stress levels (sustained load/ultimate load) below 0.69, the sustained load was removed after three months. For stress levels greater than 0.69, the sustained load was maintained until creep failure. The higher the sustained load was, the larger the displacements of anchors were (Figure 1). At a stress level of 0.35, the displacement of the inorganic-type post-installed bonded anchor remained nearly unchanged under sustained loading for 91 days. This suggests that at a stress level of 0.35, creep failure of an inorganic type post-installed bonded anchor will not occur within 50 years. At stress levels greater than 0.80, creep failure occurred within 200 days, and the displacement at the time of creep failure increased as the time to creep failure increased.

In comparing D13 and D19 bars, the displacements of the D19 bars were found to be larger than those of the D13 bars at the same stress level and the same embedment length ratio, which was seven times the nominal diameter. Because the sustained loads on the D19 bars were high, the displacements

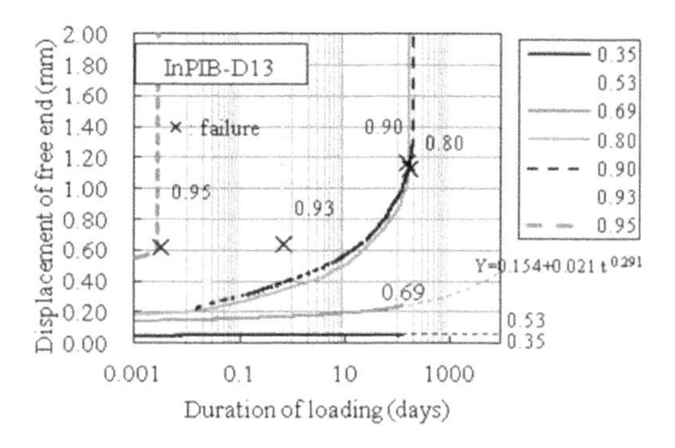

Figure 1. Creep deformation of post-installed anchors on the D13 bars.

of the D19 bars were larger than those of the D13 bars.

For organic-type anchors, the displacement increased with increasing test duration, the creep limit is considered to be approximately 0.55. The displacement at the time of creep failure was 3.5 mm, which was larger than that for the inorganic-type anchors. However, the displacement at the time of creep failure was 2–3 times that measured in the ultimate loading test.

For the inorganic anchor type, the slope of the line of the time to creep failure versus the stress level was approximately half of that for the cast-in-place anchor type. Therefore, at high stress levels, creep failure is less likely to occur in an inorganic type of anchor than in a cast-in-place anchor.

Assuming that the creep limits of the different types of anchors are represented by the inflection points of the displacement curves, the creep limit of the InPIB anchors is considered to be approximately 0.70, and that of the OrPIB anchors is approximately 0.50–0.55. It is likely that the creep limit of an inorganic-type anchor will be higher than that of an organic-type anchor.

Concrete Repair, Rehabilitation and Retrofitting IV – Dehn et al. (Eds)
© 2016 Taylor & Francis Group, London, ISBN 978-1-138-02843-2

Assessment of the use of Arcelor Mittal electric arc furnace slag as coarse aggregates in concrete production

D. Maharaj & A. Mwasha
University of the West Indies, Kingston, Jamaica

ABSTRACT

Due to the increasing cost of raw materials and the continuous reduction of natural resources, the use of industrial by products could be a potential alternative in the construction industry. In concrete production, the continued and expanding extraction of natural aggregate causes serious environmental problems which lead to irremediable deterioration of rural areas. Quarrying of aggregates alters land topography and causes potential environmental problems such as erosion. Aggregates occupy approximately 70% of the volume of concrete and the properties of the concrete produced are dependent on the characteristics of the aggregates used.

One type of industrial by product that has the potential for aggregate replacement in concrete production is electric arc furnace slag.

Electric arc furnace slag is a by product from the steel making process. Locally, electric arc furnace slag is generated at 300,000 tonnes annually and when produced, it is left in stockpiles at the production site. The use of this industrial by product in the construction industry could be beneficial in minimizing natural resource depletion and the high environmental impacts caused by concrete production. As such, this study focuses on the effectiveness of using electric arc furnace slag as coarse aggregates in the making of concrete. However, since the electric arc furnace slag is an industrial by product, a comprehensive knowledge of its characteristics is required. Therefore, this paper examines the physical, mineralogical and chemical properties of electric arc furnace slag. The feasibility of using electric arc furnace slag in the concrete was examined by conducting compressive and tensile strengths of concrete mixtures using replacement ratios of 0%, 15%, 30%, 45%, 60%,75% and 100% of the slag as coarse aggregates.

The findings indicate that the compressive and tensile strengths of the concrete achieved using the different ratios of the slag is comparable to that using purely natural aggregates. Therefore, the utilization of electric arc furnace slag in concrete production could be an efficient means of reducing the costs and environmental impacts.

For each ratio of the electric arc furnace slag, ten samples were prepared and tested. The highest average compressive strength of the concrete achieved was 64.6 MPa using 75% electric arc furnace slag and the lowest average compressive strength achieved was 50.4 MPa using 30% electric arc furnace slag. Additionally, the highest average tensile strength achieved was 3.6 MPa and the lowest average tensile strength attained was 2.95 MPa using 15% electric arc furnace slag. These results did not indicate any increase in the strengths of concrete as the percentage ratio of the electric arc furnace slag increased but it generated comparable strengths of concrete when natural aggregates were used. Therefore, it could be deduced that the tensile and compressive strengths of concrete are enhanced through the utilization of electric arc furnace slag in concrete production.

Concrete Repair, Rehabilitation and Retrofitting IV – Dehn et al. (Eds)
© 2016 Taylor & Francis Group, London, ISBN 978-1-138-02843-2

Self compacting grout and concrete: How it is produced and why it is needed

H.S. Abdelgader & A.S. El-Baden
Department of Civil Engineering, Faculty of Engineering, University of Tripoli, Tripoli, Libya

ABSTRACT

Using non-traditional concrete in engineering applications such as the construction of nuclear reactor shields, dams, massive under water bridge piers and repairs of building foundations ... etc., have been considered as an efficient solution to overcome challenges of limitations of the use of normal conventional concrete. Such new types of concretes which have been developed and produced are completely dissimilar from the conventional concrete in the method of mixing, handling, pouring, consolidation, behaviors, cost ....etc. Based on the technology of ready-mixed Self-Compacting Concrete (SCC), two types of concrete been introduced and named as: Two-Stage Concrete (TSC) and Rock-Filled Concrete (RFC), where a Self-Compacted Grout (SCG) injected or poured to fill the void space of preplaced or Self-Compacted Aggregate (SCA) or rocks. By other words, TSC (Pre-placed Aggregate) unlike Normal Concrete (NC), it is made by first placing the coarse aggregate in the formwork and then injecting a grout consisted of sand, cement and water to fill the voids between the aggregate particles. The main benefits of the method are widely appreciated as Low heats of hydration, high compressive strengths and density, economic savings, practically no mass shrinkage, low coefficient of thermal expansion, excellent bond to existing structures. Generally, the properties of two-stage concrete are thus influenced by the properties of the coarse aggregate, the properties of the grout, and the effectiveness of the grouting process. This paper illustrates the importance, advantages and special requirements of introducing TSC in the concrete industry field. Also, it demonstrates results from research work conducted in the last decade to address TSC in terms of grout property requirements and some important mechanical properties. From the results it appears that there is a good correlation between the compressive strength and tensile strength of TSC. As the compressive strength increased with grout, the tensile strength was also found to increase in the same manner. Relationship between compressive strength (fc.′) and tensile strength (ft.) of different grout proportions was developed by regression analysis.

$$ft = A + (B) \times fc' + (C) \times (fc')^D \quad (1)$$

Table 1 shows the values of the regression coefficients.

Table 1. Regression results for equation (1).

Type of grout	A	B	C	D	Correlation Coefficient
Without Adm.	−50	−0.4	38	0.1	0.724
Super-palsticier	40	0.4	−32	0.1	0.8
Expanding Admixture	−4.3	−0.3	1.82	0.6	0.721
Expanding Adm. & Super-plasticiers	163	1.2	−132	0.1	0.68

Concrete Repair, Rehabilitation and Retrofitting IV – Dehn et al. (Eds)
© 2016 Taylor & Francis Group, London, ISBN 978-1-138-02843-2

Research and development of polymer modified self compacting concrete used for replacement of large area deterioration concrete

Xiangzhi Kong, Gaixin Chen, Shuguang Li, Guojin Ji & Sijia Zhang
State Key Laboratory of Simulation and Regulation of Water Cycle in River Basin, China Institute of Water Recourses and Hydropower Research, Beijing, P.R. China

ABSTRACT

A lot of concrete dams and their ancillary buildings need to be repaired because of freezing and thawing, scour and erosion. For large area concrete repair, replacement of concrete is a good choice, especially when the depth of the repairing area exceeds 6 inches. But this method has some disadvantages in some aspects, compared with the cement or polymer mortar. For example, the durability, interface bond strength, and the complicated repair procedure. In this study we developed a Polymer Modified Self Compacting Concrete (PMSCC) which can achieve self compaction without any vibration. Experimental results show that the tensile strength of the PMSCC is 3.63 MPa at 28 days, the ultimate tension strain is 179 microstrain increased by 54.3% compared with Portland cement self compacting concrete, and the relative dynamic modulus of elasticity is 97% after 300 times freezing and thawing circles. Simulation test shows that the interfacial bond strength between new and old concrete is higher than 1.56 MPa without any adhesives. In addition, finite element analysis results show that appropriate setting joints could effectively avoid thermal cracks during the operation stage. PMSCC has the features of "high anti-crack ability, good durability and high bond strength" which is suitable for the replacement of large area deterioration concrete, especially for the dam upstream concrete.

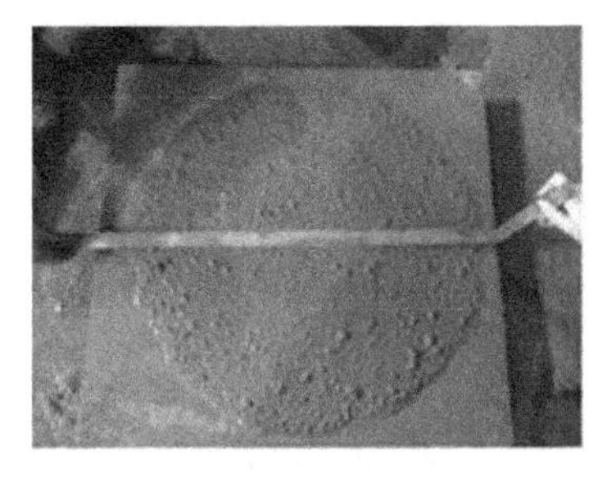

Figure 1. Polymer modified self-compacting concrete.

Table 1. Hardened concrete properties.

	Compressive strength (MPa)		Tensile strength (MPa)		Modulus of elasticity (GPa)		Ultimate tension strain (millionths)		Permeability coefficient ($m/s \times 10^{-12}$)	DF	Bond strength (MPa)
No.	7d	28d	7d	28d	7d	28d	7d	28d	28d	28d	28d
SCC	25.0	44.6	2.57	3.21	25.5	30.6	111	116	0.5	90	
PMSCC-1	22.1	31.8	2.27	3.19	18.1	23.1	137	147	0.7	95	
PMSCC-2	29.0	37.0	2.90	3.63	20.1	23.5	170	179	0.3	97	2.27

(a1) (a2)

(b)

Figure 2. Concrete samples before and after repair (a1) and (a2) before repair, (b) after repair.

Concrete Repair, Rehabilitation and Retrofitting IV – Dehn et al. (Eds)
© 2016 Taylor & Francis Group, London, ISBN 978-1-138-02843-2

Using steel fibered high strength concrete for repairing continuous normal strength concrete bending elements

K. Holschemacher
University of Applied Sciences, Leipzig, Germany

I. Iskhakov & Y. Ribakov
Ariel University, Ariel, Israel

ABSTRACT

Using Steel Fibered High Strength Concrete (SFHSC) for repairing normal strength concrete bending elements has been previously proved to be effective for statically determined beams (Iskhakov et al., 2011).

Bruhwiler & Denarie (2008) investigated methods for rehabilitation of existing structures using Ultra High Performance Fiber Reinforced Concrete (UHPFRC). The rehabilitated full-scale structures have also significantly improved durability.

Simple supported full-scale Two-Layer Beams (TLB) were tested (Iskhakov et al., 2014). A logical extension of these tests is experimental investigation of a Continuous Two-Layer Beam (CTLB), which is a focus of the current study.

The present study is focused on experimental investigation of statically non-determined TLB. A two-span TLB was tested under concentrated loads acting in the middle of each span.

The concrete classes for NSC and SFHSC were selected as in the previous studies. An optimal steel fibers' content for the SFHSC was 40 kg/m^3. The tests were focused on studying the load-deflection relationships, bonding between the SFHSC and NSC layers, transfer of shear between the layers and Poisson deformations in the SFHSC layer.

Experimentally obtained deflections at the mid spans of the beam and above the middle support are shown in Figure 1. The graph demonstrates an elastic– plastic behavior of the beam at the post-cracking range. Thus, it experimentally confirms that, as for simple supported TLB (Iskhakov et al., 2014) also for the investigated continuous beam, proper selection of steel fiber content allows achieving the desired ductile behavior, in spite that the beam's compressed zone is made of brittle HSC. The force-deflection lines on the graph are smooth. It demonstrates that there is proper interaction between the SFHSC and NSC layers.

The tested beam exhibited proper behavior under the design load. After the design load was achieved the forces were further increased. The test has demonstrated that when the load was about 150% of the design value cracks propagated through the whole NSC layer, but not appeared at the SFHSC one.

Solution for the problem of the SFHSC layer thickness and its location along the continuous beam was found: the thickness of the layer is equal to the section compressed zone depth and length of the layer corresponds to the distance between the zero values in the bending moments' diagram.

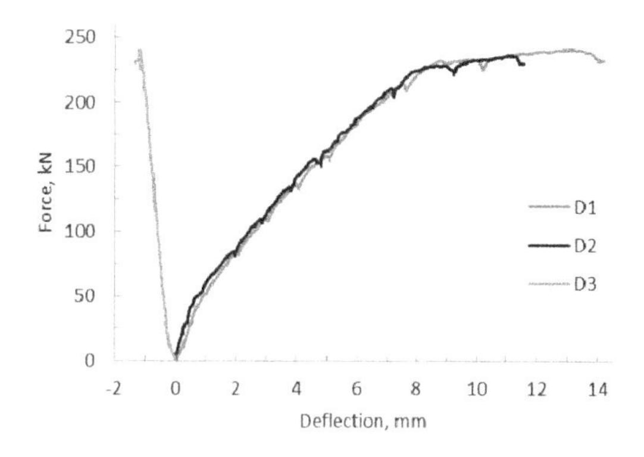

Figure 1. Load – deflection curves in the middle of the spans (D1 and D2) and the middle support uplift (D3).

Concrete Repair, Rehabilitation and Retrofitting IV – Dehn et al. (Eds)
© 2016 Taylor & Francis Group, London, ISBN 978-1-138-02843-2

Investigations on the suitability of technical textiles for cathodic corrosion protection

Amir Asgharzadeh & Michael Raupach
Institute of Building Materials Research, RWTH-Aachen University, Germany

Detlef Koch
Koch GmbH, Kreuztal, Germany

ABSTRACT

Cathodic Protection (CP) is a widely used method to protect steel reinforcements against corrosion. In the course of the last half century, it has been established as a proven system for repairing corrosion-affected reinforced concrete structures, which have mainly been damaged by chloride-induced corrosion. The impressed current anode system for the protection of steel in concrete is latest state of technology. The CP anodes can be embedded in mortar, as coating or distinct anode on the repair structure surface and exposed to external current. In this way, the potential of carbon steel is shifted in cathodic direction and the anodic dissolution of carbon steel is suppressed. The current densities on the surface of the reinforcement play a key role in the shifting of the potential in cathodic direction. Nowadays, Mixed Metal Oxide coated Titanium (MMO) is used as an anode material for CP due to its high durability under anodic polarization. Also other materials such as carbon fibers are being studied.

Carbon textiles in combination with mortar, which provide high mechanical properties and are also conductive, have not been studied systematically so far. In this paper investigations are described, which have been carried out in order to evaluate the capabilities of different carbon-textile anodes and different mortar mixtures for the cathodic protection of steel in concrete. In order to evaluate the polarization behavior of carbon-textile in mortar, galvanostatic experiments were performed. Based on these experiments, current density-potential-curves were derived.

The test setup presented in figure 2 was selected to simulate a component surface. The undermost layer of concrete containing reinforcement (here: MMO coated titanium), while the subsequent

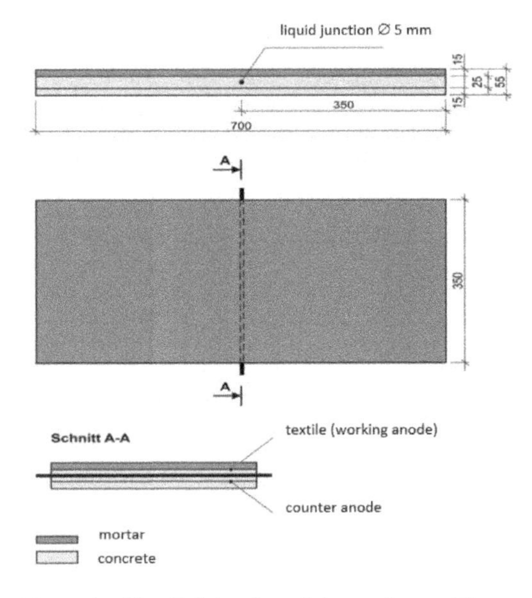

Figure 1. Detailed drawing of the specimen—Dimension [mm] according to Henkel (2014).

Figure 2. Installed liquid junction (left), concrete base plate (right) according to Henkel (2014).

Table 1. Properties of the textiles.

Textile	Material/ structure	Coating	Mesh width 0°/90°
			mm
1	Carbon/ Single mesh	styrene-butadiene-rubber	14/8
2	Carbon/ Single mesh	epoxy	20/20
3	Carbon/ Double mesh	styrene-butadiene-rubber	15/15
4	Carbon/ Single mesh	styrene-butadiene-rubber	14/13

layer of concrete simulates the concrete cover. To simulate the reinforcement the MMO coated titanium was used. Carbon Textile is applied onto the component surface by means of a specially developed mortar. Electrolyte bridges were placed in the specimens to be able to connect external reference electrodes to the specimen (Fig. 3).

The following chart, (Tab 1), provides an overview of the textile meshes.

Concrete Repair, Rehabilitation and Retrofitting IV – Dehn et al. (Eds)
© 2016 Taylor & Francis Group, London, ISBN 978-1-138-02843-2

The evaluation of concomitant use of Metakaolin and Limestone Portland Cement to durability of HPC Concrete

A.A. Ramezanianpour
Head of Concrete Technology and Durability Research Center, Amirkabir University, Tehran, Iran

Nima Afzali
Master of Engineering of Construction Management, Amirkabir University of Technology, Tehran, Iran

ABSTRACT

Nowadays, the importance of the durability in concrete structures is more than the other factors like the compressive strength. Reinforced concrete structures, which are exposed to harsh environments are expected to last over long periods of time. For that important reason, a durable structure needs to be produced. Especially for reinforced concrete structures, one of the major forms of environmental attack is chloride penetration, which leads to corrosion of the reinforcing steel and a subsequent reduction in the strength, serviceability of the structure. This may lead to early repair or replacement of the structure. The ability of chloride ions to penetrate the concrete must then be known for design and quality control purposes.

A common method of preventing such deterioration is to prevent chlorides from penetrating the structure by using additives like High Reactivity Metakaolin (HRM) in the system. Metakaolin is highly processed kaolinite clay that has been heat-treated under controlled conditions. Based on the important researches, also using appropriate portions of limestone contributes to improve compressive strength and permeability of concrete, But using 15% of limestone in concrete has severe effects on durability.

In this investigation after implementation of comprehensive experiments for evaluation of physical, chemical, and mechanical characteristics of applied materials, concomitant use of Metakaolin and Limestone Portland Cement to durability of HPC Concretes containing low water-cement ratio 0.31 and excessive cement paste has been evaluated. The maximum size of aggregate 12 mm selected, and proportion of sand to coarse aggregate was 3/1. This selection applied due to strengthening the appearance of cement paste and pozzolans' role and reduction of aggregate influences in probable weakness of concrete micro-structure. On the other hand, researchers reported conflicting views due to application of limestone powder in concrete. Since the reported conclusions in this article show the high rate of water absorption of applied limestone powder, then use of metakaolin in such HPC concrete caused a reduction of compressive strength and also durability at early ages in all the experiments. However, pozzolanic activity of Metakaolin is obvious during 7–90 days, the proper effect of this material observed in 90 days in comparison with control specimens.

Which is unique In this paper is, the simultaneous effects of different percentages of Metakaolin (10%, 12.5%, 15%) with Type 1-425 Portland Cement containing 15% limestone at a high performance concrete (>60 MPa) have been evaluated.

In order to produce appropriate limestone cement, additional tests like Alpine Sieve, Bending test, and Compressive Strength, etc. have been carried out in this investigation. Results of using a high reactivity Metakaolin in high-performance concrete in adjacent with excessive amount of limestone, and the relative influences on the Compressive Strength, Impermeability, Electrical Resistance, and RCMT indexes (Rapid Chloride Migration Test) are just some important factors presented in this paper.

Calculation of the none-steady-state migration coefficient, as shown in the Figure 1, was implemented by (RCMT) apparatus based on (NT Build 492) on cylindrical concrete specimens with 100 mm diameter and 50 mm thickness.

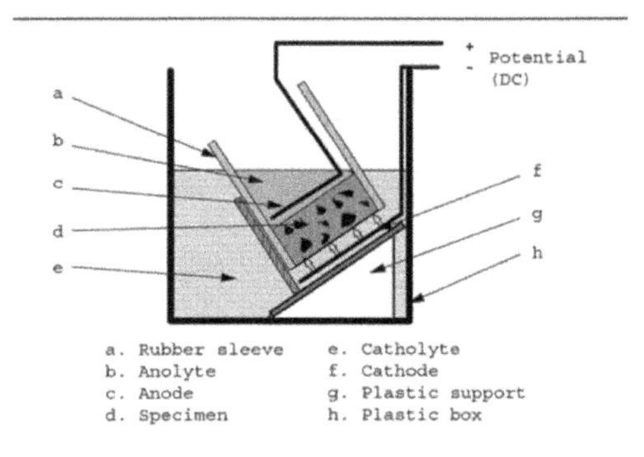

Figure 1. Schematic picture of (RCMT) apparatus.

Concrete Repair, Rehabilitation and Retrofitting IV – Dehn et al. (Eds)
© 2016 Taylor & Francis Group, London, ISBN 978-1-138-02843-2

Polymer modified high performance concrete as structural repair material of existing structures

K.D. Zavliaris
Greece

ABSTRACT

In many instances concrete structures need to be either repaired because of degradation or strengthened for upgrading. There is a variety of available repair methods involving techniques for restoring the initial load bearing capacity such as different types of injections and patching and strengthening techniques which aim at increasing the load bearing capacity and the stiffness of the structural members, among which jacketing by means of cast in place concrete, shotcrete and placement of external steel glued plates or epoxy bonded FRPs.

Polymer Modified High Performance Concrete (PMHPC) and mortar (PMHPM) can be used as repair and strengthening materials as is stated and discussed in this paper.

PMHPC and PMHPM have mechanical properties superior to that of normal concrete, controlled workability, high adhesion to normal concrete substrates and high durability.

These properties are further enhanced by the addition of reinforcing fibers of different types (steel, carbon, glass, mineral, PVA-Polyvinyl Alcohol-polypropylene) and in different forms (chopped fibers, mats, meshes).

From the point of view of constructability PMHPC and PMHPM are treated in the same way as that of normal concrete, thus contributing to reliable and cost effective application.

The structural performance of PMHPC and PMHPM is controlled by the cement type, the presence of pozzolans in the mix, the type and size of aggregates, the W/C ratio, the type and the content of Polymer and the type and content of fibers.

From design considerations point of view the shear transfer and slip across the interface, the effect of shrinkage, and the combined action of steel bars and fibers are important and discussed.

Finally a comparison of PMHPC, HPC, normal cement concrete and shotcrete in terms of mechanical resistance, durability, bond across the interface and cost is made which can serve as selection guide for the repair materials and techniques.

Concrete Repair, Rehabilitation and Retrofitting IV – Dehn et al. (Eds)
© 2016 Taylor & Francis Group, London, ISBN 978-1-138-02843-2

A new approach for internal curing of high performance concrete to reduce early-age volume variations

P.A. Savva & M.F. Petrou
University of Cyprus, Nicosia, Cyprus

ABSTRACT

Cement hydration and particles flocculation are responsible for concretes solidification. Several physiochemical mechanisms are activated from the moment that cement comes in contact with water until the material becomes solid. Autogenous shrinkage consists of chemical shrinkage and self-desiccation. Chemical shrinkage occurs because the hydrated materials occupy less space than the unhydrated constituent materials. Shortage in water due to low w/b ratio increases self-desiccation. Self-desiccation is governed by two physical mechanisms. The reduction of disjoining pressure will cause the material to contract, whereas, the meniscus formation within the pore will cause tension in the pores fluid (Laplace's Law). The disjoining pressure will be greater when the Relative Humidity (RH) is high and therefore a drop in RH will reduce the disjoining pressure causing shrinkage. The availability of water within the pores will prevent RH drop and disjoining pressure reduction. Pores partially filled with water within the materials structure cause the creation of water-air menisci and RH drop (Kelvin's Law). The water-air menisci will result in tension stress development within the pore fluid and consequently the surrounding material will develop compressive forces to restore equilibrium. The surrounding compressive stress will lead to further contraction.

An ideal maturing period for concrete would be a stress-free situation. However, due to the aforementioned mechanisms it is almost impossible for concrete to maintain the same volume from its very first moment until solidification. These mechanisms will develop differential volume variations within the structure of the material which will yield in stress development. Since the material in the early-age will not be able to develop adequate strength, it will crack due to these stresses.

The objective of this research is to develop an Internal Curing (IC) approach which will focus on reducing these volume variations in order to prevent early-age cracking. Attention is given on delivering additional water within the materials' structure using existing constituents of the mixture. The technique of Self Curing Concrete (SCUC) or internal curing as it is widely known has arisen at the beginning of the last decade and much advancement has been achieved since. It is an on-going research topic and it attracts great attention since the advantages of establishing an IC technique are both short and long term. Improved performance during the structures service life as well as extended life-cycle is only two of the main benefits that contribute to a less expensive construction and maintenance cost ensuring environmental sustainability. The ideal IC method should be able to reduce the autogenous shrinkage of High Performance Concrete (HPC) without compromising any other properties. The effort of this research has been focus on establishing IC in HPC without reducing any other properties of the material.

High Absorptive Normal Weight Aggregates (HANWA) have been employed to deliver water for Internal Curing (IC) within the materials' structure. HANWA exhibit high absorption capacity that benefits IC and sound mechanical properties that ensure the mixtures strength and durability. In order to evaluate the results, concrete mixtures including dry and saturated HANWA have been compared with control (CTRL) concrete mixtures prepared with ordinary (low-absorptive) Normal Weight Aggregates (NWA). Furthermore, mixtures including NWA and SuperAbsorbent Polymers (SAP) have been tested. The mixtures have been compared in terms of autogenous shrinkage, compressive strength, chloride diffusivity (Rapid Chloride Permeability-RCP) and vacuum porosity. Finally, two curing methods have been employed in order to assess the effectiveness of external curing on such dense concrete mixtures.

The HANWA have entered the mixtures in two different states, namely, Saturated Surface Dry (SSD) and oven dried. It has been shown that the HANWA-SSD mixture has performed better than the HANWA-DRY mixture in terms of compressive strength, porosity and RCP. Both HANWA and SAP mixtures have significantly reduced the mixtures autogenous shrinkage. However, the SAP mixtures have observed increased porosity and reduced compressive strength in comparison with the HANWA-SSD mixture. The SAP mixtures have observed significant reduction in RCP. This was attributed to the polymers dispersion within the matrix that enhances the pore network depercolation. Finally, in comparison with the CTRL mixture, the HANWA-SSD mixture was the only mixture that had recovered its initial loss in compressive strength.

Concrete Repair, Rehabilitation and Retrofitting IV – Dehn et al. (Eds)
 ISBN 978-1-138-02843-2

Real-scale testing of the efficiency of self-healing concrete

K. Van Tittelboom, D. Snoeck, E. Gruyaert, B. Debbaut & N. De Belie
Magnel Laboratory for Concrete Research, Ghent, Belgium

J. Wang
Magnel Laboratory for Concrete Research, Ghent, Belgium
Laboratory of Microbial Ecology and Technology, Ghent, Belgium

A. De Araújo
Magnel Laboratory for Concrete Research, Ghent, Belgium
Polymer Chemistry and Biomaterials Group, Ghent, Belgium

ABSTRACT

After several years of research in the Magnel Laboratory (Belgium) to obtain concrete with self-healing properties, the most promising self-healing approaches were tested on a larger scale (150 mm × 250 mm × 3000 mm beams). The first self-healing approach, consisted of the incorporation of encapsulated healing agent (SHC-PU). A one-component polyurethane was encapsulated by tubular glass capsules, which were embedded inside the concrete matrix. In total 350 capsules, filled with polyurethane, were prepared and positioned within the mould by means of a network of wires attached through the walls of the mould 10 mm below the top. As soon as cracks appear in the hardened concrete matrix, the brittle glass capsules, which are crossed by one of these cracks, will break. Due to capillary action, the healing agent will be released from the capsules and will be drawn into the crack. There, the agent will contact moisture inside the pores of the cementitious matrix causing a foaming reaction and hardening of the agent, finally leading to autonomous crack repair. The second self-healing approach which was ready to be tested on a larger scale, consisted of the use of hydrogels or superabsorbent polymers (SAPs, SHC-SAP). These were added to the concrete upon mixing. This self-healing mechanism is not immediately activated at the moment cracks appear but requires water ingress into the cracks to become active. When water enters into the cracks, contact with the SAP particles near the crack surface will result in immediate swelling of the particles and thus blockage of the crack. However, over time the particles will release their water content again so the crack blocking effect will be gone but on the other hand the slowly released water will become available to the cementitious matrix and will result in permanent crack closure by further hydration of unhydrated cement particles and calcium carbonate precipitation of leached calcium hydroxide if also carbon dioxide is available within the crack. To evaluate the autonomous crack healing efficiency, a comparison was made with a reference beam (REF), containing no embedded self-healing approach.

In order to create multiple cracks in the concrete beams, they were loaded in four-point bending. The force was exerted in upward direction and measured with a load cell. Next to the exerted load, the curvature of the beam was measured by means of five linear variable differential transformers positioned at the bottom side of the beams. As the concrete strength varies for the different beams under investigation, not the load or the curvature, but the average crack width was used to control the four-point bending test. Therefore, a measurement frame was positioned at the top of the beams, symmetrically with respect to the middle. The total displacement in horizontal direction within the area covered by this measurement frame was measured by an LVDT. Supposing that the contribution of concrete elongation is rather limited, the value measured by this LVDT represents the total crack opening within the zone covered by the measurement frame. Division by the total amount of cracks seen within this zone, results in the average crack width. Once an average crack width of 250 μm was reached, the average crack width and the deformation of the beams was fixed and the jack was taken away during the seven weeks period of crack healing.

For the beam with encapsulated polyurethane (SHC-PU), crack formation triggered breakage of the capsules, release of the healing agent and subsequent crack repair when the healing agent came into contact with moisture in the concrete matrix. For the other self-healing approach under investigation (SHC-SAP) contact with water is needed in order to activate the mechanism. As contact with water also induces autogenous crack healing, the natural mechanism of crack healing which is inherent to concrete, it was decided to bring all beams (REF, SHC-PU and SHC-SAP) in contact with water. In this way all beams would exhibit, to some extent, autogenous crack healing and the influence of this effect is filtered out when results are compared. Bringing the beams into contact with water in order to obtain autogenous healing and to activate the self-healing approach with SAP, was done by giving the beams a shower with water four times a day during one minute for a time span of six weeks (within the seven weeks healing period).

In order to evaluate the self-healing efficiency, the reduction in water ingress due to crack healing was measured. As can be seen from Figure 1, before crack healing, the water ingress into the cracks of the beam with embedded SAPs was clearly higher compared to the ingress into the other series. This is due to the fact that the SAP particles within the matrix of this beam attract an additional amount of water. However, this will result in a beneficial effect later on, as the water, absorbed by the SAPs, will be released to the surrounding cementitious matrix and result in further hydration and calcium carbonate precipitation. When these newly formed crystals are precipitated inside the cracks this results in an increased autogenous crack healing efficiency. This improved healing efficiency is partly represented by the results shown in Figure 1. While for the REF beam and the beam with encapsulated polyurethane higher water ingress values were obtained after healing, the SHC-SAP beam showed lower water ingress. We believe this should be attributed to healing of the cracks as for the SAP series crack closure was also shown from the microscopic analysis. The fact that higher water ingress was measured for the two other test series is in contradiction with our expectations. However, we believe that this finding is due to the fact that the saturation state of the beams at the moment of the water ingress measurements was different before and after healing.

The initial widths of the cracks in the beams were measured by means of optical microscopy. For the beam with encapsulated polyurethane, it was seen immediately that part of the cracks were filled up by polyurethane. During the seven weeks healing period for all of the beams, crack widths became lower due to autogenous (and autonomous) crack healing. Therefore, crack widths were measured a second time after six weeks spraying with water and the crack width reduction over time was calculated. This evaluation procedure only takes into account the amount of healing obtained during the six weeks showering period and not the extent of immediate healing of the beam with encapsulated polyurethane. Cracks were divided into different categories, based on the original crack width, and crack closing ratios were calculated for each category of each test series. What immediately can be seen from Figure 2 is that smaller cracks are more likely to be healed compared to larger cracks. A gradual decrease of the healing ratio for each test series can be seen with increasing crack width range. A second very clear finding is that the healing ratio for the beam with embedded SAP particles is considerably higher compared to the healing ratios which are obtained for the two

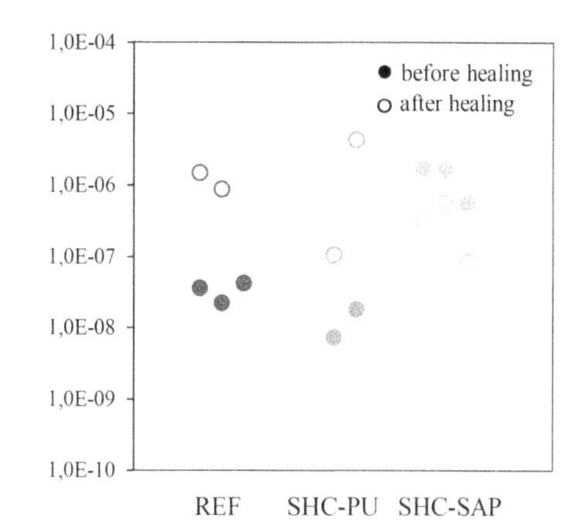

Figure 1. Water ingress into three (some results are missing) selected cracks for each test series measured before and after the healing period.

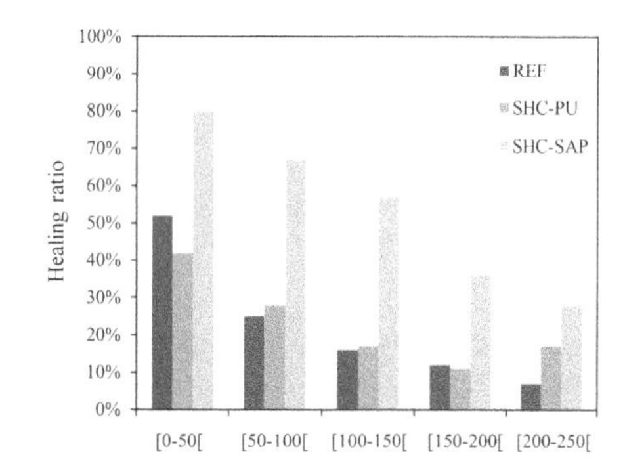

Figure 2. Crack healing ratio obtained within each crack range and for each of the test series under investigation.

other test series under investigation. This finding clearly indicates that the addition of SAPs to the concrete matrix promotes crack healing when the crack faces are exposed to water.

From this study it can be concluded that both self-healing approaches under investigation have some potential to be applied on real scale. However, the use of encapsulated polyurethane requires much more preparation to fill the tubes and position them in the moulds compared to the addition of SAPs. The approach making use of embedded SAP seems to result in the highest healing efficiency. Crack closure was obviously enhanced through the addition of SAPs to the concrete matrix. This was less clear for the beams with embedded polyurethane, however, the healing efficiency of this approach was not reflected by the crack width measurements. While it was hard to prove a good healing efficiency from the water ingress measurements, after crack formation release of polyurethane from the capsules inside the cracks was clearly noticed.

Concrete Repair, Rehabilitation and Retrofitting IV – Dehn et al. (Eds)
© 2016 Taylor & Francis Group, London, ISBN 978-1-138-02843-2

Sprayed textile reinforced concrete layers for a durable protection of waterway engineering structures

C. Morales Cruz & M. Raupach
Institute of Building Materials Research (ibac), RWTH Aachen University, Germany

ABSTRACT

The results of the inspections of hydraulic structures carried out by the Federal Office of Hydraulic Engineering (BAW) have shown that many hydraulic structures require an increased maintenance demand. In several cases, unreinforced tamped concrete hydraulic structures show not only low concrete strengths, but also cyclic opening movement joints and cracks. Considering the upcoming repair measures and taking the different concrete properties of the existing waterway engineering structures into consideration, four "old concrete" classes were created (A1–A4). These concrete substrates are classified depending on their compressive and surface strength (see Table 1).

For the surface repair of these structures anchored, steel-reinforced concrete or sprayed-concrete layers are permitted. Alternatively, non-anchored, unreinforced concrete, mortar, sprayed-concrete or sprayed-mortar layers, which are connected by adhesive strength to the concrete substrates, can be used.

A disadvantage of steel-reinforced concrete or sprayed concrete is the high layer thickness, which is required for the necessary corrosion protection of the steel reinforcement. The associated increase in the component cross-section can be problematic for the repair of hydraulic structures, such as floodgates and weir structures, since the inside space is often limited.

For the old concrete classes A2–A4, it is permissible to produce thinner ($20 \leq d \leq 60$ mm), unreinforced, non-anchored layers by using sprayed-mortar/concrete systems (ZTV-W LB 219 copy 2013). In addition to the limited availability of unreinforced surface repair systems, cyclically opening cracks and joints in the concrete substrates hinder the use of unreinforced sprayed-mortar/concrete systems by causing a "break-through" of the cracks in the repair system and a reduction of the intended service life due to water penetration. This can further result in spalling of the repair layer.

To solve this problem, a durable, watertight, and crack-bridging protective textile-reinforced concrete layer called DURTEX was developed at the Institute of Building Materials Research (ibac) of RWTH Aachen University for restoring concrete, steel-reinforced concrete or natural stone structures (see Figure 1).

A pilot application features the surface repair of a pillar of the *Weir Horkheim* in Heilbronn-Horkheim where promising first results under real conditions were achieved (Orlowsky et al. 2011, Orlowsky et al. 2012, Morales Cruz et al. 2014a). This pilot application is vital for the joint research project between public administration partners, industry partners and the research institutes of RWTH Aachen University. It seeks to further develop and create procedures for the surface repair of hydraulic structures with low strength concrete surfaces.

The paper at hand provides an overview of the functional principle of DURTEX. Furthermore, the components of the developed DURTEX-protective layer, with which a crack-bridging up to 0.3 mm has been achieved, are presented. Additionally, the practical implementation of the developed application methods on the boiler house of the power generating plant *eins energie in sachsen* is introduced. To conclude, first results of an inspection three months after the completion of the sample surfaces are presented.

Table 1. Classification of the concrete substrates according to ZTV-W LB 219.

		Old concrete class			
Characteristic values	Unit	A1	A2	A3	A4
compressive strength	N/mm²	≤10	>10	>20	>30
surface strength (arithmetic mean)	–		≥0.8	≥1.2	≥1.5
surface strength (lowest single value)	–		≥0.5	≥0.8	≥1.0

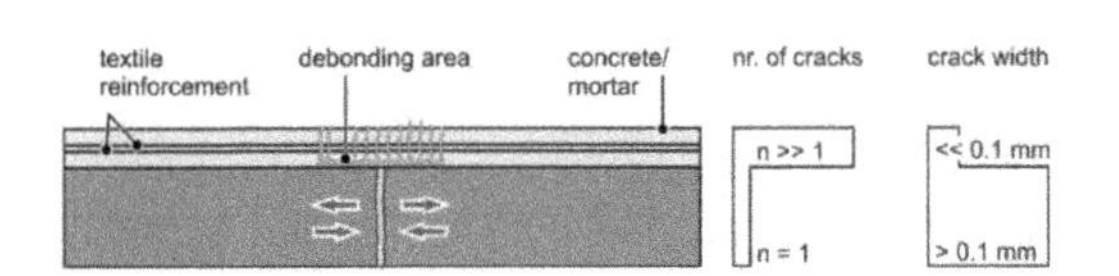

Figure 1. Schematic drawing of the textile reinforced concrete DURTEX-layer. The crack width and the number of cracks on the DURTEX-layer and supporting surface are presented—the de-lamination zone is exaggerated on purpose.

Concrete Repair, Rehabilitation and Retrofitting IV – Dehn et al. (Eds)
© 2016 Taylor & Francis Group, London, ISBN 978-1-138-02843-2

Production and fresh properties of powder type self—compacting concrete in Sudan

O.M.A. Daoud & T.M. Kabashi
Building and Roads Research Institute, University of Khartoum, Khartoum, Sudan

ABSTRACT

Sudan is one of the developing countries that face many problems in the field of construction and infrastructures; one solution of these problems is towards employment of self-compacting concrete. The application of SCC in Sudan is very rare because of lack of research and published data pertaining to locally produced SCC. Therefore, there is a need to conduct studies, further researches, and investigations to get the knowledge of SCC and to be aware about its all aspects; starting from selection of suitable constituents including super plasticizer, mineral admixtures, mix proportion optimization, assessment of the fresh state and hardened state properties to produce it successfully.

This paper presents the results of an experimental study conducted to produce powder type self-compacting concrete in Sudan. The primary objective of the study is to conduct an exploratory work towards the development of a suitable SCC mixes in the hot climate of Sudan, with special consideration focus on solving problems related to satisfactory segregation resistance, filling, and passing abilities thorough conducting a series of trial mixes. In addition, the paper investigates different fresh properties of SCC and their interrelations at the fresh state. Different tests were carried out to characterize raw materials, such as sieve analysis for coarse and fine aggregate, physical properties of aggregates, physical and chemical composition as well as the XRD test for cement and fly ash. The fresh state properties of the SCC trial mixes were assessed using very special tests; such as the slump flow test, V-funnel test, J-ring test and screen stability test. All necessary modifications in proportioning and mixing are mentioned. Mixing procedure adopted in the study is well explained. Different successful SCC mixes were produced and their fresh properties were studied. Some reliable statistical relations between fresh properties of SCC were developed and presented. Comparisons of these relations with previously published relations showed that in spite of the different material type, proportions, humidity, environmental and climatic conditions, the overall performance of the fresh self-compacting concrete mixes achieved, demonstrated same pattern. The effects of fly ash content and super plasticizer dose on fresh properties of SCC are also highlighted. Results also showed that marginal high strength concrete could be produced with same constituent proportions.

The significance of this work lies in its attempt to provide some performance data of powder type SCC containing Class F fly ash; In order to obtain a more complete understanding of these special types of concretes, and to get the practical knowledge towards the commercialization of such concrete in Sudan.

Concrete Repair, Rehabilitation and Retrofitting IV – Dehn et al. (Eds)
 ISBN 978-1-138-02843-2

Utilization of polypropylene fibre reinforced cement composites as a repair material: A review

A. Baricevic, M. Pezer & N. Stirmer
Faculty of Civil Engineering, University of Zagreb, Zagreb, Croatia

ABSTRACT

Environmental and structural loads cause stresses, deformations, and displacements in structures. According to the building codes, they have to be designed and built to safety resist all those actions during entire service life. When deteriorated, structures must be repaired and strengthen so that during usage they do not represent a risk for the users. Due to the low expenses and good mechanical characteristics, cement-based composites are often used for repairs. In structures susceptible to shrinkage and/or exposed to specific load conditions, used cement-based composites are improved with polypropylene fibres. These composites have many beneficial characteristics; like as reduced occurrence of microcracks, positive effect on autogenous, plastic and restrained shrinkage, increased resistance to spalling during the fire, accessible price etc. Due to those characteristics, they are often used for repair of slabs, pavements and concrete overlays.

Detailed state-of-the-art is presented and encompasses both fresh and hardened state properties. Also, a possibility for further improvement of repair materials is discussed.

INTRODUCTION

Environmental and structural loads cause stresses, deformations, and displacements in structures. According to the building codes, they have to be de-signed and built to safely resist all those actions during entire service life. When deteriorated, structures must be repaired and strengthen so that during usage they do not represent a risk for the users. Major issues during repair are the bond quality between substrate concrete and overlay, stresses caused by differential shrinkage and durability of the repaired structure (Silfwerbrand 2006).

When choosing a repair material basic requirement is its compatibility to the substrate concrete. Repair material and substrate should be similar in terms of stiffness or flexibility, measured by modulus of elasticity, in terms of thermal expansion, and other mechanical and durability properties. Due to the low expenses and good mechanical characteristics, cement-based composites are often used for repairs. In structures susceptible to shrinkage and/or exposed to specific load conditions, used cement-based composites are improved with PP fibres. In such cases cement-based composites are improved with PP fibres.

Detailed state-of-the-art analysis on properties of such repair materials is presented in this paper. A short discussion on their possible improvements with ecologically sustainable materials is shown.

Why polypropylene fibres?

Today different types of fibres are available on market, from steel and polymer as well as fibres produced from different recycled materials. Their role is to distribute stresses between damaged and undamaged cross sections during entire service life of structures. During design phase, depending on their properties, fibres are selected based on their future application.

Polymer fibres from different origin; polypropylene, nylon, polyethylene, carbon or aramid fibres have found their application in fibre reinforced concrete (Hannant 1978). Within this paper, properties of cement composites with PP fibres are discussed as these fibres are most frequently used due to their low price, inert behaviour in alkaline environment and ease dispersion in cement matrix (ACI 544.5R-10 2010).

CEMENT COMPOSITES WITH POLYPROPYLENE FIBRES

Plastic shrinkage presents a major concern during repair of deteriorated structures (Ghoddousi & Javid 2010). At lower dosage rates, PP fibres are very effective in controlling plastic shrinkage cracking. By increasing diameter and fibre volume an adequate control of plastic cracks is achieved (ACI 544.5R-10 2010; Naaman et al. 2005; Sanjuhn & Moragues 1997). In such way is possible to reduce the occurrence of cracks by 10% if an appropriate volume fraction (0.2%) and diameter of fibres is used (ACI 544.5R-10 2010).

Positive influence is also recorded for the autogenous and restrained shrinkage (Saje et al. 2012; Saje et al. 2011, ACI 544.5R-10 2010), therefore is possible to establish a relationship between the amount of PP fibres and shrinkage values. Investigation on influence of different PP fibre volumes shown that for 0.25, 0.5 and 0.75% of fibres, autogenous shrinkage decreases progressively for 5, 15 and 26% respectively in relation to the plain concrete after 24 hours (Saje et al. 2012). The value of their contribution to restrained shrinkage depends on type, diameter and shape of used fibres. Monofilament fibres with diameter 18 μm showed higher influence on restrained shrinkage than fibres with lager diameters (30 μm and 36 μm) (Lamour et al. 2005).

Increase of tensile and flexural strength, together with toughness (Song et al. 2005, Ramezanianpour et al. 2013; Patel et al. 2012, Gencel et al. 2011, ACI 544.1R-96 2009; Skazlic 2003; Sadrmomtazi & Fasihi 2010) is achieved with addition of micro and macro PP fibres. ACI 544.1R-96 2009; Skazlic 2003; Sadrmomtazi & Fasihi 2010.

Presence of PP micro fibres has also positive impact on the behaviour of concrete during elevated temperatures. Due to their chemical composition, polypropylene fibres melt at temperatures above 165°C. It is this melting that is believed to facilitate the reduction in the internal stresses in the concrete that cause the explosive spalling (Smith & Atkinson 2010, Jelcic Rukavina et al. 2014). Investigation of influence of different volumes of PP fibres on the values of compressive strength showed that increase in PP fibre volume has minor effect on the compressive strength values (Seferovic 2002; Bayasi & Zeng 1994; Sadrmomtazi & Fasihi 2010; Skazlic 2003; Kumar et al. 2013; ACI 544.1R-96 2009).

Due to their low modulus of elasticity addition of fibrillated PP fibres in quantities varying from 0.1 to 2.0 percent by volume have no effect on the modulus of elasticity compared to the ordinary concrete (ACI 544.1R-96 2009). If high stress levels are maintained, polymer fibers behave viscously and creep, even exhibiting creep rupture (ACI 544.5R-10 2010). Performed tests indicated that creep of macro synthetic fibre concrete is not only considerable, but also leads to creep failure at service loads (Lambrechts 2009).

POSSIBILITY OF REPAIR MATERIALS FURTHER IMPROVEMENT

Using PP fibers in concrete has many benefits. However, surface smoothness, chemical structure and hydrophobic properties, results in poor adhesion characteristics between fibers and cement matrix (Wang et al. 2006). All that, presents a severe limitation of their usage in high performance applications so further improvement of their interfacial strength with matrix is need. Modifying fibre surface by roughening assures improved adhesion in fibre/ matrix interface presents one of the improvements (Lopez-Buendia et al. 2013).

Other possibilities for improvements include usage of secondary raw materials such as Recycled Tyre Polymer Fibres (RTPF). Based on the limited literature data (Serdar et. al 2013; Serdar et al. 2014; Bjegovic et al. 2010; Mavridou 2011, Project Anagennisi 2014), RTPF can be used as substitution of PP fibres, as they do not induce negative effects on concrete mechanical properties, but do enhance concrete behaviour regarding autogenous and restrained shrinkage. Further optimization is needed for their usage as repair material.

CONCLUSIONS

The addition of PP fibres in general has significant influence on specific concrete properties. The main contribution of PP fibres is their beneficial effect on autogenous, plastic and restrained shrinkage of cement composites together with increased resistance to spalling during the fire. Due to those characteristic they are often used in repair materials.

By improving interfacial strength between PP fibres and the cement matrices, further improvement of such materials is possible. Surface treatments like wet chemical treatment, flame treatment, mechanical micro-pitting, cold plasma treatment and alkaline surface treatment and alkaline precursors on PP fibres contribute to the improvement of fibre/concrete adhesion in fibre-reinforced concrete. Other possibilities for improvements include usage of secondary raw materials such as recycled tyre polymer fibres. Their usage has positive effects on shrinkage behaviour, especially autogenous and restrained shrinkage, but further optimization is needed.

ACKNOWLEDGEMENTS

The research presented is part of the project "Anagennisi—Innovative Reuse of all Tyre Components in Concrete" funded by the European Commission under the 7th Framework Programme Environment topic.

Concrete Repair, Rehabilitation and Retrofitting IV – Dehn et al. (Eds)
© 2016 Taylor & Francis Group, London, ISBN 978-1-138-02843-2

A novel technique for self-repair of cracks of reinforced concrete structures

S. Pareek
College of Engineering, Nihon University, Koriyama, Fukushima-ken, Japan

ABSTRACT

Cracking in Reinforced Concrete (RC) structures is an inevitable phenomenon, which may occur due to tensile loads, shrinkage and or thermal loads etc. These cracks in RC structures are one of the most prominent reason for the durability loss by corrosion of reinforcing steel by forming channels for CO_2, water and Cl^- ions to penetrate into the concrete easily. The service life span of RC structures is drastically shortened by these cracks in concrete. The objective of this research work is to develop a self-repair system for cracks of reinforced concrete structures by using a network system. In this study, the regain of flexural strength of ordinary cement mortars by the self-repair system using a repair material in the network system and the storage period of the repair material in the network was taken into consideration. In addition to this, the influence of viscosity of the repair material injected in the network on the regain of flexural strength by self-repair system was also evaluated. In order to prolong the storage period of the repair material in the network, several types of surface treatment techniques were applied and tested.

Figure 1 shows the autonomic self-healing or self-repair system used in this to repair the cracks and the strength regain was obtained at the first (before crack repair) and the second loading (after crack repair). Mortar specimens were prepared having a hollow network in the specimens and loaded to develop by using a fully-automatic load controlled machine to generate cracks of 0.05~0.20 mm. Photo 1 shows the mortar specimen with crack injection system into the network. Table 1 shows the one-component epoxy resin used as crack repair material through the network system. Figure 2 shows the test program to investigate the effect on epoxy resin as repair material by pre- and post injection, and curing conditions on the regain of flexural strength at 2nd loading(after crack repair). Furthermore, in order to prolong the storage of epoxy resin in the network, several types of surface treatments were adopted to prevent the hardening in the network.

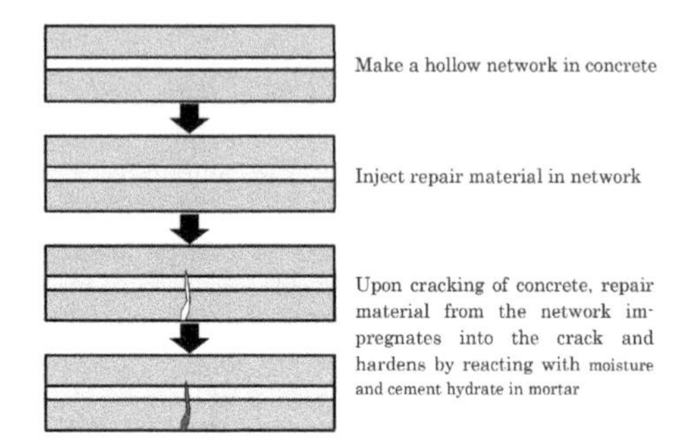

Figure 1. Concrete self-repair system.

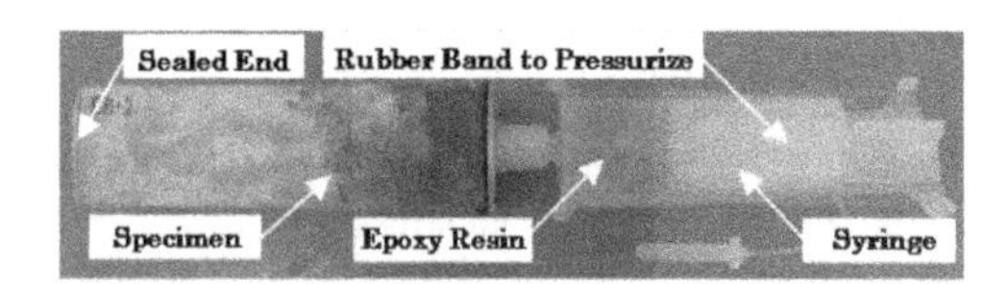

Photo 1. View of crack injection system.

Table 1. Properties of epoxy resin.

Type of epoxy resin	Hardening mechanism	Thiro-tropic index	Specific gravity	Viscosity (mPa · $s_2$23°C)
A	Moisture Sensitive	6.5	1.07	14000
B		2.2	1.07	1900
C		1.0	1.15 ± 0.05	150 ± 100

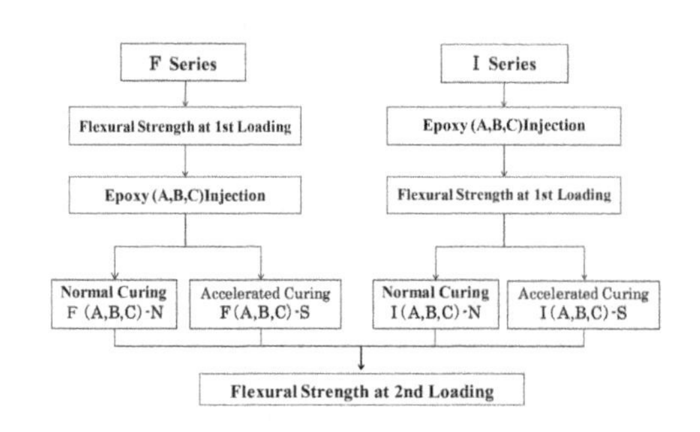

Figure 2. Test program to investigate the basic condition of repair material and curing conditions.

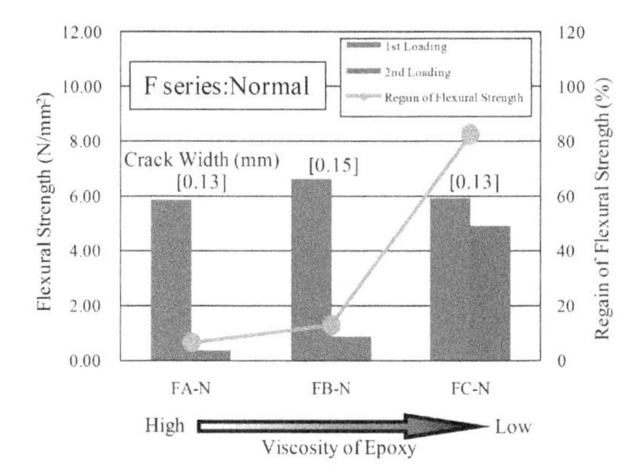

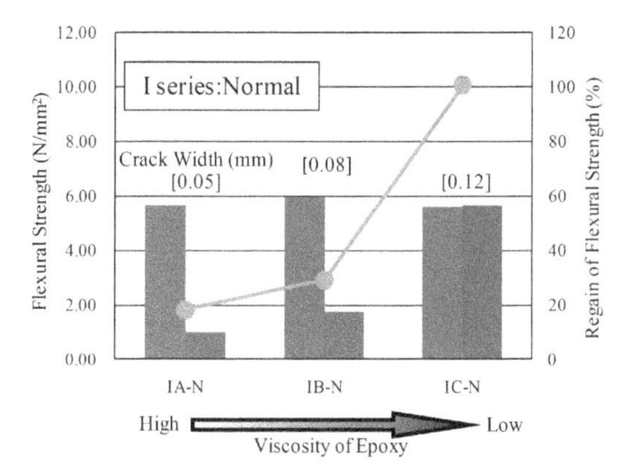

Figure 3. Influence of viscosity and curing condition on regain of flexural strength.

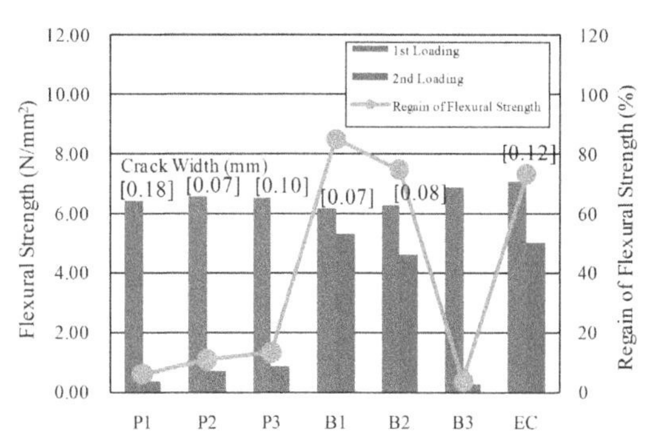

Figure 4. Influence of surface treatment of network on regain of strength after 28d storage on regain of flexural strength.

Figure 3 shows the influence of viscosity of epoxy resin and crack width of the mortar specimens in consideration to the regain of flexural strength of mortars at second loading. Figure 4 shows the influence of surface treatment of network on the regain of flexural strength of mortars at second loading. From the test results, it was found that the repair material with a lower viscosity was more effective as a repair material and showed a remarkable regain of flexural strength of mortar specimens.

Furthermore, the internal network surface treatment techniques helped to increase the storage period of the repair material in the networks.

Concrete Repair, Rehabilitation and Retrofitting IV – Dehn et al. (Eds)
© 2016 Taylor & Francis Group, London, ISBN 978-1-138-02843-2

The effectiveness of corrosion inhibitors in reducing corrosion in chloride contaminated RC structures

P.A. Arito & H.-D. Beushausen
Concrete Materials and Structural Integrity Research Unit, Department of Civil Engineering, University of Cape Town, South Africa

ABSTRACT

This paper presents the results of an investigation into the effectiveness of a corrosion inhibitor (Protectosil CIT®) using reinforced concrete beam specimens with three different binder types. Active corrosion in the specimens was induced by cyclic (2 weeks) drying and immersion of test specimens in a salt solution. Three levels of corrosion, namely: low (<0.5 μA/cm^2), moderate (0.5–1.0 μA/cm^2) and high (>1.0 μA/cm^2) were induced in the specimens. The effectiveness of the corrosion inhibitor was monitored over a duration of 134 weeks from their time of application. The assessment of the effectiveness of the corrosion inhibitor was done with respect to corrosion rate, Half-Cell Potential (HCP) and concrete resistivity measurements. The test results show that corrosion rates and HCPs decreased significantly in all the specimens in which the corrosion inhibitor was applied, indicating that the inhibitor was successful in lowering the risk of corrosion damage to the specimens.

INTRODUCTION

Corrosion inhibitors are chemical substances that decrease corrosion rate when present in the corrosion system at suitable concentration, without significantly changing the concentration of any other corrosion agent (Söylev and Richardson, 2008). They control corrosion in Reinforced Concrete (RC) structures through a combination of the following mechanisms: delaying the onset of corrosion by reducing the rate of ingress of chlorides, increasing the chloride threshold value for corrosion initiation, reducing the rate of corrosion once it is initiated, influencing the degree to which chlorides are chemically bound in the concrete cover, and by influencing the electrical resistance and chemical composition of the concrete system. This paper presents the results of an investigation into the effectiveness of a chloride inhibitor in reducing corrosion of steel in RC beams specimen exposed to a chloride contaminated environment.

EXPERIMENTAL SET UP

Test beam specimens conformed to the ASTM G-109 specifications. They were exposed to a controlled laboratory environment—average temperature and relative humidity ranging between 23 ± 2°C and 50 ± 4% respectively. Corrosion rates, half-cell potential and concrete resistivity were measured bi-weekly over 134 weeks except for the period during the application of the corrosion inhibitor (between weeks 62–66).

Test materials, mix design, casting and curing

The following materials were used to make the beam specimens:

i. Binders: CEM I 42.5 N, Fly ash (class F), Ground Granulated Blastfurnace Slag (GGBS),
ii. Klipheuwel sand—(F.M = 2.7; rel. density = 2.65),
iii. Greywacke stone—(CBD = 1570 kg/m^3; rel. density = 2.7).

A summary of the concrete mix proportions used to make the specimens, and some selected concrete properties, are presented in Table 1.

For each binder type, 6 beams (3 with the corrosion inhibitor applied after inducing corrosion and 3 without the corrosion inhibitor after inducing corrosion) were cast resulting in a total of 18 beams. The specimens were cast and de-moulded after 24 hours. De-moulded specimens were thereafter water-cured in a bath maintained at a temperature of 23 ± 2°C for 28 days.

Inducing active corrosion, application of corrosion inhibitor and corrosion monitoring

After 28 days of water-curing, active corrosion (i_{corr} $ 0.1 μA/cm^2) was induced in the specimens by exposing them to 2 cycles of 2-week drying at 70°C followed by a 2-week submersion in a 3% NaCl solution. After the induction of active corrosion, specimens were subjected to a cyclic 2-week wetting with 3% NaCl solution followed by a 2-week drying until steady corrosion rates were recorded.

Table 1. Summary of concrete mix proportions and selected concrete properties.

Binder composition		100% PC	50/50 PC/GGBS	70/30 PC/FA
Material (kg/m^3)	w/b ratio	0.50	0.50	0.50
	Mix label	PC-50	SL-50	FA-50
PC (CEM I 42.5 N)		370	185	259
GGBS		–	185	–
Fly Ash		–	–	111
Klipheuwel sand:	2 mm max.	827	827	827
Greywacke stone:	13 mm	1040	1040	1040
Water content		185	185	185
28-day compr. strength (MPa)		49.3	45	39
Slump (mm)		58	72	94

Table 2. Average corrosion rates in test specimens.

Specimen ID		Corrosion rate ($\mu A/cm^2$)	
		Week 62	Week 134
PC-50	Without inhibitor	1.22	1.53
	With inhibitor	1.25	0.76
FA-50	Without inhibitor	0.83	1.10
	With inhibitor	0.83	0.48
SL-50	Without inhibitor	0.50	0.80
	With inhibitor	0.52	0.21

The corrosion inhibitor was applied to the specimens, according to the manufacturer's specifications, when corrosion rates had stabilised. The stabilisation of corrosion rates was achieved at 62 weeks.

Corrosion monitoring comprised bi-weekly measurements of corrosion rate, Half-Cell Potential (HCP) and concrete resistivity. Corrosion rate was measured using the Gecor 6™. HCP was measured against a silver/silver chloride (Ag/AgCl) reference electrode as specified in the ASTM C 876. Concrete resistivity was measured using the 4-point Wenner probe technique. Corrosion rate, HCP and concrete resistivity were not measured during the period of application of the corrosion inhibitor (week 62–66).

RESULTS AND DISCUSSIONS

A summary of average corrosion rates in the specimens is presented in Table 2.

From Table 2, it is evident that corrosion rates increased in specimens without an inhibitor. Thus, it can be inferred that the inhibitor reduced corrosion rates. The corrosion rates in specimens without an inhibitor, however, increased by: 25% in PC specimens, 32% in PC/FA specimens and 60% in PC/GGBS specimens. Whereas the experimental setup could not facilitate the direct comparison of corrosion rates across specimens made using different binders, owing to the fact that different corrosion rates were induced in each series of test specimens initially, it can be observed that PC/GGBS specimens experienced the greatest percentage reduction in corrosion rates. The inhibitors in this study seem to be more effective in mitigating corrosion in specimens whose levels of corrosion are in the following order: low, moderate and high respectively. An explanation of this phenomenon could not be obtained based on the tests that were undertaken. Nevertheless, a more comprehensive and accurate explanation of this phenomenon could be obtained by investigating various specimens made using different binder types and exposed to similar initial corrosion conditions. HCP values became less negative after the application of the inhibitor. This observation implies a reduction in corrosion rates – a phenomenon that corroborated results from corrosion rate measurements. The application of the corrosion inhibitor did not affect concrete resistivity.

CONCLUSIONS

The effectiveness of a commercial corrosion inhibitor, was investigated by carrying out accelerated corrosion experiments in the laboratory. A modified version of ASTM G109 test method was used. The test results show that corrosion rates not only decreased in the specimens with low corrosion rates (<0.5 $\mu A/cm^2$) but also in those with moderate (0.5–1.0 $\mu A/cm^2$) and high (>1.0 $\mu A/cm^2$) corrosion rates. Corrosion rate and half-cell potential measurements showed very good agreement in determining the effect of the corrosion inhibitor, confirming that both methods yielded valid results. The corrosion inhibitors used in this study therefore show good potential for application in the prevention of damage related to chloride-induced reinforcement corrosion.

ACKNOWLEDGEMENTS

The authors would like to express their sincere acknowledgements to Mr. Peter Horn (of Evonik Industries, Germany), Mr. Stephen De Klerk (of BASF South Africa (Pty) Ltd) and Dr. Mike Otieno (of the University of the Witwatersrand) for their support and contribution towards the success of this project.

Concrete Repair, Rehabilitation and Retrofitting IV – Dehn et al. (Eds)
© 2016 Taylor & Francis Group, London, ISBN 978-1-138-02843-2

Protection against biogenic sulfuric acid corrosion—development of a sprayed polymer concrete

Robert Schulte Holthausen & Michael Raupach
Institute of Building Materials Research, RWTH-Aachen University, Germany

ABSTRACT

Concrete sewage systems can be severely damaged when attacked by biogenic sulfuric acid corrosion. While minor aggressive exposures can be accounted by using a high sulfate resistant cement, higher concentrations of acid (pH 3.5) require the protection of concrete surfaces. (Kampen et al. 2011) A number of different linings and coatings have been used in sewage systems throughout the decades (Orlowsky et al. 2010), such as Polymer Concrete (PC) and or manually applied ceramics and inorganic glass. However, these have been limited to newly constructed pipes and manholes as well as local repair.

In cooperation with Massenberg GmbH, the Institute of Building Materials Research of RWTH Aachen University developed a new way of applying PC for the rehabilitation of damaged concrete surfaces of sewage systems; by dry spraying an epoxy mortar on concrete, layers with an average thickness of 20 mm on vertical and overhead surfaces were successfully produced. (Schulte et al. 2014) The dry sprayed PC was examined for its overall performance such as strength, adhesion, thermal expansion, and wear. A special focus has been put on the diffusion properties and the durability in an acidic environment.

The results in Table 1 (excerpt) show the good mechanical properties and the technical impermeability. The sprayed PC shows very good adhesion to the concrete substrate on vertical and overhead surfaces, even on partly wetted substrates. The measurements by nuclear magnetic resonance to valuate the diffusion properties, shown in Figure 1, indicate, that an ingress of sulfur can be monitored non-destructively. As seen in Figure 1, a reduction of T_1-Relaxation, a material property indicating molecular mobility as well as the existence of magnetic impurities in the material, is detected at the specimens surfaces.

Table 1. Material properties of the sprayed PC.

Property	Unit	Value
Mortar density	[g/cm³]	2.02
Compressive strength	[N/mm²]	90 to 105
Tensile Strength		12 to 16
Young's-modulus		13 500 to 18 000
Coefficient of sorption	[kg/(m² d$^{0.5}$)]	0.001
Pull off strength	[N/mm²]	> 3.5 average > 2.6 minimum

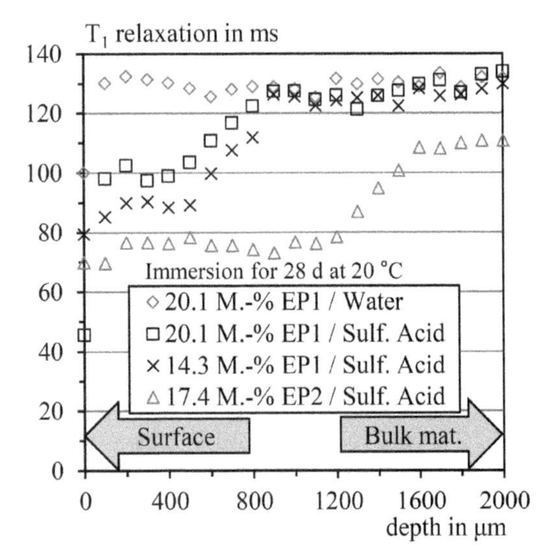

Figure 1. NMR spin-lattice-relaxation profiles through the first 2 mm from surface of SPC-specimens after the immersion in water and sulfuric acid for 28 days.

Based on these positive laboratory results, a first practical application in a waste water structure is currently underway.

Concrete Repair, Rehabilitation and Retrofitting IV – Dehn et al. (Eds)
© 2016 Taylor & Francis Group, London, ISBN 978-1-138-02843-2

Colloidal nanosilica application to improve the durability of damaged hardened concrete

M. Sánchez & M.C. Alonso
Institute of Construction Sciences "Eduardo Torroja" (IETcc–CSIC), Madrid, Spain

I. Díaz & R. González
Autonomous University of Nuevo León, Monterrey, Mexico

ABSTRACT

Nowadays, ageing of existing concrete structures exposed to highly aggressive environments is one the highest concerns for the construction industry. In this context, developing sustainable, efficient and secure technologies for the repair and maintenance of such structures rises as a challenge to assure their service life. In present paper, a non-invasive repair method, based on the application of colloidal nanosilica on the concrete surface, is proposed. Two mechanisms to force the nanosilica penetration have been considered: capillary suction and migration. The ability of colloidal nanosilica to consolidate the cementitious matrix has been confirmed; nanosilica promotes the decrease of capillary pores not only filling the pores but also interacting chemically with the substrate to form new gels, which seem to evolve with time in presence of the appropriate humidity conditions.

INTRODUCTION

The migration of silica nanoparticles covered with nanoalumina as further stage of a ECE treatment to seal the concrete surface protecting it against new penetration of aggressive has been proposed (Cárdenas, 2006). Also the viability of promote the migration of colloidal nanosilica under the action of an electric field, showing a compatible interaction between the nanosilica and the cementitious substrate, has been confirmed (Sánchez-Moreno, 2013). A decrease of the total porosity, with a refinement of the capillary pores is observed with this type of treatment. The reactivity of the nanosilica is proposed by these authors (Sánchez, 2014) forming C-S-H gels with high silicon content. A maximum penetration depth of 2 mm was reached after the treatment of nanosilica migration.

In presence study the influence of the transport mechanism on the efficiency of superficial treatment with colloidal nanosilica is assessed. Two different treatments are proposed: migration under the action of the electric field (M) and capillary suction through a previously dried mortar surface (CS).

EXPERIMENTAL PROCEDURE

Cylindrical mortar samples of 7.5 in diameter and 15 cm were prepared. The samples were cured for 7 days within a humid chamber under controlled conditions (98 ± 2% relative humidity and 21 ± 2°C). After curing, the cylindrical samples were cut in thinner slices, 5 cm thick, which were submitted to the nanosilica penetration treatments (15% by weight).

Two different methods have been assessed to promote the transport of the colloidal nanosilica through the concrete pores: Migration (M) and Capillary Suction (CS).

Migration treatment

Before treatment, mortar samples were saturated in water under vacuum conditions. During testing, an external anode and cathode were located on both surfaces of the mortar slice. Stainless steel meshes were used as electrodes. A tank with the colloidal nanosilica solution was located on the mortar surface in contact with the cathode; at the opposite side of the mortar sample, connected as anode, a tank with distilled water were located. After finishing the treatment, the handled samples were cured in an environment with high humidity, stored in a closed box with water in the bottom.

Capillary suction treatment

Before capillary suction of colloidal nanosilica diluted solution, mortar samples were dried at controlled conditions (25°C, 30% RH) until constant weight in order to favor the transport by suction of the nanosilica solution. The arrangement was similar than in the case of migration but without the connection of the electric field. The treatment was carried out in 5 cycles defined by 2 days of nanosilica penetration (tank filled with the solution) and 5 days of drying at the laboratory atmosphere (tank empty).

Sealing efficiency characterization

Indirect sealing parameters

Resistivity measurements were carried out following the arrangement defined in the Spanish standard for direct measurements (UNE 83988-1:2008). After the migration treatment, the resistivity was monitored during the 28 days of storage under high humid conditions. In the case of capillary suction treatment, resistivity measurements were carried out during the 28 days of treatment; the measurements were obtained in high humid conditions, just before empty the tank for starting each drying stage.

Direct sealing parameters

After the migration treatment, the effective chloride diffusion coefficient was determined at the age of 28 days after finishing the treatment, using an accelerated test defined in the Spanish standard (UNE 83987:2009).

After the capillary suction treatment with colloidal nanosilica, the coefficient of water absorption (K) and the mortar effective porosity (ε_e) were estimated following the procedure described in the Spanish standard for determining the absorption of water by capillarity (UNE 83982:2008). This test was carried at the end of the treatment (30 days total length).

RESULTS AND DISCUSSION

Efficiency of migration treatment

The efficiency of the migration treatment in protecting the mortar samples against the further penetration of chloride was analyzed and the value of the effective chloride diffusion coefficient was estimated at the age of 28 days after finishing the treatment.

In Figure 1 the evolution of chloride in the anolyte during the accelerated migration test for estimating the effective chloride diffusion coefficient is included. A non-treated mortar sample, stored under the same humid conditions than the treated ones has been included as the reference case.

The estimated values for D_{eff} are included in Table 1 for both treated and reference mortar samples; a significant decrease in D_{eff} is registered for treated samples (>65%).

Efficiency of capillary suction treatment

In Figure 2 the increase of mass (ΔQ) the two tested samples after the capillary suction treatment with nanosilica penetration has been represented. Two non-treated mortar samples at the same age were also tested as reference cases.

In Table 2 the values estimated for the coefficient of water absorption (K) are resumed for both treated and reference cases. A significant decrease on this coefficient is obtained for mortar samples after the treatment with penetration of colloidal nanosilica by capillary suction.

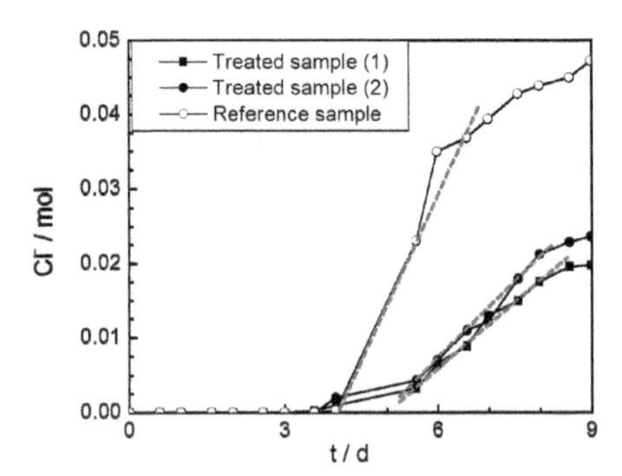

Figure 1. Evolution of chloride in the anolyte during chloride migration test. Colored lines: stationary-state period.

Table 1. Effective chloride diffusion coefficient in treated samples after the colloidal nanosilica migration treatment.

	D_{eff} (10^{-8} cm^2/s)	% Decrease
Treated sample 1	1.33	76
Treated sample 2	1.99	65
Reference sample	5.62	

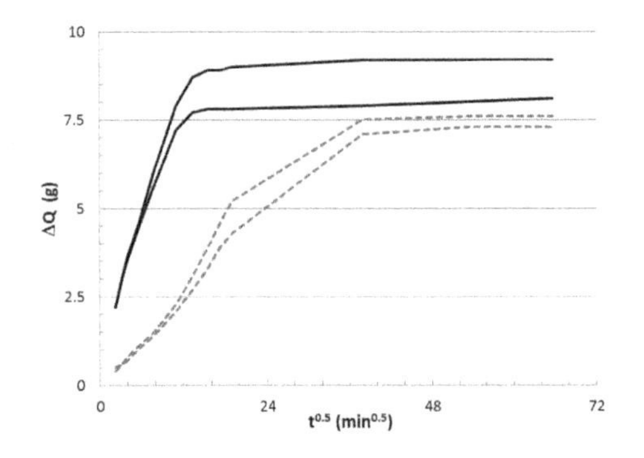

Figure 2. Increase of mass during test for characterizing the capillary water absorption coefficient after the nanosilica penetration by capillary suction.

Table 2. Coefficient of water absorption (K) and mortar effective porosity (ε_e) after capillary suction.

	K (kg/m^2 $min^{0.5}$)
Treated sample 1	$6 \cdot 10^{-4}$
Treated sample 2	$5 \cdot 10^{-4}$
Reference sample	$2 \cdot 10^{-3}$
Reference sample	$2 \cdot 10^{-3}$

ACKNOWLEDGEMENTS

M. Sánchez also acknowledges the Spanish Ministry of Economy and Competitiveness (MINECO) her "Juan de la Cierva" postdoctoral position (JCI-2011-09572).

Concrete Repair, Rehabilitation and Retrofitting IV – Dehn et al. (Eds)
© 2016 Taylor & Francis Group, London, ISBN 978-1-138-02843-2

Combined influence of slag composition and temperature on the performance of slag blends

O.R. Ogirigbo & L. Black
Institute of Resilient Infrastructure, University of Leeds, West Yorkshire, England

ABSTRACT

Ground Granulated Blast Furnace Slag (GGBS) is known to improve the chloride ingress resistance of concrete when used in combination with Portland cement. However, the nature of the ore, composition of the limestone flux, coke consumption and the type of iron being made are factors which affect its chemical composition. There are standards which regulate the use of GGBS. For example, BS EN 197-1:2011stipulates that for GGBS the $(CaO + MgO)/SiO_2$ ratio by mass must exceed 1 (EN197-1:2011). However, work done by several authors has shown that these oxide (basicity) ratios may not be an accurate predictor of slag reactivity and performance. For example, a study carried out by (Mantel, 1994) investigating the hydraulic activity of five different slags concluded that there was no clear correlation between the basicity ratios and the properties of the slags.

This work focused on the impact a change in chemical composition of GGBS will have on its performance in terms of strength and transport properties in different temperature conditions.

Two slags (1 & 2) having Ca/Si ratios of 1.05 and 0.94 respectively were used to partially replace a CEM I 52.5R at 30%. Various tests including compressive strength, sorptivity, chloride ingress and chloride binding were carried out on mortar specimens to measure the performance of the slag blends against a CEM I 42.5R at 20 and 38°C. The mortar specimens were cured for 28 days before exposure to a 3% sodium chloride solution.

The results obtained showed similar strength performance for the slag blends at 20°C especially at later ages. The early age strength of the slag blends was lower than that of the neat system but this trend reversed at later ages. The higher temperature curing of 38°C improved the early age strength of the slag blends but not the later strength with slag 1 performing better than slag 2.

Similar trend of results was obtained for the transport properties studied. The slag blends were much more resistant to the penetration of water and aggressive ions. The high temperature exposure increased the sorption of water and the ingress of chloride ions for all the mixes. The performance of slag 2 was affected more by the high temperature exposure as compared to that of slag 1.

The results from the chloride binding test showed chloride binding capacity in the order of slag 1 > slag 2 > CEM I 42.5R, which correlated with the results obtained for the transport properties. Slag 1 had a higher chloride binding capacity and better resistance to chloride ingress than slag 2 and this was attributed to its higher alumina content (Table 1).

The results obtained showed that temperature had more influence on the performance of the slag blends than the variation in chemical composition.

Table 1. Properties of cementitious materials.

Property	Unit	C42.5R	C52.5R	S1	S2
LOI 950°C	%	2.20	2.54	(+1.66)	(+0.40)*
SiO_2	%	19.71	19.10	36.58	40.14
Al_2O_3	%	5.08	5.35	12.23	7.77
MnO	%	0.03	0.03	0.64	0.64
Fe_2O_3	%	2.97	2.95	0.48	0.78
CaO	%	63.16	62.38	38.24	37.90
MgO	%	2.19	2.37	8.55	9.51
SO_3	%	2.97	3.34	1.00	1.47
Glass	%	na	na	99.3	97.1
Blaine	cm^2/g	3510	5710	4490	4090
Density	g/cm^3	3.23	3.18	2.94	2.95

*The sample was oxidized with HNO_3 before the determination of LOI.

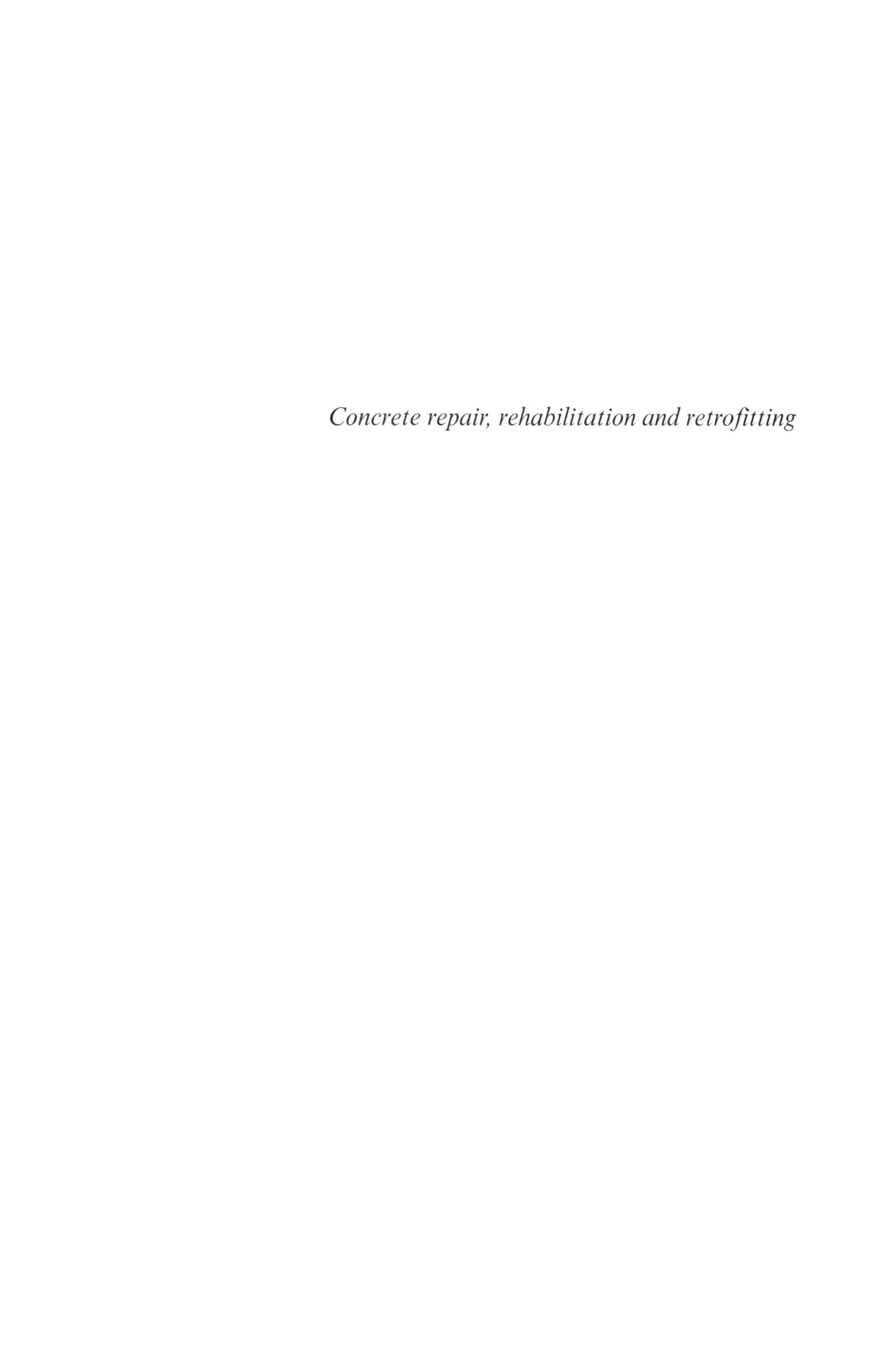

Concrete repair, rehabilitation and retrofitting

Concrete Repair, Rehabilitation and Retrofitting IV – Dehn et al. (Eds)
© 2016 Taylor & Francis Group, London, ISBN 978-1-138-02843-2

The construction of the bridge over the Vaal River in Warrenton—a case study

Tiago Massingue
The South African National Roads Agency Limited SOC (SANRAL), South Africa

ABSTRACT

The construction of the bridge over the Vaal River in Warrenton replaces an existing low level bridge that had been in operation for 83 years. From an economic perspective, the new bridge offers significant improvements in terms of living conditions to the people of Mahareng as during the rainy season they were deprived from crossing the Vaal River for approximately six weeks. Specific objectives outlined during the design stage, highlighted to the following improvements over the existing low level bridge:

- providing an extra lane capacity, improving traffic flow in both directions,
- providing new pedestrian footpaths and thus enhancing safety,
- providing safer road shoulders for bicycles crossing the bridge,
- improved discharge capacity during high floods of the Vaal River,
- reducing the risk of floods which have in the past affected the local community in Warrenton,
- improving the economic development for Warrenton,
- Enhancing the overall traffic flow and safety on the N18 as a National Road.

The existing low level bridge was to be demolished in order to accommodate the new one. However the existing legislation in South Africa [Section 34 Act 25,1999—South African Heritage Resources Agency] outlines that any structure that is in existence for over 60 years must be protected in terms of heritage requirements. Therefore, while the above objectives were still justified, Structural Retrofitting for altered service conditions was to be given to the old structure in order to preserve its historical significance. The Project has shown at the end that new and old can come together in a proactive bridge development interface. Critics argue that the old bridge has enhanced the significance of the new one. Additional benefits were gained by means of job creation, training programs and the SMME development in the area. This is a positive South African story about the development of bridge infrastructure projects as part of the South African National Roads Agency's mandate to promote and develop road network of the country.

INTRODUCTION

Historically, the original bridge over the Vaal River in Warrenton opened for traffic on the 29 October 1931 and has the following dimensions: 55 spans of 16 m each and 2.5 m height. To accommodate pedestrians on the single carriageway bridge of 3 m width, 1 m × 0.6 m precast concrete side sidewalks were provided and painted in yellow for aesthetics. Although it contained several constraints in terms limited bridge cross section allowing one direction of traffic flow at a time and overtopping by floods causing the isolation of local Municipality up to six weeks, it has been successfully in use for 83 years.

The National Roads Agency initiated the process to revert this situation by defining a design team for the new bridge in 2007. Ravi Rooney and Edwin Kruger were responsible for the Drainage and Structural Design. Petronella Theron was assigned to work on access roads design while T Massingue would be responsible for the overall project management and implementation in the Western Cape. Limited scope was later assigned to expert consultancy firm, SNA to oversee some of the SANRAL functions and later the supervision of works. The new design criterion is based on the South African design requirements in terms of the Technical Recommendations for Highways THR 4 Road Category, THR 7 for Bridges and the National Water Act 1998, Section 21(c): Impeding or diverting the flow of water in a watercourse and Section 21(i): Altering the bed, banks, course or characteristics of a watercourse.

This paper deals with the construction of a new bridge, Bridge No. B0067 and the associated road works for the realignment of National Route 18 Section 1, N18-1 across the Vaal River at Warrenton, in proximity to and downstream the old existing single lane, low level river bridge, Bridge No. B2284 on N18-1. The new designed outlined a total length of realignment of N18-1, inclusive of the new bridge to was 1.48 km and the bridge length is 368 m.

Work undertaken in terms of the contract includes:

- Establishment of the contractor;
- The provision of offices, laboratories and housing for the engineer's site personnel,
- The accommodation of traffic. The existing low level river bridge remained in use for the accommodation of traffic throughout the construction of the project.
- The relocation of existing electrical (Eskom and Municipal) and telecommunication services,
- The construction of roads works,
- The construction of surface drainage,
- The construction of subsoil drainage
- The construction of the new Bridge B0067 across the Vaal River.
- The rehabilitation of the existing low level bridge, which was retained due to the historic significance thereof,
- Ancillary work consisting of:
  - The installation of service ducts;
  - The construction of segmental block paved sidewalks
  - The construction of gabion erosion protection, including the protection of approach fills to the new bridge across the Vaal River;
  - The erection of guardrails;
  - The erection of fencing,
  - The relocation of existing road signs and the erecting of new road signs
  - The application of road markings,
  - The establishment of grassing,
  - Finishing of the road and road reserve and the treatment of the old road;
- The erecting of the following at the existing low level bridge across the Vaal River;
  - Bollards to prevent vehicular access to the bridge
  - Information signs regarding the use of the bridge
- The construction of a segmental block paved rest area on the Warrenton side of the Project. A commemorative plaque providing a brief background of the existing low level bridge and new bridge was erected at the rest area. A blocked paved walkway was constructed from the rest area to the existing low level bridge.
  - The erecting of the overhead lighting over an 800 m length of the project, including across Bridge no. B0067.

PROJECT OUTPUTS

At the end of the 22 months construction period, the following road and bridge construction outputs were achieved:

- 144 Precast Beams
- 27.2 tonnes per beam;
- 16 Prestressing Strands,
- Total length presstressing Strand in Beams approximately 59.04 km,
- Total Length Rebar in Beams: approximately 209.4 km
- Elastomeric Bearings: 288
- Total Volume Concrete in Bridge: approximately 6295 m^3
- Total Area Formwork (Temporary and Permanently): Approx 10350 m^2
- Mass of Total Structure: Approx 16000 tonne

In line with the South African Government's Policy to address poverty the following aspects were also included in the project:

- Labour Intensive Methods,
- The Contract Participation Goals (CPG)
- Training & Skills transfer Programs, and
- The Socio-Economic Development,
- Environmental Control & Mitigation Actions

As a result, the proportion of persons-hours in terms of employment throughout the project duration was: Male: 255884 and Females: 38392. The value of employment of local persons was in the order of R5.6 million. The value of employment of all persons was: Male: R10.9 million and for Female was R1.1 million, totaling overall R12 million.

The overall project cost was R86 million which is approximately EURO 8 million.

Concrete Repair, Rehabilitation and Retrofitting IV – Dehn et al. (Eds)
 ISBN 978-1-138-02843-2

Numerical study on effect of ductility in the flexural capacity enhancement of RC beam strengthened with FRP

A. Kashi & M.Z. Kabir
Department of Civil and Environmental Engineering, AmirKabir University of Technology, Iran

F. Moodi
Concrete Technology and Durability Research Center, AmirKabir University of Technology, Iran

ABSTRACT

The repair, rehabilitation and strengthening of Reinforced Concrete (RC) beams is a challenging engineering problem. Flexural strengthening of these members is achieved by attaching FRP strengthening system to the tension face of a flexural member. Previous studies have demonstrated by attaching externally bonded fabric of Carbon Fiber Reinforced Polymer (CFRP) laminate to the tension side of reinforced concrete beams can increase stiffness and maximum load of the beams.

In this report, a finite element model was posted out to examine the effect of ductility on the flexural capacity enhancement of RC beams after attaching CFRP. Two cases of RC beam (552 mm × 305 mm × 6840 mm) were analyzed by FE modelling (Figure 1). The field of longitudinal flexural rebar has been dissimilar in these two types of the beam. The total area of flexural rebar in series A beams is more than series B. Modelling of three point bending experiment for control beams (without attaching CFRP) showed that beam in type B is more ductile than type A. By applying CFRP on the tension face of the beams, the gain in maximum flexural load of the specimens reached values of approximately 8.8% and 13.4% in type A beam and 35% and 46% for type B beam respectively, for 1 mm and 2 mm thickness of CFRP. By comparison, between series A and B, it was found that the effect of CFRP in increasing of flexural capacities, was higher in a more ductile beams. On the other hand by reduction of ductility the effect of CFRP becomes lower.

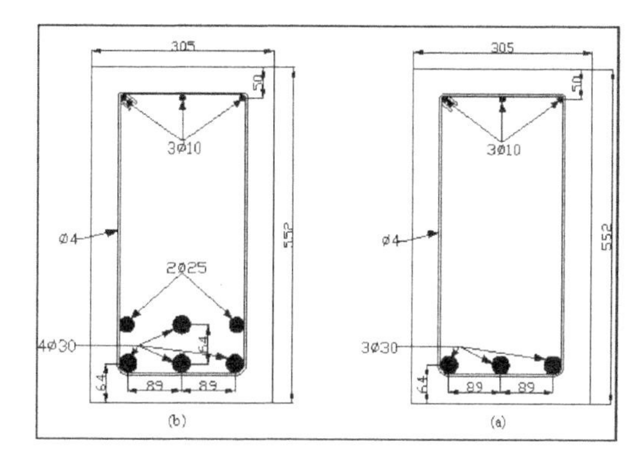

Figure 1. Beams cross section (a) Group B (b) Group A.

Concrete Repair, Rehabilitation and Retrofitting IV – Dehn et al. (Eds)
© 2016 Taylor & Francis Group, London, ISBN 978-1-138-02843-2

Study of the crack pattern and its evolution by DIC of RC beams externally reinforced with TRC and CFRP

S. Verbruggen, T. Tysmans, J. Wastiels & S. De Sutter
Department of Mechanics of Materials and Constructions, Vrije Universiteit Brussel, Brussel, Belgium

INTRODUCTION

Textile Reinforced Cements (TRC) are emerging as an external reinforcing material for concrete structures. Especially the high fibre volume fraction TRCs are investigated by researchers as an alternative for the already established Carbon Fibre Reinforced Polymer (CFRP) strips because of their high mechanical performances, their advantages related to fire safety and their relatively low cost. The application of an external bending reinforcement on a reinforced concrete beam will considerably influence the cracking behaviour of this beam, with potential benefits for the concerning serviceability limit state.

This paper presents an extensive experimental study of this influence and analyses the difference between the different reinforcing materials (CFRP and TRC). This paper proposes the use of the Digital Image Correlation (DIC) technique to measure the crack pattern and its evolution. DIC is an optical, non-contacting measuring technique, enabling the possibility to determine the displacement- and deformation field of the specimen's surface under any type of loading condition. Based on these DIC measurements the bending cracking behaviour of reinforced concrete beams with spans of 0.6 and 2.3 meters is studied. For the TRC reinforced beams this external reinforcement is applied over the full bottom surface, while for the CFRP alternative a smaller strip, still applied over the full beam length is necessary to achieve a similar ultimate load. This is the consequence of the larger material strength and stiffness compared to TRC.

0.6 METER SPAN SPECIMENS

For the small scale beams, the paper shows that the TRC-reinforced beams exhibit a similar crack pattern and evolution to the CFRP-reinforced beams, with the same number of cracks and a comparable crack width. Both external reinforcements cause a retention of the initial high stiffness in the load-deflection curve, but for both materials cracks initiate at a similar load level as for the reference non-externally reinforced beam. This is the consequence of the crack bridging capacity of the external reinforcement.

2.3 METER SPAN SPECIMENS

To exclude scale effects of the smaller scale beams, beams with a 2.3 meter span are tested. Due to the much smaller CFRP strip compared to the TRC, a less efficient crack bridging is obtained. This results in a smaller amount of cracks which grow wider in the CFRP reinforced specimen. Moreover, these CFRP strengthened beams are not able to retain the initial high stiffness compared to a reference non-strengthened beam.

CONCLUSIONS

Compared to CFRP strips and depending on the set-up and the failure modes, IPC TRC as a reinforcing material can offer several advantages related to the cracking and damage behaviour of a concrete beam. These advantages originate mainly from the lower strength and stiffness of the IPC TRC, which leads to a larger cross section (and thus contact width) than the one necessary with CFRP to meet a similar ultimate load.

The DIC measuring technique proved a valuable tool to measure the cracking behaviour of externally reinforced concrete beams. This is the consequence of the broad spectrum of possible quantitative and qualitative results, obtained by performing these DIC measurements on the side and the bottom surfaces of the specimens.

Concrete Repair, Rehabilitation and Retrofitting IV – Dehn et al. (Eds)
 ISBN 978-1-138-02843-2

Intelligent, multifunctional textile reinforced concrete interlayer for bridges

C. Driessen & M. Raupach
Institute of Building Materials Research, RWTH Aachen University, Aachen, Germany

ABSTRACT

In bridges which show defective sealing chloride ions migrate into the concrete, catalyse the corrosion of the reinforcement and thus threaten the durability of the construction.

The corrosion of the reinforcement can be seen on the surface of the bridge only when a high degree of impairment is already reached. Consequently, at that state, comprehensive building operations are necessary which lead to traffic obstructions and thus to economic losses. By now, sensor systems for an all-over monitoring of the effectivity of sealings do not exist.

In the context of the joint research project "Intelligent, multifunctional textile reinforced concrete surface for bridges—SMART-DECK "—an innovative, multifunctional and thin (approx. 30 mm) interlayer will be developed, which will provide significant benefits in comparison to the state of the art for new bridge constructions as well as for existing bridges.

SMART-DECK is based on textile reinforced concrete and will provide three functions: the system will combine an all-over real time humidity-monitoring; a preventive cathodic corrosion protection, which can be adjusted by sections and a reinforcement of bridges with an insufficient load-bearing capacity. By means of the monitoring system an early recognition of damages in the sealing layer and, combined with the cathodic corrosion protection, the prevention of traffic obstructions are possible. The renewing of the bridge deck sealing must not be carried out immediately but can be postponed to periods with little traffic. All three functions - the monitoring system, the preventive cathodic corrosion protection and the reinforcement—are realised by textile reinforcement, which consists of carbon, and a new developed mortar. Thus it is possible that Smart—Deck does not increase the dead load of the structure significantly. The aim is that Smart-Deck extends the durability of bridge constructions and reduces economic losses which emerge through traffic delays because of comprehensive building operations. Furthermore despite the higher initial investments the life cycle costs of bridges can be reduced because of fewer maintenance intervals.

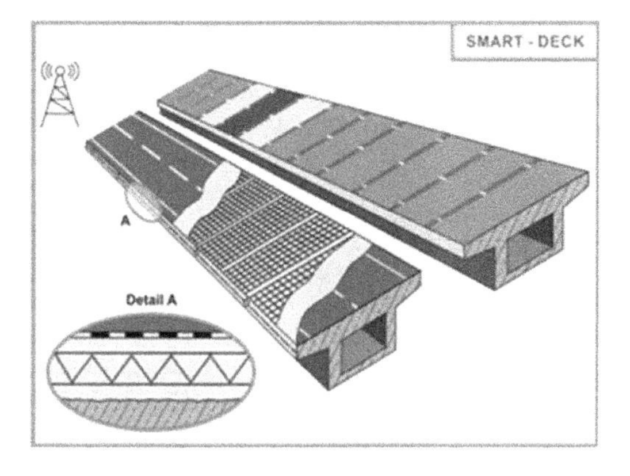

Figure 1. The intended structure of Smart-Deck.

Concrete Repair, Rehabilitation and Retrofitting IV – Dehn et al. (Eds)
© 2016 Taylor & Francis Group, London, ISBN 978-1-138-02843-2

Preliminary investigation of flexural strengthening of RC beams using NSM iron-based shape memory alloys bars

H.N. Rojob & R. El-Hacha
University of Calgary, Calgary, Alberta, Canada

ABSTRACT

The applications of Shape Memory Alloys (SMA) in structural engineering field are limited to small-scale projects and not adapted in practical application due to its high cost. Recently, the iron-based SMA (Fe-SAM) is being developed. The inexpensive constitutive materials of the Fe-SMA and the availability of mass production facilities of steel products makes this material more suitable for large-scale structural engineering applications than the most common SMAs (i.e. NiTi) (Awaji, 2014). SMA is mainly characterized by the Shape Memory Effect (SME) phenomenon. The SME represents the ability of the SMA to recover part of the inelastic strain through heating. In the current research project the SME phenomenon is utilized for flexural strengthening of RC beams by using the Fe-SMA as prestressing reinforcement using the NSM technique. The pre-deformed Fe-SMA bar was embedded in a groove cut on the tension side of the beam. Because the bar was restrained through steel anchors at both ends, the application of heat above the activation temperature will result in a permanent prestressing force in the bar that will remain even after the bar cools down. This paper reports on the flexural performance of 150 × 305 × 2000 mm RC beam strengthened with NSM Fe-SMA bar tested under four point bending setup. 6% strain was initially applied to the Fe-SMA bar. The ends of the bar were then anchored inside a groove cut on the tension side of the RC beam. The bar was then heated using flexible heating tapes up to 350°C, at which the transformation from martensite to austenite phase occurred causing a prestressing force developed in the Fe-SMA bar that counteracts the applied loads. The load mid-span deflection curve of the strengthened beam (Beam-SMA) was compared to a control unstrengthened beam (Beam-C) tested by Hadiseraji & El-Hacha (2014) as presented in Figure 1. The yielding and ultimate load capacities were increased by 22% and 32% over the control beam, respectively. Furthermore, the ductility of the beam was significantly improved. That is reflected by the 43% increase in the deformability index and 146% increase in the total dissipated energy as presented in Table 1. The deformability index was calculated as the ratio of deflection at ultimate load to the deflection at yield load, and the total dissipated energy was calculated as the total area under the load deflection curve up to the ultimate load.

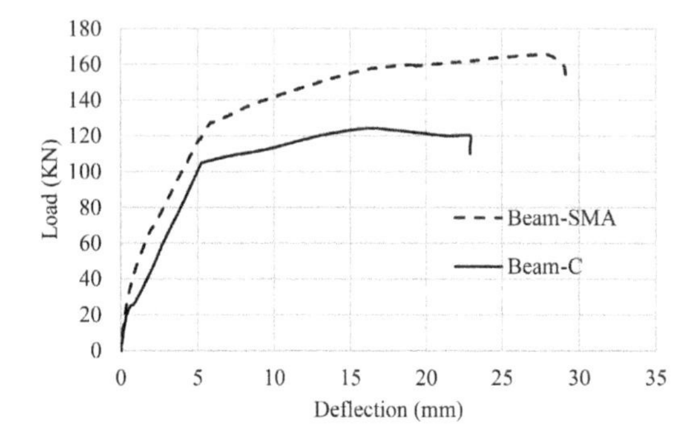

Figure 1. Load-deflection curves of Beam-C (Hadiseraji & El-Hacha, 2014), and Beam-SMA.

Table 1. Summary of the test results.

	Beam-C	Beam-SMA	%Δ
Cracking load: P_{cr} (kN)	19	40	111
Yielding load: P_y (kN)	104	126.8	22
Ultimate load: P_u (kN)	124	165.4	33
Deformability index	3.4	4.8	43
Energy dissipated (kN · mm)	1534	3777	146

REFERENCES

Awaji. 2014. Shape Memory Alloy. (Awaji Materia Co. LTD) Retrieved Dec 27, 2014, from http://www.awaji-m.jp/english/r_and_d/about.html#alloy

Hadiseraji, M., & El-Hacha, R. 2014. Strengthening RC beams with prestressed near-surface mounted CFRP stip. CICE 2014 Conference, 20–22 August 2014. Vancouver, BC, Canada.

Concrete Repair, Rehabilitation and Retrofitting IV – Dehn et al. (Eds)
© 2016 Taylor & Francis Group, London, ISBN 978-1-138-02843-2

Preliminary experimental investigation of reinforced concrete columns confined with NiTi SMA wires

K. Abdelrahman & R. El-Hacha
University of Calgary, Calgary, Alberta, Canada

ABSTRACT

Strengthening concrete columns through confinement techniques is commonly used to enhance the structural performance of the concrete. Recently, an innovative strengthening technique that utilizes smart materials termed Shape Memory Alloys (SMA) have been introduced. The SMA possesses unique characteristic properties that lie in their ability to undergo large deformations and return back to their original shape through stress removal or heating. This paper aimed to investigate the use of SMA in the form of wires to actively confine the concrete columns. The preliminary investigation included characterizing the material properties of the SMA wire relevant to the confinement application. In this case, namely the tensile properties and the recovery stress test was conducted. The experimental testing conducted to determine appropriate splicing connection for the SMA wire was also presented. Finally, a preliminary experimental study was conducted to determine the effectiveness of utilizing SMA wires to actively confine the concrete columns.

Once the SMA material was characterized and the test parameters were defined in this study, the preliminary investigation was conducted to assess the validity of the proposed test parameters and to evaluate the innovative active confinement technique of the concrete. The concrete cylinders used in this study measured 100 mm in diameter and 200 mm in height, and were subjected to uni-axial compression loading. One cylinder was confined with the SMA wire at a pitch spacing of 10 mm and the other cylinder was left to act as a control specimen. The success of the confinement technique can be determined preliminarily by quantifying the strength and ductility gain of the SMA confined concrete columns. A summary of the experimental test data is graphically presented in Figure 1.

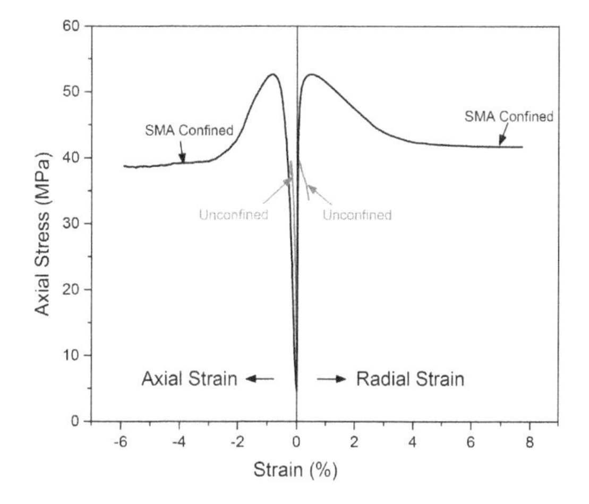

Figure 1. Test results of unconfined and SMA confined concrete cylinders.

The experimental data showed significant enhancement in the strength and ductility of the SMA confined concrete cylinder, when compared to the control cylinder. The actively confined SMA wrapped cylinder experienced an increase in the ultimate strength, axial strain and radial strain by factors of 1.4, 28, and 20, respectively, compared to the unconfined concrete cylinder.

The test results of the conducted preliminary experimental investigation clearly showed the enhanced structural performance of SMA confined concrete over the unconfined concrete specimen. The outcome of this study verified the success of actively confining the concrete specimens with the SMA wires. Thus, the research study undertaken by the authors can be further progressed towards its main objective of investigating the behaviour of eccentrically loaded reinforced concrete columns actively confined using SMA wires.

Concrete Repair, Rehabilitation and Retrofitting IV – Dehn et al. (Eds)
© 2016 Taylor & Francis Group, London, ISBN 978-1-138-02843-2

Chillon Viaduct deck slab strengthening using reinforced UHPFRC: Full-scale tests

D. Zwicky
School of Engineering and Architecture Fribourg (HEIA-FR), Institute of Construction and Environmental Technologies (iTEC), University of Applied Sciences of Western Switzerland (HES-SO), Fribourg, Switzerland

E. Brühwiler
Laboratory of Maintenance and Safety of Structures (MCS), Civil Engineering Institute (IIC), Swiss Federal Institute of Technology Lausanne (EPFL), Lausanne, Switzerland

BACKGROUND

The twin Chillon Viaduct, situated next to world-famous Chillon Castle on the A9/E27, was conceived by Jean Muller, designed by Jean-Claude Piguet, and built in the late 1960s. The variable height box girder structure, spanning between 92 m and 104 m over a total length of 2,210 m, was built by prestressed segmental construction with epoxy-glued joints which was a world novelty at the time.

Recent examination of the structural performance showed that the prevailing failure mode at Ultimate Limit State is punching of wheel loads through the only 18 cm thick deck slab, though structural safety requirements currently can be fulfilled. Further investigations revealed, however, that the concrete is prone to Alkali Silica Reaction (ASR). The latter is expected to lead to concrete shear strength reduction in the future, with an adverse impact on punching resistance.

As the replacement of the waterproofing on the deck slabs was planned for 2014–15, its combination with a strengthening intervention was investigated. The chosen strengthening consists in casting a 40 to 50 mm thin layer of ultra-high performance fiber reinforced concrete, additionally reinforced with steel rebars (R-UHPFRC), on the top surface of the deck slab. The R-UHPFRC layer also serves as waterproofing layer, leading to reduction of the concrete humidity and thus of further ASR development.

The full paper reports on full-scale failure tests on specimens representing zones between and above the webs of the box girder, respectively. The primary goal was to verify the expected increase in ultimate resistance due to the R-UHPFRC layer, as expected through analytical models. This contribution is complemented by an additional paper on finite element analyses for reproduction of test results (Sadouki et al. 2015).

FULL-SCALE TESTS

Failure tests on full-scale specimens representing potentially prevailing zones of the unstrengthened and the strengthened deck slab were performed and evaluated, in order to determine the impact of the strengthening layer on ultimate resistance and failure modes, in particular with regard to punching failure. Figure 1 shows the concreting of the R-UHPFRC strengthening layer during specimen fabrication.

Interaction of punching shear and bending

According to Swiss code SIA 262 (2013) and other international codes (e.g. fib MC 2010) as well as underlying research (Muttoni et al. 2013), the resistance of a concrete slab without transverse reinforcement to punching interacts with the exploitation of the available bending resistance.

Two types of specimens were tested, as the deck slab of the Chillon Viaduct is reinforced with different rebar diameters in the zones over the webs and in the zones between webs. One test series consisted of three unstrengthened specimens and three strengthened specimens.

Figure 1. Concreting the R-UHPFRC strengthening layer.

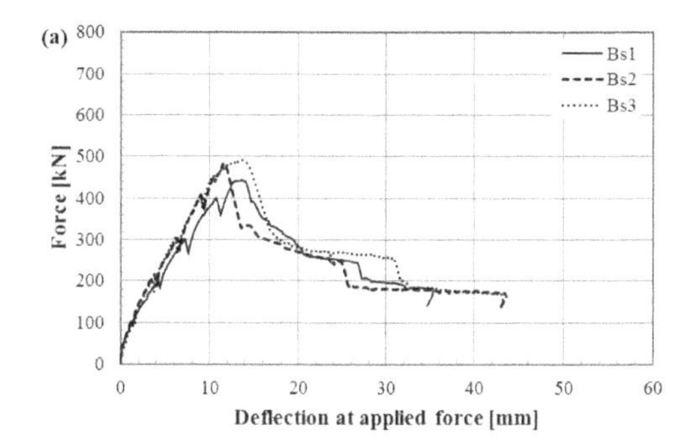

Figure 2. Overall force—deflection response of test specimens B—(a) unstrengthened and (b) strengthened specimens.

Test results evaluation

The empirical evaluation of test results consisted in determining the influence of the R-UHPFRC layer through comparison of the behaviors of strengthened and unstrengthened specimens. Figure 2 shows examples of the force – deflection behavior of specimens submitted to negative bending, representing deck slab zones near girder webs.

The full paper shows more details w.r.t. test setups and measurements and provides comprehensive evaluations of ultimate resistances, failure modes, residual resistances and deformation capacities of strengthened and unstrengthened specimens.

RESULTS AND CONCLUSIONS

- Punching failure could not be observed in any of the tests. Failure of the unstrengthened specimens was attained at forces between 400 kN and 480 kN, being about three times the characteristic value of wheel loads as given by codes.
- Casting a 40 mm layer of reinforced UHPFRC increases ultimate resistance by at least 50%, provided that the interface is properly prepared and that the stiffness of the fresh UHPFRC mix is appropriately chosen.
- Inner slab zones between the box girder webs (submitted to positive bending moments) fail with moderate to normal ductility, i.e. announcing failure by major deflections. Residual resistance capacity of strengthened zones is still considerably higher than that of unstrengthened.
- Slab zones adjacent to girder webs (submitted to negative bending moments) show limited deformation capacity, strengthened or unstrengthened. In unstrengthened slab zones, failure is associated with shear failure in the substrate concrete. In strengthened slab zones, failure is related to R-UHPFRC layer delamination. No sign of shear failure of the substrate slab could be observed at this stage. Deformation capacity (ductility) is slightly improved by the presence of an R-UHPFRC strengthening layer.

APPLICATION TO CHILLON VIADUCT

The full-scale tests confirmed that also an unstrengthened slab undergoing strength reduction due to ASR would not fail by punching of a wheel load. Nevertheless, the client decided to strengthen the deck slabs of the Chillon Viaduct by means of an R-UHPFRC layer due the following advantages:

- The strengthening layer also serves as waterproofing.
- The R-UHPFRC strengthening layer considerably increases the flexural stiffness of the deck slab, thereby reducing fatigue stresses in the (potentially ASR-affected) concrete and the reinforcing bars.
- The strengthening layer provides a reserve for future traffic development and axle loads.
- In view of future interventions to rehabilitate and improve further highway bridges, the UHPFRC works on the Chillon Viaduct allow to gain important experience.

Casting of the R-UHPFRC strengthening on the 2.1 km long Chillon Viaduct was performed during five weeks, ending in September 2014. Structural design of the strengthening layer was performed using analytical resistance models which were validated by non-linear finite element modelling (Sadouki et al. 2015), calibrated to the present test results.

REFERENCES

fib MC. 2010. *Model Code 2010—Final draft, Volume 2*, bulletin no. 66, Lausanne: International Federation for Structural Concrete (fib).

Muttoni, A.; Fernández Ruiz, M.; Bentz, E.; Foster, S. & Sigrist, V. 2013. Background to fib Model Code 2010shear provisions – part II: punching shear. *Structural Concrete* 14(3):204–214.

Sadouki, H.; Brühwiler E. & Zwicky, D. 2015. Chillon Viaduct deck slab strengthening using reinforced UHPFRC: Numerical simulation of full-scale tests. *Proc. International Conference on Concrete Repair, Rehabilitation and Retrofitting (ICCRRR 2015), Leipzig, 5–7 October 2015.*

Concrete Repair, Rehabilitation and Retrofitting IV – Dehn et al. (Eds)
© 2016 Taylor & Francis Group, London, ISBN 978-1-138-02843-2

Chillon Viaduct deck slab strengthening using reinforced UHPFRC: Numerical simulation of full-scale tests

H. Sadouki & E. Brühwiler
Laboratory of Maintenance and Safety of Structures (MCS), Civil Engineering Institute (IIC), Swiss Federal Institute of Technology Lausanne (EPFL), Lausanne, Switzerland

D. Zwicky
School of Engineering and Architecture Fribourg (HEIA-FR), Institute of Construction and Environmental Technologies (iTEC), University of Applied Sciences of Western Switzerland (HES-SO), Fribourg, Switzerland

BACKGROUND

The deck slab of the Chillon Viaduct located on the Swiss National Highway on the East end of the Lake of Geneva, was strengthened by adding, on the previously hydro-jetted top surfaces of deck slabs, a 40 to 50 mm thick layer of Ultra High Performance Fiber Reinforced cement-based Composite material (UHPFRC) which is additionally reinforced by steel rebars (Figure 1).

Figure 1. Deck slab strengthening using UHPFRC.

The choice of UHPFRC as a strengthening material, was motivated (1) by its outstanding mechanical properties, namely high tensile and compressive strengths and a its important deformation capacity due to the high amount of incorporated steel fibres in the cement-based matrix of the material, as well as (2) by its very low porosity implying minimized moisture exchange and ingress of aggressive chemical substances such as chloride ions from the surrounding atmosphere (Brühwiler & Denarié, 2008).

To validate this novel strengthening method, an experimental program has been carried out including a series of full-scale tests on instrumented composite RC—R-UHPFRC slab elements representing zones between and above the girder webs of the viaduct (Zwicky D. et al., 2013).

The present paper reports mainly on the results of numerical simulation of the tests as obtained by non-linear Finite-Element Analysis (FEA) targeting at verifying the effectiveness of the strengthening method using R-UHPFRC and at calibrating the FEA model with regard to the verification of the ultimate resistance and structural safety of the strengthened deck slab of the viaduct.

Figure 2. Test slab Series B (negative bending over web of box girder).

FINITE ELEMENT MODELLING

A three dimensional non-linear finite element was used for analyzing the structural behavior of the tested slabs. Finite volume elements with 20 nodes were employed to represent the geometry of the structure, consisting of RC monolithic slab elements and composite R-UHPFRC—RC slab elements, steel loading plates and supporting beams (Figure 2). The model was elaborated using the non-linear finite element analysis software DIANA (DIANA, 2010) with its graphical pre- and post-processors FEMGEM and FEMVIEW, respectively. The validation of the FE model was accomplished by comparison of numerical results with those recorded from instrumented test specimens (Zwicky & Brühwiler, 2015).

NUMERICAL RESULTS

Figure 3 gives a comparison between the numerical and experimental force-deflection responses of the RC/UHPFRC composite slabs of series B.

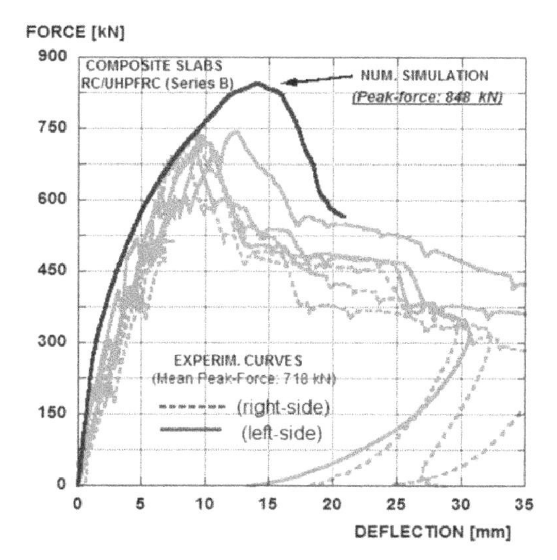

Figure 3. Comparison between experimental force-deflection curves of the three RC/UHPFRC composite slabs (series B) (grey solid curves for the left part of slabs and grey dashed for the right part) and the predicted one by the numerical model (black solid curve).

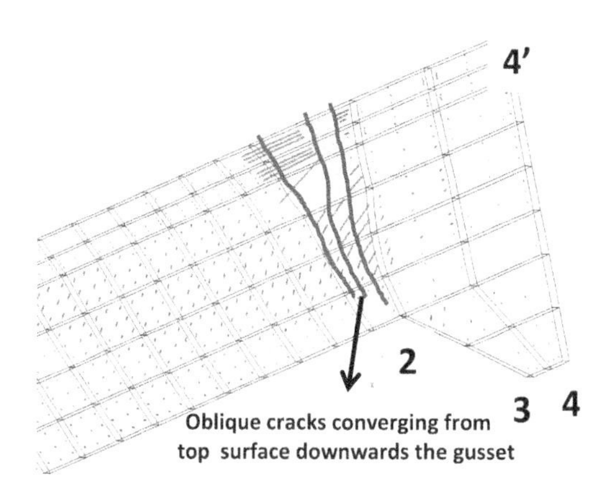

Figure 4. Visualisation of crack pattern of the corner of the composite slab (series B).

The computed peak-force (848 kN) is found to be 16% higher than the mean experimental value (728 kN). Compared to the experimental results, the computed force-deflection curve overestimates the force values while the deflection values are well predicted.

Figure 4 gives a side view of the crack opening distribution of the corner (2-3-4–4′) of the composite slab. Three dominant oblique cracks are distinguished; they are highlighted by solid blue lines. It must be outlined that these cracks started to form already in pre-peak regime.

Overall, numerical results show that strengthening of the slab elements by means of UHPFRC increases the peak force by 45%, while experimental test revealed an increase of 47% (based on the mean value of the three specimens). The increase in resistance due to the UHPFRC-reinforcing layer is significant.

CONCLUSIONS

The strengthening method of bridge deck slabs by means of an UHPFRC overlay has been validated by a specific experimental program which is presented in a companion contribution (Zwicky & Brühwiler, 2015). This paper presents results of a non-linear finite element analysis of the structural and cracking behavior of composite UHPFRC/concrete slabs subjected to loading similar to loading of real bridge decks.

The results of the non-linear 3D-numerical modelling of three monolithic concrete slabs and three UHPFRC/concrete composite slabs show good agreement between experimental findings and the slab response as obtained by the numerical model, both in terms of force-deflection responses and cracking patterns. The model confirms that the failure is essentially governed by the flexural mode, as observed experimentally.

In accordance with the experimental results, the computed results show that the UHPFRC layer increases significantly the resistance of concrete slabs.

In summary, 3D FE-analysis has proven to effectively simulate with sufficient accuracy the structural behavior of the tested slab elements.

This numerical model can be further used to explore the structural behavior of structures like bridges or industrial buildings subjected to different loadings and environmental conditions like temperature variation, creep/relaxation, drying shrinkage, alkali silica reaction-induced swelling or ageing effects. This validated model was used to design the strengthening of the deck slab of the Chillon Viaducts.

REFERENCES

Brühwiler E. & Denarié E. (2013), Rehabilitation and strengthening of concrete structures using Ultra-High Performance Fibre Reinforced Concrete, Structural Engineering International, Volume 23, Number 4, November 2013, pp. 450–457.

Zwicky D. & Brühwiler, E. (2015), Chillon Viaduct deck slab strengthening using reinforced UHPFRC: Full-scale tests, Proc. International Conference on Concrete Repair, Rehabilitation and Retrofitting (ICCRRR 2015), Leipzig, 5–7 October 2015.

Concrete Repair, Rehabilitation and Retrofitting IV – Dehn et al. (Eds)
© 2016 Taylor & Francis Group, London, ISBN 978-1-138-02843-2

Strengthening of existing reinforced concrete beams using ultra high performance fibre reinforced concrete

Andreas P. Lampropoulos & Spyridon A. Paschalis
University of Brighton, UK

Ourania T. Tsioulou & Stephanos E. Dritsos
University of Patras, Greece

ABSTRACT

The addition of reinforced concrete layers and jackets for the earthquake strengthening of existing beams and columns is one of the most commonly used techniques in earthquake prone areas.

In this study the efficiency of the technique of strengthening existing structures using Ultra High Performance Fibre Reinforced Concrete (UHPFRC) has been investigated. The uniaxial stress-strain relationship in tension and the compressive strength characteristics have been determined by direct tensile (Figure 1) and compressive tests respectively.

These results (Figure 1) have been used for the constitutive modelling of UHPFRC.

Finite element method has been used for the modelling of strengthened reinforced concrete beams with UHPFRC layers, and a parametric study has been conducted on beams strengthened with UHPFRC layers and jackets. Three different types of models have been examined, one with an additional layer in the compressive zone (SCL_UHPFRC), one in the tensile (STL_UHPFRC), and another one with a three side jacket (S3SJ_UHPFRC) (Figure 2). In all the examined cases a 50 mm thick layer or jacket has been used. The technique of strengthening with UHPFRC layer in the tensile and in the compressive zone are compared to respective results of strengthened elements with additional RC layer in the tensile side (STL_RC) (Tsioulou et al. 2013), and normal plain concrete with 40 MPa compressive strength in the compressive zone (SCL_NC).

The results of the various strengthening techniques are compared to the initial beam's results and to the results are presented in figure 3.

The results presented in figure 3 indicate that the addition of UHPFRC layers/jackets can considerably improve the strength and the stiffness of the existing structures. In case of beam strengthened with additional UHPFRC layer in the tensile side (STL_UHPFRC), the stiffness is significantly

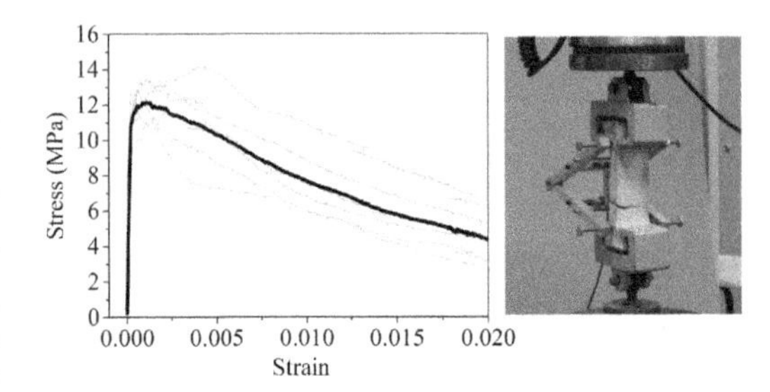

Figure 1. Stress-strain experimental results and setup of direct tensile tests.

Figure 2. Finite element models for strengthened beams with additional layer in the compressive side, in the tensile side, and with three side jacket.

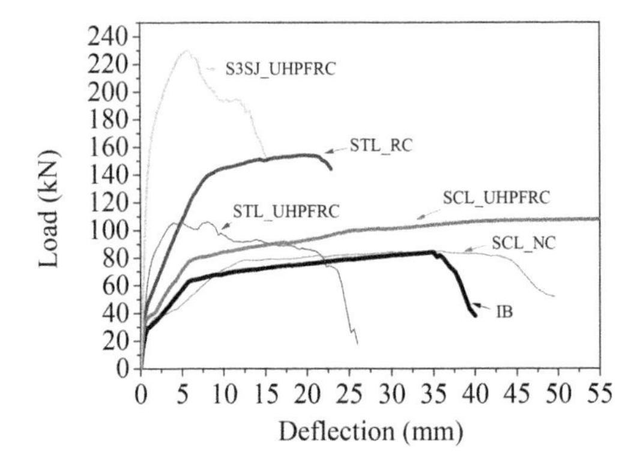

Figure 3. Numerical load-deflection results for beams strengthened with various techniques.

Table 1. Yield and the ultimate bending moment values of the examined specimens.

Specimen	M_y (10^3 Nm)	M_u (10^3 Nm)
IB	24	32
SCL_UHPFRC	29	41
STL_UHPFRC	30	40
SCL_NC	29	32
STL_RC	52	58
S3SJ_UHPFRC	63	86

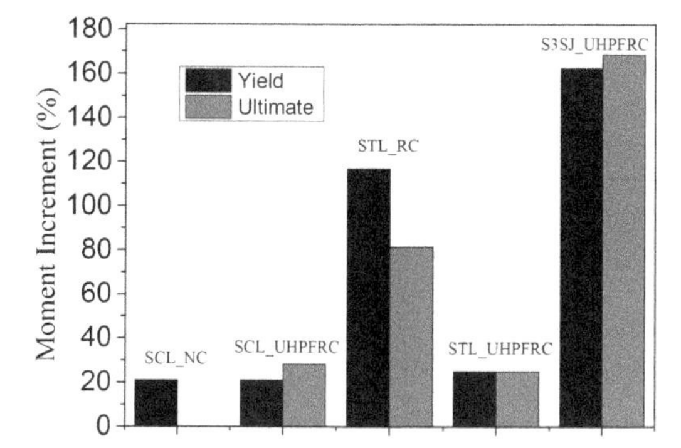

Figure 4. Moment increment for the examined techniques.

increased but the yield and the maximum load values are lower compared to the respective load values of the beam with additional RC layer (STL_RC). In case of beam strengthened with UHPFRC in the compressive zone (SCL_UHPFRC), both strength and stiffness are considerably improved compared to the respective specimen with additional normal strength concrete layer (SCL_NC). The results of the beam strengthened with a three side UHPFRC jacket (S3SJ_UHPFRC) are characterized by superior performance compared to all the other examined specimens.

The yield and the ultimate bending moment (M_y and M_u) for all the examined cases have been calculated using the results of figure 3 (Table 1).

The yield and ultimate moment increment for all the examined techniques as percentage of the respective values of the initial beam (100*ΔM/M) is presented in figure 4.

The addition of UHPFRC layer in the compressive or in the tensile side can contribute to 20–30% increment of yield and ultimate moment, while in case of a three side UHPFRC jacket the respective increment is 160–170%. When comparing the specimens with the additional layer in the tensile zone with RC and UHPFRC, it can be observed

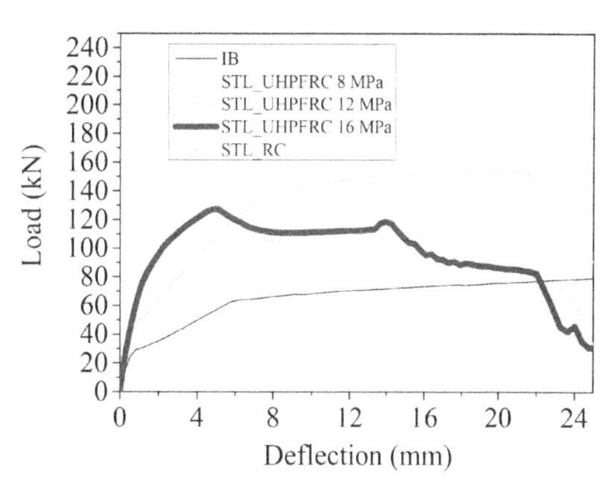

Figure 5. Parametric study results using different values for the tensile strength of UHPFRC.

that the specimen with the additional RC layer shows better performance compared to the respective strengthened beam with UHPFRC.

A further parametric study has been conducted for the specimens with additional UHPFRC layer in the tensile side (STL_UHPFRC) using various UHPFRC tensile strength values. In the current study 8 MPa, 12 MPa and 16 MPa tensile strength values have been used and the numerical results are presented in figure 5 together with the results of STL _RC specimen.

The results of figure 5 indicate that there is a significant increment of the ultimate moment M_u when higher tensile strength values are used. As the tensile strength of UHPFRC is increased, the maximum strength of STL_UHPFRC is approaching the ultimate strength of STL_RC. However, it may worth mentioning that in all the examined specimens with UHPFRC, the stiffness is higher compared to the respective specimens strengthened with conventional strengthening techniques with normal plain and reinforced concrete layers.

The overall conclusion of this study is that the application of UHPFRC can considerably enhance the performance of existing reinforced concrete beams in terms of stiffness and yield & maximum strength. Further experimental and numerical investigation is required in order to take into account concrete shrinkage effect and investigate the durability of the examined technique.

REFERENCE

Tsioulou, O., Lampropoulos, A. & Dritsos, S. 2013. *Experimental investigation of interface behaviour of RC beams strengthened with concrete layers*, *Construction and Building Materials* 40: 50–59.

Concrete Repair, Rehabilitation and Retrofitting IV – Dehn et al. (Eds)
© 2016 Taylor & Francis Group, London, ISBN 978-1-138-02843-2

"Integralization" with new UHPC decks for existing motorway bridges

P. Hadl & R. della Pietra
Institute of Structural Concrete, Graz University of Technology, Graz, Austria

M. Reichel
König und Heunisch Planungsgesellschaft, Leipzig, Germany

N.V. Tue
Institute of Structural Concrete, Graz University of Technology, Graz, Austria

ABSTRACT

Ultra-High Performance Concrete (UHPC) has a high potential for the repair and retrofitting of existing bridges. Due to its excellent durability properties a longer service life of bridge decks made of UHPC can be expected. Therefore, the life-cycle costs decrease. Because of the high strength of UHPC furthermore the bearing capacity of the structure will be enhanced.

Bridge decks generally provided with a replaceable layer made of asphalt, which require an additional waterproof arrangement and have a limited service life. The basic idea of using UHPC for the retrofitting of existing bridges is to create a monolithic connected and maintenance-free integral bridge for conventional concrete bridges by replacing the waterproofing as well as the asphalt pavement with a thin UHPC layer of a few centimeters. In addition to a durable and waterproof surface, the UHPC also participates in the load transfer, and enhances the bearing capacity of the bridge.

This paper presents the application of an UHPC mixture with micro steel fibers for the rehabilitation of Steinbach-bridge, a motorway bridge in Austria. The special focus is on the concrete technology, related experimental testing of bearing behavior and the construction work.

For this pilot project more than 40 m³ UHPC were produced in a ready-mix concrete plant and cast on site. Figure 2 illustrates the monolithic connected UHPC topping with a thickness of 7 cm. The principle of the "integralization" is shown in figure 3.

Within this project a successful transfer of laboratory findings into practical productions under site conditions was achieved. Currently, the behavior of the structure is observed by a monitoring system. This innovative solution can thus make a significant contribution to the reduction of the life-cycle costs for the renovation and retrofitting existing bridge structures. The pilot project Steinbach-bridge illustrates the potential of this UHPC application regarding its material benefits and its practicability. Finally, this project benefited from an innovative cooperation between concrete technologists, structural designers and construction site.

Figure 1. Steinbach-bridge.

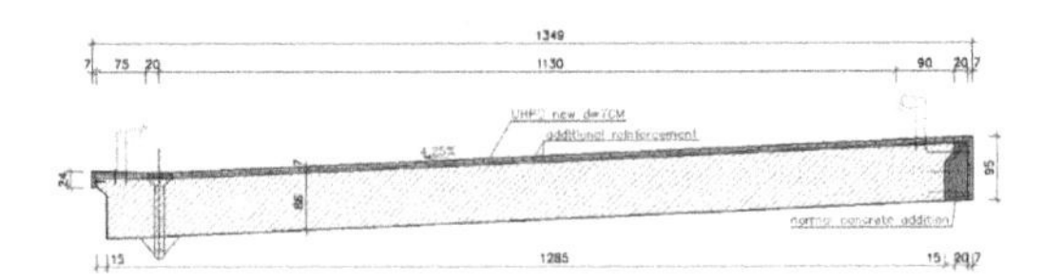

Figure 2. Cross-section after renovation.

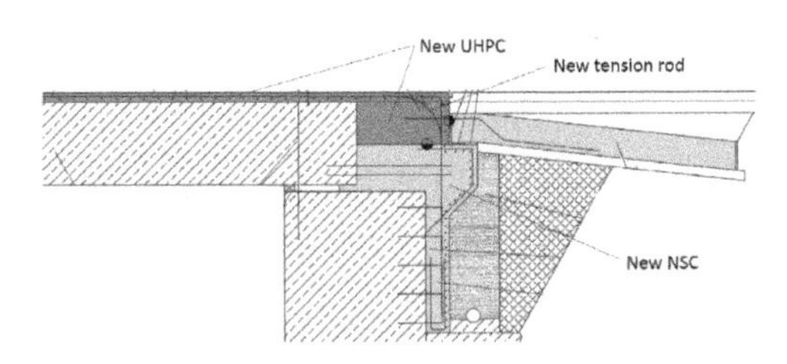

Figure 3. Principle of "integralization".

Concrete Repair, Rehabilitation and Retrofitting IV – Dehn et al. (Eds)
© 2016 Taylor & Francis Group, London, ISBN 978-1-138-02843-2

Structural strengthening of a shop-house for use as a medical centre

C.C. Lim & Y.C. Chua
Public Works Department, Malaysia

ABSTRACT

A 5-storey shop-house unit was procured by the Ministry of Health Malaysia for conversion into a Medical Centre. The 8-year old building was formerly built and completed by a private developer but was never in use before it was procured by the Ministry of Health Malaysia. The building was inspected, assessed and re-evaluated structurally to determine the design adequacy and functionality for use as a medical centre with respect to design code BS8110 and loading criteria to BS6399. Unfortunately, as-built drawings, which are crucial for structural evaluation, were not available except for some as-tendered drawings. Hence, a detailed investigation, comprising a comprehensive material test on the structure including verification of as-built structural details, was necessary.

A general visual survey was first made on the building before developing a detailed material tests plan for the structure. The objectives of the material tests plan are to evaluate the in-situ concrete strength and quality, to verify the as-built structural details and to assess the level of concrete deteriorations. The test results were subsequently used for structural integrity and capacity evaluations.

Information from the as-built details was reliably used to develop a 3-D model of the building structure. Structural analysis was done by Staadpro computer software. Design was checked against two limit state conditions, i.e., Serviceability Limit State (SLS) and Ultimate Limit State (ULS). Structural capacity was computed and checked against the maximum load effects under ULS. Crack widths and deflections were computed from maximum load effects under service load condition. Structural design, load combinations and load intensities are in accordance with BS8110 and BS6399.

Live load and superimposed dead load intensities are generally higher for building used as medical centre than for shop or residential unit. For the medical centre, a typical load intensity between 3.0 to 5.0 kPa was considered for public areas, file-storage rooms and designated areas for heavy equipment installation.

From the visual inspection and material test results, the overall structure is found to be in good condition generally. The building was never used since completion. All structures are fully plastered except for the soffit of the slabs. No sign of serious concrete defects (spalling, delamination or structural crack) except for some hairline cracks on slabs and plaster due to shrinkage. The structure was slightly carbonated but no sign of reinforcement corrosion on many parts of the structure. Field exposure of reinforcements at selected locations in the building, confirmed the reinforcements are in good conditions. The average in-situ concrete compressive strength is within an acceptable limit of BS6089 for Grade 25 concrete.

The as-constructed structural details are generally in compliance with the as-tendered drawings. However, the slab is found to be 100 mm thick. It is deemed relatively "thin" compared to normal slab thickness of about 125–150 mm. The material test also reveals that some slabs do not have tension reinforcements to resist hogging moment at the support. Discrepancies between the as-constructed beam layout plan and the as-tendered drawing are also detected.

Structural strengthening is essentially needed to fulfil both the ultimate and serviceability requirements of BS8110. The slab thickness is increased to 160 mm by an additional 60 mm thick reinforced concrete overlay. It increases the effective depth of the slab. CFRP laminates, 50 mm × 1.2 mm thick, are installed to the slab soffit at 500 mm centres. Both CFRP strengthening and 60 mm thick overlay are necessary to address three issues pertaining to design inadequacy of the slab i.e., to enhance flexural capacity, to keep deflection within limit and to provide a sufficient 25 mm cover to reinforcements. Beams do not have deflection issue. They are strengthened to enhance moment and shear capacities by means of CFRP laminate and wrap respectively.

CFRP is particularly susceptible to damage caused by high temperatures that may take place in the event of fire. A 40 mm thick vermiculite fire protection mortar was sprayed over the entire CFRP area.

Concrete Repair, Rehabilitation and Retrofitting IV – Dehn et al. (Eds)
© 2016 Taylor & Francis Group, London, ISBN 978-1-138-02843-2

Overall precision uplift rehabilitation technology for uneven settlement of concrete structure of slab ballastless track

Xinguo Zheng, Jing Liu & Shuming Li
Railway Engineering Research Institute of China Academy of Railway Sciences, Beijing, China

Weichang Xu, Zhiyuan Zhang & Youneng Wang
Shanghai Railway Bureau, Shanghai, China

Fei Cheng & Xihua Wen
Shanghai-Hangzhou Railway Passenger Dedicated Line Co. Ltd., Shanghai, China

ABSTRACT

The concrete slab ballastless track is an advanced track structure with high stability and integrity. It has been widely used in many high speed railway lines all over the world. However, when the foundation of ballastless track appears large uneven settlement deformation especially in roadbed section, compared with ballast track which can be restored by increasing the ballast under the sleepers simply, the smoothness of concrete slab ballastless track can only be restored by adjusting the thickness of the base plates of fastener system accordingly. But if continuous settlement deformation exceeds the specified adjustable range of fastener system, the smoothness of concrete slab ballastless track cannot be restored timely and completely, and then the speed of passing trains has to be limited to ensure the safety. To solve this problem only just using daily skylight time (only 4 hours every day), we systematically studied a complete set of new overall precision uplift rehabilitation technology including special grouting materials, dedicated miniaturization equipment and construction processes through a series of indoor reduced scale tests and field full scale simulations with real concrete slabs. And combined with field practices, the overall precision uplift rehabilitation technology for concrete slab ballastless track had been studied and developed.

Results indicated during the overall grouting uplifting of ballastless track concrete structure, the grouting materials was filled into the space between the concrete bottom layer and the surface layer of graded broken stone foundation bed (see Figure 1) through the special grouting equipment. The uplifting was realized by the expansibility of the grouting materials and the hydraulic transmission effect resulting from the grouting materials and grouting pressure. this technology could not only restore the smoothness of settlement concrete slab

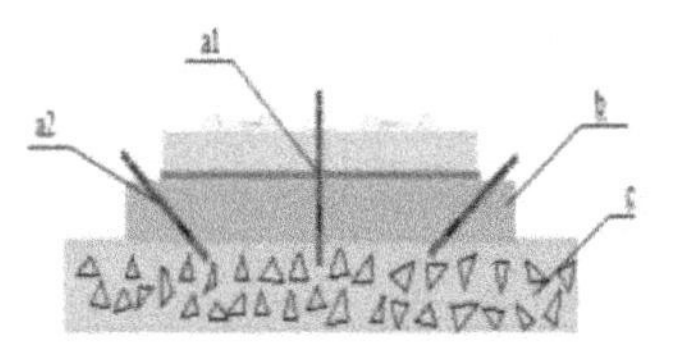

Before uplifting rehabilitation

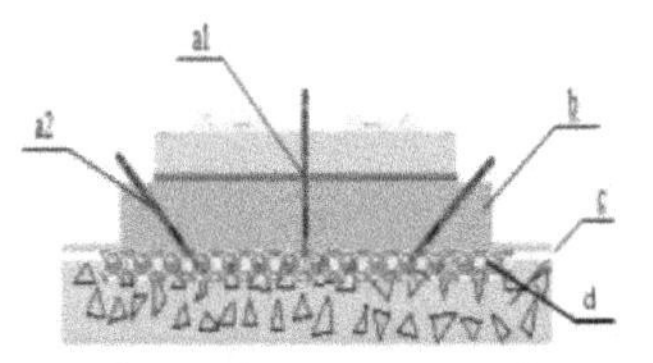

After uplifting rehabilitation

Figure 1. Uplifting mechanism and comparison of ballastless track structure before and after uplifting rehabilitation (Note: a1/a2—grouting pipes, b—concrete bottom slab, c—graded broken stone layer, d—polyurethane solid foam layer).

ballastless track effectively and accurately, but also recover the adjustable range of the fasteners just in daily skylight time without disturbing the following operation of the railway lines. The smoothness of the track was improved significantly (see Figure 2 for the track elevation comparison before and after the grouting uplifting rehabilitation). Moreover, the adjustable range of the fasteners was restored. The maximum uplifting height of the concrete structure was up to 82 millimeters (No cracking during uplifting rehabilitation). The railway speed in this section had recovered to the design speed of 250 km/h and the dynamic response of the ballastless track in this section showed normally.

Based on the field practices, the uplifting mechanism of this overall precision uplift rehabilitation was also analyzed. The uplifting could be divided

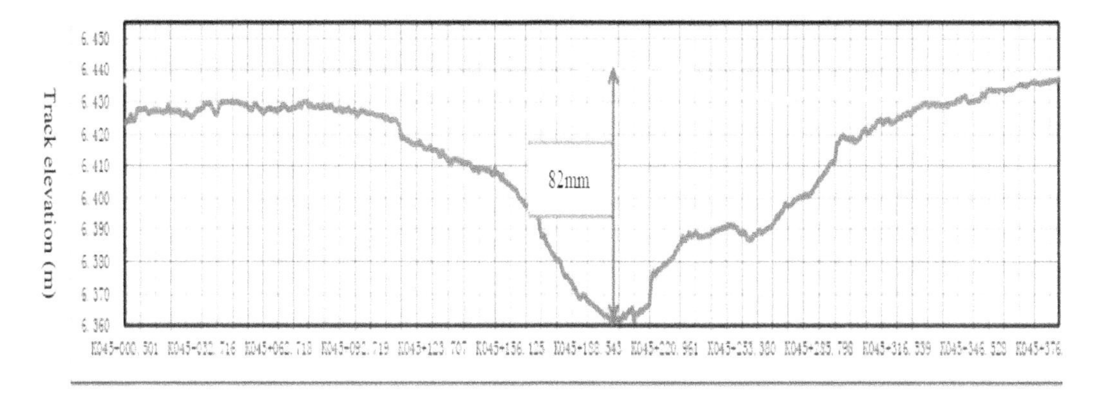

Figure 2. Comparison of the elevation of the track before and after uplifting rehabilitation (Note: the below pink line represents the track elevation before rehabilitation, the above yellow line represents the track elevation after rehabilitation).

into two stages, namely the filling & compacting stage and the uplifting & filling stage.

In the filling & compacting stage, the grouting materials will fill and compact the gaps such as the ones between the supporting layer and graded broken stone layer or those inside the graded broken stone layer. And the track structure will not be uplifted at this stage.

In the uplifting & filling stage, the filled grouting materials and the grouting pressure from grouting equipment will cause hydraulic transmission effect. Meanwhile, the volume expansion from the grouting material solidification will cause expansive uplifting forces. With the effects of these two factors, the concrete structure can be uplifted continuously and stably, and the gaps between concrete structure and graded broken stone layer due to uplifting can also be filled completely and densely.

The research will provide valuable and useful references for the rehabilitation of concrete slab structure occurred differential settlement.

Concrete Repair, Rehabilitation and Retrofitting IV – Dehn et al. (Eds)

Axial compression testing of an emergency-retrofitted shear-damaged RC column

K. Nakada & M. Kochi
University of the Ryukyus, Okinawa, Japan

S. Arakaki
Kozo Factory Co. Ltd., Fukuoka, Japan

H. Karwand & M.Z. Noori
Former PEACE Participant of Japan International Corporation Agency, Tokyo, Japan

ABSTRACT

The aim of this study is to develop a manual emergency retrofitting technique for damaged Reinforced Concrete (RC) buildings. In this study, lashing belts comprising a ratchet buckle and an aramid fiber belt were used as the emergency retrofitting method of choice. As one of the vertical load carrying members, it is important that the emergency retrofitted RC column sustains the vertical load as well as the lateral force. In this study, the residual axial compression capacity of shear-damaged RC columns and the recovered axial compression capacity of shear-damaged RC columns emergency-retrofitted by lashing belts was investigated and discussed.

The retrofitting details of the test specimens and test results are shown in Table 1. The RC column specimens used in this test were square with cross-sectional dimensions of 250 × 250 mm² with a total height of 900 mm. The height of the test section of the RC column was 500 mm, with a shear span-to-depth ratio (M/VD) of 1.0. The experimental procedures were as follows. 1) shear-critical RC columns were damaged by cyclic lateral forces under a constant axial force ratio of 0.2 on the assumption that earthquakes cause shear-damage to RC columns. The damage level was categorized as between II and IV, as defined by the Japan Building Disaster Prevention Association. 2) a universal testing machine was used in the axial compression tests of the shear-damaged RC column and emergency-retrofitted RC column. The test parameters of these specimens included damage level (II–IV), the lateral confinement by the lashing belts, and the conditions wherein epoxy resin repair was and was not applied.

Table 1. Details of specimens and test results.

No.	Series	Specimen (AC13)	$_c\sigma_B$ (MPa)	Damage level	W_{cr} (mm)	epoxy resin	$_As$ (mm)	ε_{pt} ( )	σ_r (MPa)	N_{max} (kN)	N_{max}/N_0
1	1	D3	21.2	III	1.5	-	-	-	-	1128	0.75
2		D3e			1.4	apply				1435	0.95
3		D4-1		IV	5.0	-				384	0.25
4		D4-2			3.5					455	0.30
5		D4e-1			3.5	apply				1011	0.67
6	2	N1	26.5	-	-	-	-	-	-	1849	-
7		N2								1785	-
8		εL'					63	800	0.69	1930	1.06
9		D2		II	0.3		-	-	-	1658	0.91
10		εLD4		IV	4.0		63	800	0.69	1319	0.73
11		εLD4e			5.0	apply				1167	0.64
12		D4e-2			5.0		-	-	-	1613	0.89

Notes: $_c\sigma_B$ = compressive strength of concrete cylinder, W_{cr} = residual crack width after shear failure test, $_As$ = interval of lashing belt, ε_{pt} = initial strain of lashing belt, σ_r = lateral confining pressure, N_{max} = max axial load, N_0 = max axial load of non-damaged RC column.

Figure 1. After the loading test.

The main results of this study are summarized in the following points: 1) The residual axial compression load of the shear-damaged RC column was 91% of that of the compression capacity of the non-damaged RC column at damage level II, and 75% and 28% at III and IV, respectively; 2) at damage levels III and IV, a residual axial compression load N/N_0 of 0.67–0.95 was obtained by epoxy resin repair; 3) when only the lashing belt was applied, the axial load gradually increased during the loading test and low axial rigidity was confirmed; and 4) when repairing using both epoxy resin and the lashing belts for lateral confinement, the recovery of the axial compression capacity, axial rigidity, and compressive ductility was found to be significant.

Concrete Repair, Rehabilitation and Retrofitting IV – Dehn et al. (Eds)
© 2016 Taylor & Francis Group, London, ISBN 978-1-138-02843-2

Rehabilitation of corrosion-aged concrete T-girders with textile-reinforced mortar

T. El-Maaddawy
United Arab Emirates University, Al Ain, United Arab Emirates

A. El Refai
Laval University, Quebec City, Canada

ABSTRACT

Reinforced Concrete (RC) structures in the Arabian Gulf and worldwide suffer from extensive corrosion due to the harsh environment and/or use of deicing-salt. The present study proposes a new technique for rehabilitation of corrosion-damaged RC T-girders, namely, Textile-Reinforced Mortar (TRM), that utilizes newly produced carbon and basalt fiber textile grids embedded between cement-based mortar layers. The viability of the (TRM) system to improve the flexural response of undamaged RC beams has been demonstrated in few recent studies (D'Ambrisi & Focacci 2011, Elsanadedy et al. 2013, Hashemi & Al-Mahaidi 2012). Nevertheless, the effectiveness of the TRM system to restore the load capacity of corrosion-damaged RC beams has not been investigated. This research work aims at filling this gap and providing insight into the subject. The experimental program comprised testing of six specimens as shown in Table 1. The repair scheme consisted of TRM layers internally embedded within the corroded region of the beam or a combination of internally embedded and externally bonded TRM layers.

The load-deflection responses of the tested specimens are depicted in Figure 1. Corrosion damage and cracking significantly reduced the load capacity and ductility of the unrepaired beam. The basalt TRM repair system was not effective in improving the flexural response whereas the internally embedded carbon TRM layers fully restored the load capacity and ductility, expect for specimen D-4Ci where only 90% of the ductility was restored. The use of carbon TRM layers internally embedded within the corroded region together with TRM layers externally bonded along the beam span was more effective in improving the flexural response than the use of internally embedded TRM layers alone.

Table 1. Test matrix.

	TRM repair scheme		
Specimen	Material	Internally-embedded	Externally-bonded
Control	–	–	–
D-NR	–	–	–
D-2Ci	Carbon	2	–
D-4Ci	Carbon	4	–
D-4Bi	Basalt	4	–
D-2Ci-2Ce	Carbon	2	2

*D = damaged; NR = not repaired; 1, 2, and 4 = number of TRM layers; C and B = Carbon and Basalt textiles, respectively; i = internally embedded within the corroded zone; e = externally bonded on tension face along the beam span.

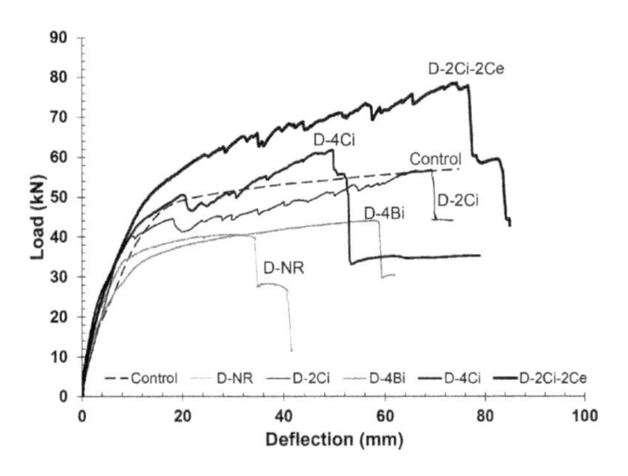

Figure 1. Load-deflection response.

REFERENCES

D'Ambrisi, A. & Focacci, F. (2011). Flexural strengthening of RC beams with cement-based composites. *Journal of Composites for Construction*. 15(5): 707–720.

Elsanadedy, H. M., Almusallam, H. M., Alsayed, S. H., and Al-Salloum, Y. A. (2013). Flexural strengthening of RC beams using textile reinforced mortar – experimental and numerical study. *Composite Structures*. 97: 40–55.

Hashemi, S. & Al-Mahaidi, R. (2012). Flexural performance of CFRP textile-retrofitted RC beams using cement-based adhesives at high temperature. Construction and Building Materials. 28: 791–797.

Concrete Repair, Rehabilitation and Retrofitting IV – Dehn et al. (Eds)
© 2016 Taylor & Francis Group, London, ISBN 978-1-138-02843-2

Dismantling of damaged PSC damaged suspended span of Varsova Bridge across Vasai Creek on NH-8, Mumbai, India

M.L. Gupta
IRB Infrastructure Developers Ltd., Mumbai, India

D.A. Bhide
Design, IRB Infrastructure Developers Ltd., Mumbai, India

P.B. Dongre
IRB Infrastructure Developers Ltd., Mumbai, India

ABSTRACT

The existing Varsova Bridge across Vasai Creek, called as Bassein Creek Bridge when constructed, was opened to traffic in 1968. It is in Mumbai Ahmedabad section of NH-8, about 35 Km from Mumbai. The Bridge is 555.32 m long with 8 spans. The span arrangement is 48.46 + 2 × 57.3 + 2 × 114.6 + 2 × 57.3 + 48.46 m. The main spans from central three piers were built with balanced cantilever construction m. These are with single cell box girder. The cantilever construction was continued in adjacent penultimate spans with overhangs of 8.84 m each. These overhangs supported suspended spans of 48.46 m. These spans are with two nos., pre-cast post tensioned girders with deck slab. Thus the bridge has 4 spans that constitute a continuous module for a length of 458.4 m. All the simply supported spans are strengthened with 20 nos. of single strand external tendons.

Major cracks were noticed on 12th December 2013 in the West side girder of the penultimate span from Mumbai end.

The main crack was 150 mm wide at the bottom flange on one side of central diaphragm. It progressed towards deck slabs with branching. The external strengthening done for the girder prevented a total collapse. Some damage was seen at the deviator blocks provided for external tendons.

The site was inspected for assessment of the damage. It was quite obvious that such damage has rendered the said girder as completely failed. A complete replacement of span was a necessity. It was decided to replace the said span with composite steel girder. The specific span arrangement and construction details dictated various constraints on the repair scheme. The 4 span continuous structure with overhangs meant the span moments in its end san as well as support moments of the end support with over hangs have to be maintained, neither increase or decrease of weight of replaced structure was possible. Similar condition existed at other end of damaged span as the balance loads on pier could not be disturbed.

Figure 1. Crack in the girder.

Figure 2. Girder with external prestressed tendons.

Figure 3. Steel girder launched over damaged span.

Figure 5. Damaged span removed.

Figure 4. Cut segment lowering in progress.

The bridge is across creek with marine clay for a depth of about 6 to 8 m overlaying rock. The region is with tidal variations. The clearance between deck bottom and water level is very low. During high tide hardly any room is available for operations. The bridge is about 58 years old. The code provisions have undergone drastic changes since then. The same structure, as existed, will simply be not allowed now.

The minimum dimensions of various components now required are quite higher than those of adopted for old structure. The bearings were fully embedded in prestressed articulations (halved joint) that are prestressed, for suspended spans. These had to be retained as the prestressing cables were part of main prestress at support.

The real task was to dismantle the existing PSC span without any debris falling in the creek from environment aspects and working in very restricted location in short period.

Various options for dismantling were studied, such as supporting the structure from bed, use of launching girders, barge mounted cranes to lift span as a whole etc. But none was found feasible. The use of composite structure, steel girders with concrete deck slabs probably only feasible option and was adopted.

Steel girders were found to be light enough to roll across the existing intact girder and robust enough to support the deck of damaged span. Thus the same girder facilitated removal of the damaged span as well could be used for new permanent structure.

The total scheme was then formulated with two nos. steel girders as main longitudinal elements and concrete deck slab. The weight of concrete deck slab could be easily adjusted to ensure that the total dead load of new structure and that of earlier structure are exactly matched.

The scheme comprised of dismantling the deck slab, partly from existing structure and balance with support from new steel girders; rolling/side shifting and erection of steel girders across damaged span, raising of the girders for making working head room and lowering them in final position;, specific sequence of cutting the concrete girders in segments for removal, supporting arrangement for segments for various conditions resulting from the cutting sequence; arrangements for holding the cut segments and lowering them in barge stationed in creek for removal to dump yard and completing of deck slab and other components.

The paper describes all the foregoing in details.

Concrete Repair, Rehabilitation and Retrofitting IV – Dehn et al. (Eds)
 ISBN 978-1-138-02843-2

Repair and retrofitting of two mega liter post tensioned, precast concrete tanks for molasses storage

J.H. Strydom
SMEC South Africa, Pretoria, South Africa
University of the Witwatersrand, Johannesburg, South Africa

B.H. Schlebusch
SMEC South Africa, Pretoria, South Africa

ABSTRACT

In this paper, retrofitting and repair of post tensioned, precast concrete molasses storage tanks, constructed in Durban, South Africa, is presented.

Tanks constructed by using precast elements generally have two shapes; rectangular or circular. The use of the precast concrete panels instead of conventional in-situ concrete, considerably increase the speed and precision of construction (Karbaschi and Goumdarzizadeh 2011).

The two mega liter precast concrete tanks were constructed to store molasses for a yeast manufacturing plant. The tank floor consists of a post tensioned concrete slab. Vertically pre-stressed concrete panels, post tensioned horizontally, form the walls of the tank. A tarpaulin roof is used to protect the molasses from the environment.

Both tanks constructed had significant defects and could not be certified to hold molasses. Defects included, insufficient grout strength, backward bending, omitted post tensioning cables, sliding of precast panels and structural defects in precast panels.

The repair of the tank was vital due to the high demand for molasses storage in South Africa. SMEC was appointed by the yeast company to design a solution that would allow the tank to be used for storage.

The tanks were repaired by retrofitting external stressing cables to withstand the pressure of the molasses during the daily use of the tanks as indicated in Figure 1. A curtain wall was used as a composite membrane to prevent leaks and to act as a structural support. The external cables were anchored with X-Anchors supplied by Freyssinet. A Sika combiflex bandage provided a leak proof movement joint on the inside of the tanks. Retrofitting and repair of the tanks were completed in two months. A water test indicated that the tanks were watertight and ready for molasses storage.

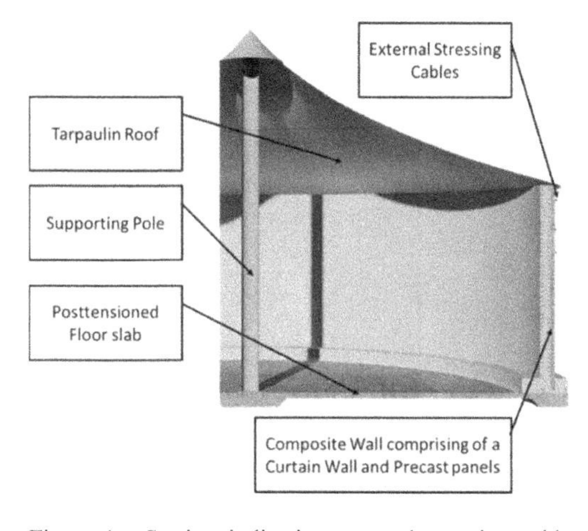

Figure 1. Section indicating external stressing cables and tented roof.

REFERENCES

373, ACI Committee. Design and Construction of Circular Prestressed Concrete Structures with Circumferential Tendons. MANUAL OF CONCRETE PRACTICE, Reported by ACI Committee, n.d.

Bijan O, A. Unbonded and Bonded Post-Tentioning Systems in Building Construction. A Design and Performance Review, Pos-Tentioning Institute, Arizona: PTI Technical Notes, 1994.

Bladon, S. "Repair Techniques for Prestressed Concrete Tanks." Maintenance Tips, Micham, n.d.

Hjorteset, K, M, W Wernli, M, W Lanier, K, A Hoyle, and H, O William. "Development of large-scale precast, prestressed concrete liquefied natural gas storage tanks." PCI Journal, 2013.

Karbaschi, M E, and R and Hedayat, N Goumdarzizadeh. "Efficiency of Post-Tensioning Method for Seismic Retrofitting of Precast Cylindrical Concrete Reservoirs." World Academy of Science, Engineering and Technology, 2011.

Concrete Repair, Rehabilitation and Retrofitting IV – Dehn et al. (Eds)
 ISBN 978-1-138-02843-2

Repair and widening of the Nels River Bridge on road R37 in South Africa

R.G. Miller & R. Nel
ARQ (Pty) Ltd., South Africa

E.J. Kruger
South African National Roads Agency (SOC) Ltd. (SANRAL), South Africa

ABSTRACT

This paper deals with the repair, strengthening, improvement and widening of the Nels River Bridge situated in the Mpumalanga province of South Africa. The paper outlines the design complexities and discusses the solutions and the construction techniques followed that resulted in a successful project.

INTRODUCTION

The Nels River Bridge is located on Road R37 between Sabie and Nelspruit in South Africa (GPS co-ordinates 25° 17′ 26.32″S 30° 46′ 1.74″E). This bridge was unique in that it comprised two arched balanced cantilevers connected with a mechanical hinge in the centre as depicted in figure 1. The hinge mechanism was on the verge of failure for a second time prompting the bridge owner and client to implement a major rehabilitation of the bridge.

This bridge, a post-tensioned box girder as depicted in figure 2, is critical for access across the river and complete failure would force detours of about 30 km.

In the final solution adopted the mechanical hinge connection was fully eliminated and the deck made continuous. When the bridge was repaired the roadway over the bridge was widened from 9.1 m to 12.4 m.

This paper presents details of the repair, strengthening, improvement and widening of the bridge. The design challenges, the solutions implemented and the construction techniques followed are discussed.

THE MECHANICAL HINGE

The central hinge connection comprised structural steel plates with slotted holes and steel pins of 90 mm diameter. It allowed for rotations and longitudinal movements but provided a restraint to vertical differential translations.

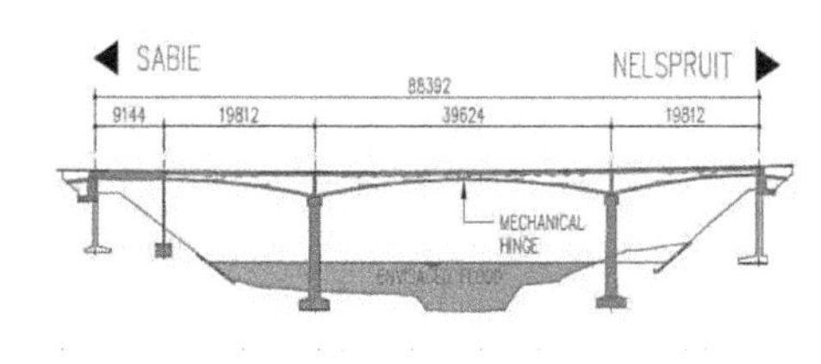

Figure 1. Section of the original Nels River Bridge.

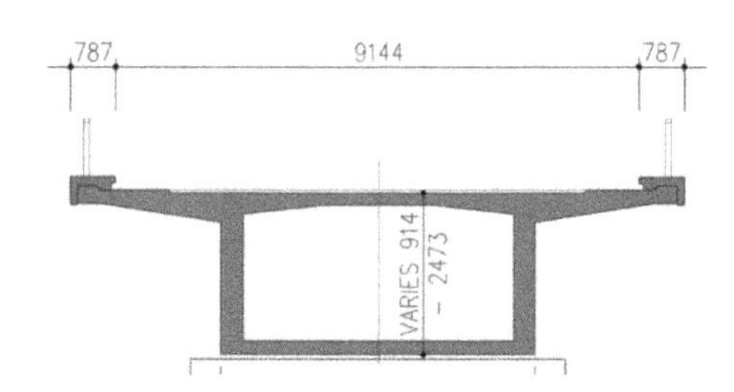

Figure 2. Cross section of the original deck.

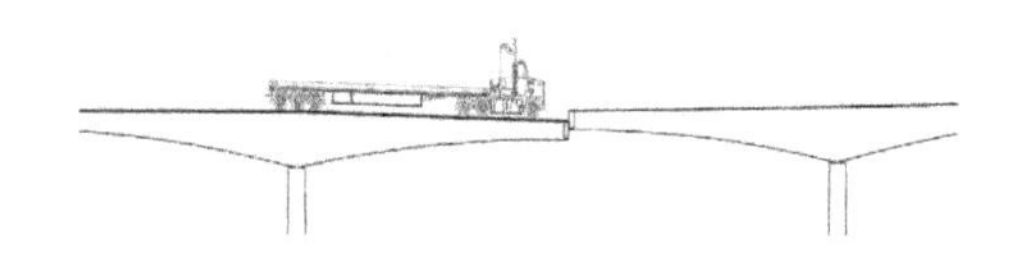

Figure 3. Vertical movements at hinge connection.

The capacity of the bridge deck would be severely compromised should the restraint to vertical differential translations fail. This is depicted in figure 3.

CLIENT REQUIREMENTS

The client's requirements for the bridge rehabilitation were the following:

- Eliminate the mechanical hinge connection by making the deck continuous.
- Widen the bridge from 9.1 m to 12.4 m.

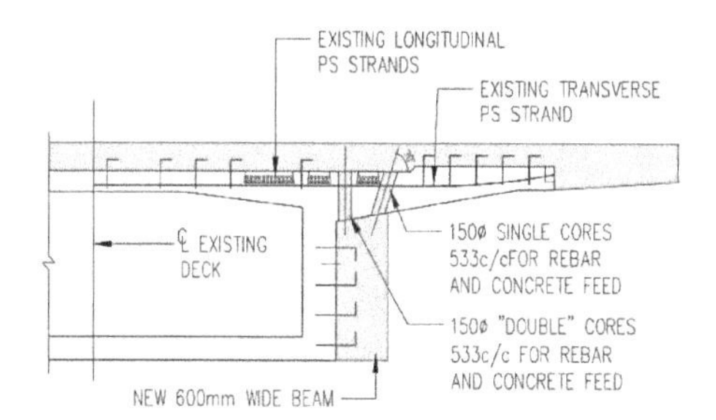

Figure 4. Cores linking new beam and slab.

- Correct the road vertical alignment over the bridge.
- Accommodate traffic over the bridge during construction.

SOLUTIONS INVESTIGATED

Several solutions were investigated to comply with the client's brief i.e.

- Widening on one side only
- Widening symmetrically on both sides
- Replacement of the old bridge

From the three options the second, widening on both sides, was adopted as it was the most cost effective and retained the form of the bridge from both an aesthetic and historic perspective.

DECK DESIGN

The existing box girder was widened with 600 mm wide post-tensioned beams on either side together with a new topping slab and cantilevers. Cores were drilled between the existing transverse stressing cables to provide links between beams and slab as depicted on figure 4. The beam and slab configuration was continuous over the full length of the bridge eliminating the problem of the central hinge.

A slab with varying thickness was constructed on top of the deck for structural purposes and in order to correct the road alignment over the bridge.

14 m high temporary concrete support columns were constructed on the sides of the river to stabilize the deck during construction as shown in figure 5.

CONSTRUCTION

Between the designer and the contractor it was decided to construct the deck in specific steps using frames attached to and cantilevering over the side of the deck as shown on figure 6.

The possibility of staging from the ground was considered undesirable, not only due to the height involved and the river below, but also because of the differential deflections of the old bridge vs the new concrete supported by "fixed" staging.

A proprietary self-levelling concrete product was used for the side beams under the cantilevers.

Continuity was achieved over the central joint by steel dowels, reinforcement continuity in the new slab and continuous post-tensioned cables.

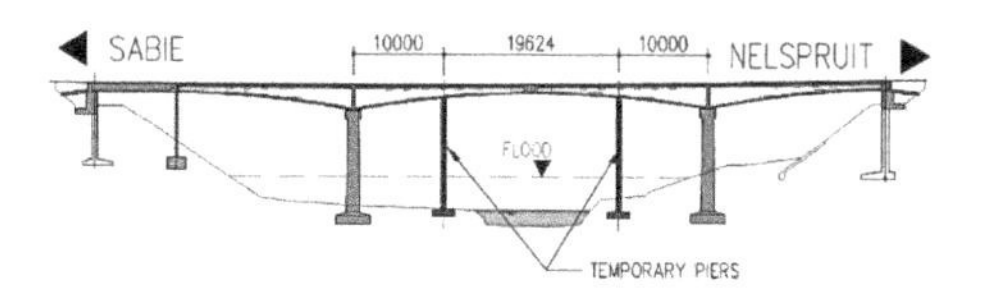

Figure 5. Temporary supports introduced on the river banks.

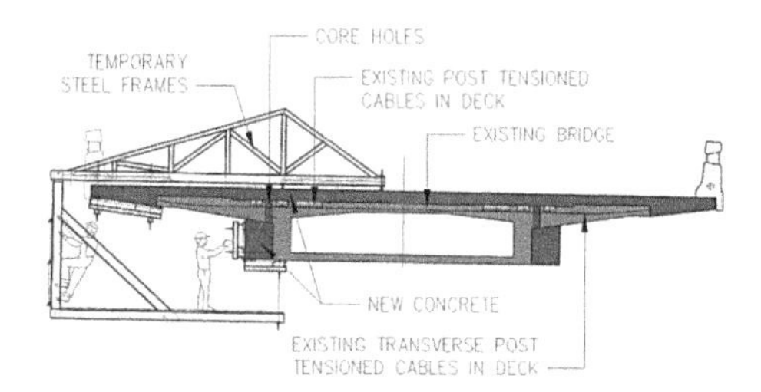

Figure 6. Deck construction methodology.

Figure 7. Roadway over the new bridge.

CONCLUSIONS

This project was challenging and through cooperation between the client, consulting engineer and contractor a successful widening of the bridge and elimination of the problem hinge joint resulted. Figure 7 shows the roadway over the new widened bridge.

DESIGN AND CONSTRUCTION TEAM

Client/Bridge Owner: SANRAL (SOC) Ltd
Consulting Engineer: ARQ (Pty) Ltd
Contractor: Stefanutti Stocks (Pty) Ltd

Concrete Repair, Rehabilitation and Retrofitting IV – Dehn et al. (Eds)
© 2016 Taylor & Francis Group, London, ISBN 978-1-138-02843-2

Evaluation of repair mortar materials for old monuments in southern India

S. Divya Rani, Madhubanti Deb, Manu Santhanam & Ravindra Gettu
Department of Civil Engineering, Indian Institute of Technology Madras, Chennai, India

ABSTRACT

Mortar samples from three historical monuments in the southern Indian state of Tamil Nadu were evaluated using X-ray diffraction, Scanning electron microscopy, and Fourier transform infrared spectroscopy for their mineralogical and morphological characteristics. The results show that the lime used is of a non-hydraulic type. Additionally, several studies were performed on lab based mortars to access the parameters of compatibility to check the suitability of different renders. From the testing of lab based samples, it was observed that, for studying the compatibility of repair mortar, parameters like modulus of elasticity and capillarity coefficient play a significant role as they help to judge the rigidity and porosity of the mortar. While cement mortar proves to be very rigid for old masonry substrate, plain lime mortar is very porous. For all properties studied, the lime mortar mix with organic additives gave optimum results.

INTRODUCTION

There are several masonry structures of historical significance in India that require conservation. Lime binders were commonly used for the construction of masonry buildings before the large scale introduction of Ordinary Portland Cement. When modern binders are used for the repair of these structures, a lot of compatibility issues arise between the repair mortar and the substrate. The characterization of mortar used in original construction is vital for formulating a new repair mortar which is compatible with the existing structure.

While designing a repair mortar, it is essential to weigh different physical factors according to their importance. For the same, several compatibility studies have been carried out in this field to assess some parameters of compatibility to check the suitability of different renders. It is widely accepted that the lime binders have greater permeability than OPC, enabling them to reduce moisture entrapment by their inherent 'breatheability'. For the lab based investigation in the present study, gallnut and jaggery are used as organic additives as these are reported to have been used widely in southern India.

MATERIALS AND METHODS

The historic mortar samples were collected from the following temple sites, (i) Veetrirundha Perumal temple at Veppathur village near Thanjavur in Tamil Nadu which is 2000 years old (ii) Sri Kolavilli Ramar Temple in Thiruvelliyankudi village, near Kumbakonam in Tamil Nadu, where the temple complex is said to be 2000 years old and (iii) Brihadisvara Temple at Gangaikondacholapuram in Tamil Nadu, which is one of the great living Chola temples and a UNESCO World heritage site, built 1000 years ago. Representative samples were collected and stored in air tight containers. Samples were air dried and powdered to a size less than 75 μm for X-Ray Diffraction analysis (XRD). For Fourier Transform Infrared Spectroscopy (FT-IR), samples were prepared by grinding a known mass of solid with dried KBr. The resulting powder was then pressed at 2000 psi for 5 min to produce a pellet for analysis. The wave number ranges analysed were from 400 cm^{-1} to 4000 cm^{-1}. Broken chunks were used for Scanning Electron Microscopy (SEM) and Energy Dispersive X-ray Analysis.

For the lab based investigation, a pure lime mortar, cement mortar and a lime mortar mix with organic additives was prepared. The lime used for the mortar preparation was procured from Pollachi, Tamil Nadu, which confirmed to class C lime upon testing. The cement used is Ultratech OPC (Ordinary Portland Cement) grade 55. Raw jaggery and gallnut powder were added as organic additives. The sand used is standard sand conforming to IS 650-1991. Modulus of Elasticity, Water Sorptivity, Compressive and Flexural strengths were tested at 14, 28 and 56 days as per the relevant standards The cement/lime to sand ratio was 1:3 and the mortar was mixed with the specified quantity of water/jaggery and gallnut juice i.e., 0.5 and 0.7 water to cement/lime ratio for cement mortar and lime mortar respectively. The cubes were demoulded after 1 day of casting for cement mortar and cured in lime solution until the required age of testing. The lime mortar cubes were demoulded after 3 days of keeping them in air. Later they were air cured at 25°C and 65% RH until the required age of testing.

RESULTS AND DISCUSSION

X-ray diffraction

X-ray diffraction indicated that there is no amorphous hump indicating the absence of any calcium silicate hydrate. Also, larnite and portlandite peaks are absent. These observations lead to a conclusion that the lime used was non-hydraulic in nature.

Fourier transform infrared spectroscopy

The bands at 712, 875 (C-O bend), 1440 (C-O stretch), 1790, 2515 and 2854 cm^{-1} correspond to calcium carbonate forms. Band at 3200–3600 cm^{-1} are due to O-H stretch bond, which could be due to the presence of moisture. The broad peak at 1000–1200 cm^{-1} can be attributed to polysaccharides or sucrose, which could possibly indicate the presence of an organic additive.

Scanning electron microscopy

The dominant presence of calcite crystals was seen in all the samples (highlighted in the images). Other minor elements present are Si, Al, Fe, Mg, Na and K. As the XRD analysis also suggests the lime to be non-hydraulic, the conclusion may be that most of the compound is in calcite form and the silica and alumina are from the aggregates. Also, the lime could be inherently impure which imparts a feeble amount of hydraulicity. The increased porosity of the matrix is another observation; dark regions in the image indicate pores or voids.

Study on lab based specimens

The results of lab based specimens showed that the compressive and flexural strength of cement mortar is much higher than the lime mortar mixes. The cement mortar can be found to be stronger than the substrate rendering it unfit for the purpose. Both the lime mixes had compressive strength in the prescribed range and can be considered suitable for repair of historic masonry. The cement mortar showed the highest risk in terms of compatibility and both the lime mixes satisfy the condition with minimum risk of incompatibility.

The capillarity coefficient of lime mortar mix with additives has values within the prescribed range that indicates an optimum amount of porosity compatible with the substrate material. Similar to the previous case, cement mortar shows high risk in terms of compatibility. In this case, the plain lime mix also shows high risks of compatibility as the value is higher than 50% of the prescribed value. However the value for the lime mix with jaggery falls under lower compatibility risk category. The dynamic modulus of cement mortar is very high suggesting that the matrix is too rigid. The values for both the lime mixes are within the prescribed limit and thus they are fit for using as repointing mortar in conservation works. The compatibility risk is the highest in case the strength is more than 50% of the prescribed value. Similar to the previous cases, in this case also cement mortar shows the highest risk in terms of compatibility and the plain lime mortar and lime mortar with additives show lower risks of incompatibility.

CONCLUSIONS

The salient conclusions drawn from the study are as follows:

- The historic samples studied are lime mortar samples and the lime is found to be non-hydraulic in nature.
- The lab based studies prove that the cement mortar is very rigid for old masonry substrate, and plain lime mortar is very porous
- Cement mortar is stronger than the two mixes of lime mortar tested. However, it was observed that for studying the compatibility of mortar with the substrate, strength parameter does not play an important role. For compatibility studies, parameters like modulus of elasticity and capillarity coefficient play a more important role as they help to judge the rigidity and porosity of the mortar
- The lime mortar mix with jaggery and gallnut powder gave optimum results and lower risk of incompatibility, which indicates that the organic additives used traditionally helped in optimising strength, flexibility and porosity of the matrix.

Concrete Repair, Rehabilitation and Retrofitting IV – Dehn et al. (Eds)
© 2016 Taylor & Francis Group, London, ISBN 978-1-138-02843-2

Selection procedures for concrete repair—patch repair and cathodic protection in atmospheric zones

F. Papworth, J. Dyson & M. Marosszeky
BCRC, Australia

ABSTRACT

Although Cathodic Protection (CP) systems for above ground structures have been used for over 30 years in Australia many designers and contractors involved in small to medium sized concrete repair projects consider it a complex and expensive method and give it no consideration for projects they manage. Hence on many projects no consideration is given to the life cycle cost savings, better structural outcome or improved performance and repair reliability that cathodic protection can provide. The paper provides some insight into these benefits.

Current codes dealing with cathodic protection of concrete above ground provide little guidance on how to select anode systems. In this paper an approach is outlined that systematically identifies what will best suit the client's overall needs.

SELECTING THE REPAIR APPROACH

When reinforcement corrodes and causes concrete to crack, spall or delaminate, the repair designer has to determine the approach to adopt for inspection and repair.

Typically the first step is to determine if the issue will only ever be localised or if it is a much wider future problem. This can be determined using visual inspection, drummy testing, electrical potential measurement and cover testing.

Where a long term repair is required and chloride ingress or carbonation is widespread then the decision becomes whether to use cathodic protection to eliminate the break out of sound but chloride contaminated/carbonated concrete or to break out around actively corroding reinforcement to remove sound but chloride contaminated/carbonated concrete.

Types of common concrete repair are:

1. Patch repairs with no electrochemical protection require breakout behind the reinforcement to remove chloride contaminated and carbonated concrete. The repair would typically include a reinforcement coating and surface coating to supplement protection.
2. Patch repair can incorporate low powered sacrificial around the edge of the patch to prevent corrosion of uncorroded reinforcement outside of the patch area, i.e. the incipient anode areas. Because the anodes do not provide much polarization of the reinforcement, i.e. they do not give cathodic protection, breakout and patching around the bar is still required with the associated issues of cost, noise and structural damage.
3. Patch Repair can have higher powered sacrificial anodes embedded within the repair patches to provide local cathodic protection. This means that breaking out behind the bars is not required. The anodes also provide incipient anode protection. The repair is still localized as the anodes only protect for a limited area around the patch. Protection can be extended by using discrete sacrificial anodes with the same high output.
4. There are various sacrificial anodes systems (surface applied, discrete and embedded) that provide high enough polarization to give cathodic protection in accordance with CP code requirements. The zinc of such anodes is consumed much faster than the lower power output anodes used for cathodic prevention. Hence a significantly higher zinc mass is required for the same life. These systems are very simple to install as they require little monitoring.
5. The typical repair approach where there is widespread corrosion activation is to patch the corrosion damaged areas and apply Impressed Current Cathodic Protection (ICCP) to all actively corroding areas. The systems are expensive to design, install and maintain but because they have a long life and proven performance they are frequently the repair system of choice for major projects.

CP SYSTEM COMPARISON

The authors have developed a procedure for CP system comparison to assist the designer in selecting the most appropriate system for a particular project.

The evaluation process has four steps:

1. Establish owner preferences.
2. Review CP systems. A five point scale is used to assess each of the acceptance criteria and a weighting applied that is specific to the project.

Table 1. Acceptance criteria weighting system.

Item	Acceptance criteria	Weighting
A	Initial Cheapness	10
B	Maintenance Cheapness	5
C	Reliability	7
D	Initial Speed Application	3
E	Customer Friendliness	4
F	Flexibility	3
G	Aesthetics	3

Table 2. Ranking system.

a) Cheapness class		b) Ranking method	
Limit ($/m)	Classification	Classification	Ranking
<300	Very High	Very Low	1
<350	High	Low	2
<400	Moderate	Moderate	3
<450	Low	High	4
>450	Very Low	Very High	5

3. Indicative pricing. The systems that best meet the client preferences are then priced to give a dual assessment.
4. Specification. The systems that are both economic and meet the client needs are incorporated into the CP specification so that they can be priced by the market.

A ranking and weighting approach is applied to the acceptance criteria. This is designed to determine how well the system meets the building owner's requirements for the structure. These criteria fall into two broad categories—suitability and cost.

Suitability factors are assessed first and any anodes not meeting the owner's preferences are discarded. Remaining systems are then assessed based on both factors.

It can be tempting for owners to focus predominantly on initial cost particularly when available budget is limited, but maintenance cost is also considered to enable whole of life (life cycle) cost to be assessed.

The system was used on a recent project where large areas of reinforced concrete structural elements required cathodic protection to achieve a future service life of 40+ years, the authors compared ICCP systems available for each type of structural element (slabs, columns, walls, beams).

To combine the various acceptance criteria a weighting system is used to describe the relative importance of each criteria. The weighting is developed for each project, and possibly each element, from the discussions with the owners. The weighting for the case study is given in Table 1.

Table 2a) shows the cheapness ranking system as developed for the columns. The cost bands are not fixed but are developed to cover the range of costs expected for each element type.

The ranking method in Table 2b) is applied to cheapness and suitability factors. A high number for both factors is positive.

The analysis is shown in Tables 3 and 4.

The analysis suggests that the expanded mesh with sprayed mortar overlay scores highest on lifecycle cost, whereas the Cassette and durAnode4 systems score were highest in terms of acceptability. CorroDisc was likely to be the most suitable system if both cost and suitability are considered equally but the higher ranking of cost made mesh the selected option.

Table 3. Weighted ranking of acceptance criteria (beams).

	Cost		Suitability				
Anode type	Init	Maint	Rel	Spd	Frnd	Flex	Asth
MMO Mesh	50	20	28	12	8	6	9
Sawtooth Rib	30	20	21	9	12	9	15
Ribbon Mesh	10	15	21	6	8	9	12
Cassette	20	15	35	15	20	9	9
CorroDisc	50	15	28	12	16	12	12
durAnode4	30	20	28	12	16	15	15

Table 4. Cheapness and suitability comparison (Beams).

	Mesh	Saw	Rib	Cass	Corro	dur4
Cheapness	70	50	25	35	65	50
Suitability	63	66	56	88	80	86

CONCLUSIONS

CP would be used more if there was a greater understanding of their benefits and method of application. SACP systems are simple and inexpensive to design and install and are suited to many small to medium sized projects. ICCP system will remain the system of choice for major projects where the capital budget is available. ICCP systems available are not being used to their best advantage because of a lack of a clear method for assessing which system is the most appropriate to a client needs. The paper provides a system for undertaking such an assessment. Use of such selection procedures is likely to increase client understanding of cathodic protection and, by removing the mystique, make it more commonly accepted.

Concrete Repair, Rehabilitation and Retrofitting IV – Dehn et al. (Eds)
© 2016 Taylor & Francis Group, London, ISBN 978-1-138-02843-2

Experience with installing an Impressed Current Cathodic Protection (ICCP) system on the multi-story car park of the Allianz Arena in Munich, Germany

Hernani Esteves
Ed. Züblin AG, Essen, Germany

Paul Chess
CPI Force, Essex, UK

Sebastian Mayer, Ronny Stöcklein & Rajko Adamovic
Ed. Züblin AG, Essen, Germany

ABSTRACT

In the end of April 2005, the new Allianz Arena in the North of Munich was opened as the home ground of the two soccer clubs FC Bayern Munich and 1860 Munich. The stadium has seats for more than 71,000 visitors.

Along with the opening of the stadium, Europe´s largest multi-story car park at a soccer stadium was completed. In 2014, concrete repair works were undertaken on approximately 800 column sockets in the car park in order to prevent further corrosion activity on the reinforcement of the concrete columns and the foundations of the columns in the 1st floor. The requirements for the repair were both to enhance the lifespan of these columns for another 50 years and to reduce mechanical and environmental impact during the repair work. After carefully evaluating several repair options, cathodic protection was chosen to be the most economic and most suitable solution /1/.

Situation

During technical inspections in the multi-story car park of the Allianz Arena in 2013, the first visual signs of corrosion in numerous column sockets were detected. Further material testing showed raised chloride contents in the concrete in a great number of column sockets.

Concept

Based on the technical inspection, the planning engineers came with a concept of Cathodic Protection (CP) with the target to protect selected column sockets in the car park in the long term from corrosion damage. Seven column types led to different CP layouts with MMO ribbon mesh anodes and MMO discrete anodes. Generally, MMO ribbon mesh anodes were installed at the concrete surface and provided current mainly to protect the reinforcement and the steel elements nearest to the concrete surface. The MMO discrete anodes provided current deeper into the structure.

INSTALLATION PROCESS

Continuity of reinforcement

When applying a cathodic protection system, it is highly important that all steel members and steel elements within the concrete are electrically connected. That way, the entire reinforcement forms one unit and can be connected with the negative pole of a DC power unit. Due to ICCP, the steel will be polarized and no single member or steel element will be exposed to stray currents.

Sensors, cathode connections and anodes

Prior to the anode installation at the columns, all required openings (e.g. for cathode connections and reference electrode installation) and slots for cables, primary anodes were established according to the ICCP layout. Cathode connections and monitoring cathode connections were welded to the exposed rebars in the openings. Further, 50 reference electrodes were placed into the concrete next to a rebar in order to monitor and to control the ICCP system in the car park of the Allianz Arena. Finally, the function of all components was controlled using multi-meter instruments and LCR meter (L = inductance, C = capacitance, R = resistance).

Depending on which type of anode will be installed at a column, the preparations on the concrete slightly varied prior the anode installation. When discrete anodes were used, holes were drilled into the concrete on exactly predefined positions and with an exact drill depth to ensure a current distribution as calculated in the design. Besides the advantage of a current distribution deep into the structure, discrete anodes represent an invisible ICCP solution on a structure as the anode rods are mounted into the concrete in holes. When using ribbon mesh anodes, the concrete surface must be cleaned thoroughly prior to anode installation. When the anode is fixed to the concrete surface, a vertically installed primary titanium strip connects the secondary anodes. Finally a layer of ICCP approved shotcrete completely covers the anode assembly around the column.

Cabling, power supply and communication

In the multi-story car park in Allianz Arena, no large surfaces are treated with ICCP. Instead of this, a number of several hundred column feet represent a huge bundle of small protection elements spread over large distances and three floors of the parking. Each column is provided separately, equipped with a control box for revision and further adjustments. Due to the long distances in this project, voltage drop in the power cabling was actively considered. With the target to reduce cable length to an acceptable extent, three separate cabinets were placed in the car park at different locations. In ICCP station 1 (master cabinet), the controller of the ICCP was hosted as well as a modem and a telephone landline for remote communication, next to a number of power supply units. In ICCP station 2 (slave cabinet 1) and ICCP station 3 (slave cabinet 2), power supply and reference electrode monitoring units were installed. All monitoring data were collected by data acquisition modules and transferred via data bus to the controller.

COMMISSIONING AND MAINTENANCE

Commissioning

In January 2015, the ICCP system was energized according to chapter 9 of the standard EN 12696. When it was clear the system was behaving well and no unexpected effects which could indicate problems, the voltage was increased stepwise with the aim to carefully polarize the steel to reach full cathodic protection according to the protection criteria of EN 12696.

Maintenance

In the multistory car park in Allianz Arena, a service contract was signed parallel with the commissioning of the system to make sure the system is maintained from the beginning. This way, the client receives a regular status record of his structure which confirms year by year the durability of a high-quality concrete repair.

CONCLUSIONS

In the car park of the Allianz Arena in Munich, a cathodic protection system has been successfully installed at the early stages of corrosion damage which was detected only 10 years after opening the car park in 2005. The system will provide a significant life enhancement to the structure.

A technically complex ICCP solution designed by the engineering office Schießl Gehlen Sodeikat GmbH in Munich with a tailor-made solution for the Allianz Arena car park was executed by specialists of the contracting company Ed. Züblin AG. An innovative power supply set up has been adopted allowing individual control of each of the columns while offering an electronic package despite the wide geographic spread of the installation.

Finally, a close cooperation between the client Allianz Arena München Stadion GmbH, the engineering office and the contractor was the key to complete this challenging project successfully, in time and with a happy client.

Concrete Repair, Rehabilitation and Retrofitting IV – Dehn et al. (Eds)
© 2016 Taylor & Francis Group, London, ISBN 978-1-138-02843-2

Application of cathodic protection on 30 concrete bridges with pre-stressing steel: Remaining service life extended with more than 20 years

A.J. van den Hondel
Vogel Cathodic Protection, Zwijndrecht, The Netherlands

E.L. Klamer
Royal HaskoningDHV, Nijmegen, The Netherlands

J. Gulikers
Rijkswaterstaat, Utrecht, The Netherlands

R.B. Polder
TNO, TU Delft, Delft, The Netherlands

ABSTRACT

In view of long term maintenance of its infrastructure facilities the Dutch Highway Administration (Rijkswaterstaat) has repaired over 1.500 heads of prestressed beams in 30 bridges and provided them with Cathodic Protection (CP). The heads of these beams showed moderate to severe concrete damage due to reinforcement corrosion caused by the penetration of chlorides from leaking joints, which required adequate intervention on the short term.

The beams are prestressed with a DYWIDAG system, which consists of 3 to 6 tendons per beam, which are anchored with a steel plate and a nut. From investigations, conducted by Royal HaskoningDHV, into the structural consequences of the concrete damage, resulting in a reduction of the concrete and reinforcement section, and of the (mathematical) failure of one of the tendons, it was concluded that further corrosion had to be stopped, in particular to maintain the concrete cover around the pre-stressing anchors and tendons. Computations showed that for the beams with a limited number of tendons (3 or 4), failure of one of the bars could already lead to the collapse of a beam.

Cathodic Protection is a successful technique to inhibit corrosion of reinforcement to a negligible rate by lowering its potential and was therefore used on the beam heads to halt the corrosion process in order to guaranty the structural integrity of the beams. In addition, the concrete repairs in the chloride contaminated parts of the beams will be durable.

Rijkswaterstaat, in collaboration with TNO, Royal HaskoningDHV and SGS Intron, designed a conceptual solution in which the leaking joints were to be replaced and an impressed current CP system (ICCP) had to be applied on the heads of selected pre-stressed beams. ICCP was chosen in order to obtain sufficient certainty about the degree of protection. A potential risk of an ICCP system is that it can result in very negative potentials at the steel surface. This so-called 'overprotection' may lead to brittle failure of prestressing steel and has to be avoided at all times.

The contract was ultimately awarded to the consortium Mourik/Salverda and CP specialist company Vogel Cathodic Protection was responsible for the concrete repairs and the application of CP.

After removal of all delaminated and spalled concrete parts, welded metallic connections to the steel were made in every beam, cable slots were milled in the concrete surface and reference electrodes (RE's; 2 or 3 in every beam) for monitoring were accurately placed by drilling holes close to either the reinforcement or close to the tendons. Over 3000 RE's were installed in the 1500 beams to assure corrosion protection and avoid hydrogen embrittlement simultaneously.

In order to measure the level of protection of the reinforcement, the pre-stressing tendons and the anchors, so-called 'decay probes' were installed in every beam to measure the changes in steel potential. In accordance with NEN-EN-ISO 12696 sufficient

Figure 1. Concrete damage and corrosion of steel reinforcement.

Figure 2. Accurately placing RE's for measuring sufficient protection and avoiding overprotection.

Figure 3. As-built CP system with decentralized power supply and monitoring units.

protection is obtained if the depolarization of the steel in the first 24 hours after the (temporary) switch-off of the CP system, is at least 100 mV.

In order to prevent hydrogen embrittlement of pre-stressing steel, the absolute potential is measured at the steel surface of the reinforcement or tendons closest to the concrete surface with the applied anode. For that purpose every beam was additionally equipped with so-called 'true reference' (Ag/AgCl) cells. NEN-EN-ISO 12696 prescribes that the steel potential must always be less negative than –900 mV.

The challenge for the CP engineer is to find a voltage at which sufficient protective current is sup-plied to all the steel in the chloride contaminated concrete section in the head of the beam in order to obtain a minimum of 100 mV depolarization in this area, but without the absolute potential at the pre-stressing steel surface to become more negative than -900 mV. At this point the limitation for the absolute potential at the pre-stressing steel surface is leading.

Next the concrete was repaired with a repair mortar suitable for the application of CP. After curing of the mortar a conductive coating (the CAST^{3+} system, a 2-component aluminum-silicate polymer filled with carbon) was applied as a (secondary) anode to the concrete surface of the beam head, by spraying it on the surface in 2 layers. In this coating a primary current carrying anode (a CuNbPt-wire) was incorporated.

Especially the restricted work space, the deep position of the beam heads at abutments, and the fact that at some locations the execution had to take place at night, complicated the repair work.

The RE's and the connections for both anode and cathode were installed and connected per 2 or 3 beam heads to a decentralized power supply and monitoring unit, which in turn was connected with a data cable and power cable to the central measuring and control unit. The decentralized units collect and store the measurement results for the connected beams and these results can be approached, read, collected and programmed remotely through the modem in the central control unit. Also the voltage applied on the CP system can be adjusted remotely as well, per every decentralized unit. The power of the CP system is, at most locations, provided by a double battery, which is fed by solar panels.

The execution phase of all 30 bridges within the project was completed in August 2014. The first bridges were completed and commissioned in December 2012.

The first CP systems are now running for over 2 years. The measurements on the CP system illustrate that the reinforcement is well protected, as is confirmed by depolarization values of the decay probes (≥ 100 mV).

Also, the pre-stressing tendons, which receives less protective current due to its deeper position compared to the steel reinforcement, are in general sufficiently protected. This is confirmed by the de-polarisation values of the 'true reference' electrodes. The risk for over-protection, and therefore hydrogen embrittlement, of the pre-stressing steel appears to be nil. In none of the measurements an absolute steel potential ('instant-off' value) is found that exceeds the criterion of –900 mV (the most negative value measured is above, i.e. less negative than, –600 mV).

Lastly, the necessary amount of protective current is found to be relatively small. After the start up of the system, initially a relatively large (protective) current was measured. After a couple of weeks or months the current consumption decreased to approximately 10–30% of the initial starting level. A further reduction is expected as the steel is protected by the CP system for a longer period of time and the conditions surrounding the steel are becoming less corrosive and gradually more steel becomes repassivated.

Severe corrosion of pre-stressing tendons and anchors could, in time, have had major adverse effects on the structural safety of the bridges. With the chosen solution of replacing the leaking joints and to protect the reinforcement and pre-stressing steel through cathodic protection, the structural safety of the beams is restored and the remaining service life of the bridges is extended by at least 20 years. With this solution the use of radical reconstruction measures could be avoided.

An aspect of most importance in this consideration is that Rijkswaterstaat has in advance introduced into the contract that the maintenance and monitoring of the CP systems will be performed by the contractor for a period of 20 years.

Concrete Repair, Rehabilitation and Retrofitting IV – Dehn et al. (Eds)
© 2016 Taylor & Francis Group, London, ISBN 978-1-138-02843-2

Towards improved cracking resistance in concrete patch repair mortars

P.A. Arito, H.-D. Beushausen & M.G. Alexander
Concrete Materials and Structural Integrity Research Unit, Department of Civil Engineering, University of Cape Town, South Africa

ABSTRACT

The failure of patch repair mortars as a result of restrained shrinkage cracking is responsible for the reduced service life in both concrete repairs and patch repair materials. Cracking in patch repair materials is dependent on important material properties such as tensile relaxation, magnitude of restrained shrinkage deformations, elastic modulus, tensile strength, etc. Many researchers have studied the effects of various mix design parameters and constituents on each of the aforementioned material properties. However, information regarding the effect of simultaneously varying multiple mix design parameters and constituents on the cracking performance of patch repair mortars is scanty. This paper reviews the effects of various mix design parameters—binder type, binder content, water content, water:binder ratio, chemical admixtures and curing—on the crack-determining material properties of patch repair materials.

INTRODUCTION

Concrete Patch Repair Mortars (PRMs) are intended to extend the service life of deteriorating concrete structures. To achieve this, PRMs are expected to possess a wide range of properties such as cracking resistance, durability and aesthetics, among other performance requirements. These properties depend, to a large extent, on mix design parameters. The selection of mix design parameters for PRMs, therefore, ought to be founded on a proper understanding of the influence of various mix design parameters, and their corresponding interactions, on the desired performance requirements. Interactions that exist among mix design parameters, however, are complex and poorly understood (Mehta and Monteiro, 2006). The subsequent subsections will attempt to elucidate the individual influence of binder-related parameters, water content, water:binder ratio, chemical admixtures and curing on the cracking of PRMs.

BINDER-CONTENT AND TYPE

Cement content affects important crack-determining material properties such as shrinkage. Increasing the cement content, at a constant water:cement ratio, results in a corresponding increase in shrinkage. However, at a given workability, shrinkage is unaffected by an increase in the cement content, or may even decrease (Neville, 2007). A reduction in cement content results in a corresponding reduction in heat of hydration and shrinkage.

The type of binder and its particle size influence important crack-determining material properties such as water requirement, tensile strength, tensile relaxation and shrinkage. Supplementary Cementitious Materials (SCMs) such as fly ash, silica fume and slag affect the following mix properties: water requirement, kinetics of hydration, heat of hydration and the performance of the mix with respect to durability, mechanical strength, elastic modulus, tensile strength and creep and relaxation (Khan et al., 2000; Langan et al., 2002; Lawrence et al., 2003). The contribution of various SMCs on specific crack-determining material properties, however, is subject to contradictory views.

WATER CONTENT AND W/B RATIO

The water requirement is an important crack-determining parameter and is dependent on the properties of materials used in a mix. For given materials and conditions, the higher the water content, the higher the drying shrinkage and the greater the tendency for mortar mixes to crack. High water contents also reduce the mechanical properties of the hardened material as well as increase shrinkage and creep deformations.

The water:binder (w/b) ratio is directly related to the pore system of the paste. The porosity of the paste, and its associated interconnectivity, influences important crack-determining properties like: tensile strength and elastic modulus. The relationship between w/b and cracking is quite complex and incorporates other factors inter alia. Increasing w/b increases the potential for shrinkage; with the effect being more pronounced with a decrease in aggregate content by volume. Increasing w/b also: decreases the total amount of heat evolved, increases the rate of evaporation as well as delay the occurrence of first crack. Lowering w/b result in: low crack areas and small crack widths, reduces creep and increases tensile strength. The extent to which the w/b can be reduced, however, ought to be considered carefully vis-à-vis the susceptibility of a material to experience autogenous shrinkage.

CHEMICAL ADMIXTURES

The extent to which an admixture affects shrinkage in a cementitious material depends on its intended use. Despite the fact that the effect of admixtures on many material properties are well established, it is difficult to predict their influence on certain material properties owing to their highly specific nature. There is no single admixture that can produce mixes of optimum quality under all conditions. It is imperative, therefore, that the influence of various types of admixtures on crack-dependent material properties such as shrinkage, elastic modulus, tensile strength, etc. be investigated.

CURING

The benefits of adequate curing, in light of the thrust of this paper, comprise: increased tensile strain capacity and volume stability (Concrete Society, 1992; Perrie, 1994; Taylor, 2014). Increasing the tensile strain capacity is a desirable material property because it implies that a material will withstand a greater tensile strains without failing. Also, volume stability minimises the possibility of shrinkage-induced stresses that could result in cracking. Adequate moist curing also delays the advent of shrinkage and increases tensile strength (Kronlöf et al., 1995; Neville, 2007; Alexander and Beushausen, 2009). The influence of the duration of curing on crack-determining material properties, however, is subject to contradictory views.

CONCLUSIONS

The influence of various mix design parameters on cracking in PRMs has been presented. For PRMs to be effective with respect to resistance to cracking, it is imperative that the relationships and synergies that exist within the aforementioned mix design parameters are investigated, and the information obtained from such investigations is harnessed and applied into the design and development of PRMs. Also, there are many contradicting views regarding the influence of various mix design parameters on crack-determining material properties. These contradicting views tend to complicate the process of selecting optimal mix design parameters that would guarantee the realisation of effective PRMs. A more holistic investigation into the combined influence of the aforementioned mix design parameters on crack-determining material properties in PRMs, therefore, is necessary. A holistic investigation will provide valuable information that would dictate the process of designing PRMs that are effective with respect to resistance to cracking.

ACKNOWLEDGEMENTS

The authors wish to acknowledge with gratitude the University of Cape Town and the Carnegie Corporation of New York for the financial support for this research. Financial support from the University of Cape Town was in the form of a Research Associateship.

REFERENCES

Alexander, M., and Beushausen, H. (2009). Deformation and volume change of concrete. Fulton's concrete technology, 9th Edition, Midrand South Africa, Gill Owens (Ed), pp. 111–154.

Concrete Society. (1992). Concrete Society Technical report No. 22-Non-structural cracks in concrete, 3rd Edition. Concrete Repair Manual, Second Edition, Volume 1. Published jointly by American Concrete Institute, Building Research Establishment, The Concrete Society and The International Concrete Repair Institute, 2000, pp. 507–551.

Khan, M.I., Lynsdale, C.J., and Waldron, P. Porosity and strength of PFA/SF/OPC ternary blended paste. Cement and Concrete Research, 30, 2000, pp. 1225–1229.

Kronlöf, A., Leivo, M., and Sipari, P. Experimental study on the basic phenomena of shrinkage and cracking of fresh mortar. Cement and Concrete Research, Vol. 25, No. 8, (1995), pp. 1747–1754.

Langan, B.W., Weng, K., and Ward, M.A. Effect of silica fume and fly ash on heat of hydration of Portland cement. Cement and Concrete Research, 32, 2002, pp. 1045–1051.

Lawrence, P., Cyr, M., and Ringot, E. Mineral admixtures in mortars Effect of inert materials on short-term hydration. Cement and Concrete Research, 33, 2003, pp. 1939–1947.

Mehta, P.K.., and Monteiro, P.J.M. (2006). Concrete: Microstructure, properties and materials. Third edition. Tata McGraw Hill, 660 pp.

Neville, A.M. (2007). Properties of concrete, 4th Edition, Pearson Prentice Hall, pp. 412–481.

Perrie, B.D. The testing of curing compounds for concrete. Department of Civil Engineering, University of the Witwatersrand, Johannesburg, 1994, MSc dissertation.

Taylor, P.C. (2014). Curing concrete. CRC Press, Taylor and Francis Group, 181 pp.

Reinforcement corrosion in separation cracks after injection with PUR

M. Kosalla & M. Raupach
RWTH Aachen University, Aachen, Germany

ABSTRACT

In separation cracks in RC elements (e. g. park decks), a short-term chloride exposure can already cause reinforcement corrosion. It is questionable, if a deep injection of the crack as an exclusive procedure is sufficient for effectively stopping the corrosion process. The crack repair methods for cracks, including the injection, are currently extremely discussed in Germany. But most available experiences derive from constructions in practice and thus, they always have various boundary conditions. So there is no scientific work available at the moment, in which specific repair methods are investigated in a scientific way, for confirming or rebutting the thesis and/or assumptions among experts in practice based on quantitative research results. So in this paper, the methods and results of the investigations at RWTH Aachen University are summarized and should provide a first step in the quantitative evaluation of the crack repair problems.

For investigating the corrosion activity during the chloride exposition and after the crack maintenance, cracked beams with crack-crossing reinforcement were fabricated. The cracks have been grout injected with PUR after a winter season of six month with a periodic chloride exposure. The anodic (element) current was recorded both before and after the injection. More than one year after the injection, the measurement stopped and the crack-crossing rebars were removed from the specimens and visually investigated.

During chloride exposure the measured current flows were very high (sometimes >200 µA). After the injection they slowly decreased and dropped nearly to zero in the next winter period. In the following summer the current flows increased just marginally, which seems to be mostly uncritical (fig. 1).

The visual inspection mostly indicates a structurally insignificant degree of damage (fig. 2), because the loss of cross-section is just in exceptional cases more than 10%. So the combination

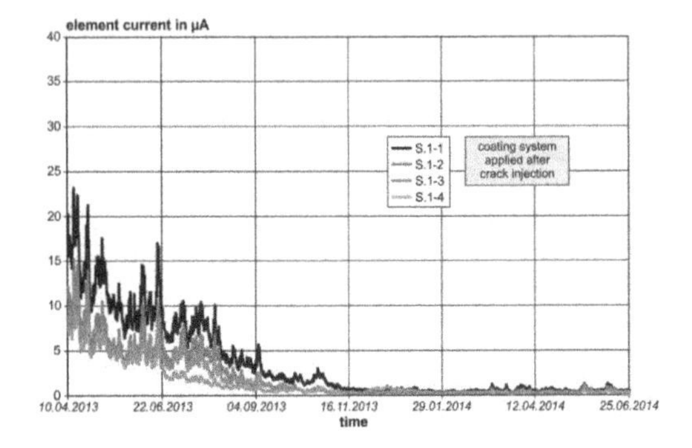

Figure 1. Element currents after the crack injection (series 1).

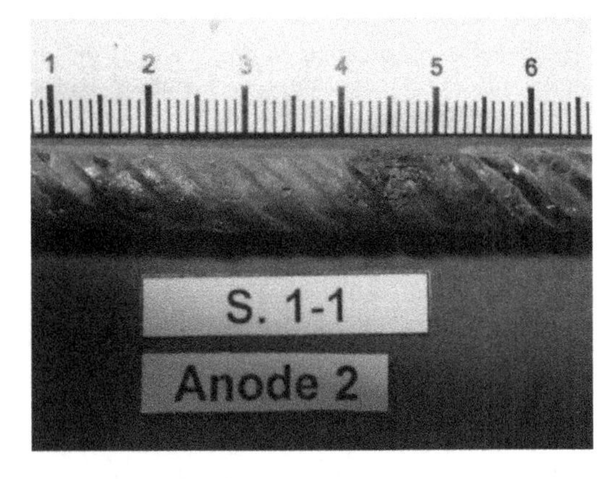

Figure 2. Representative damage of reinforcing steel.

of the previous slight damage and the low corrosion activity after the injection suggests a sufficient effectivity of the crack treatment. But since higher corrosion rates for worst case considerations are not excluded, it is recommended by the authors to install a monitoring system.

Concrete Repair, Rehabilitation and Retrofitting IV – Dehn et al. (Eds)
© 2016 Taylor & Francis Group, London, ISBN 978-1-138-02843-2

Maintenance of concrete pavements with thin-layered new concrete

Amela Cokovik & Rolf Breitenbücher
Institute for Building Materials, Ruhr-University, Bochum, Germany

ABSTRACT

Concrete pavements are exposed to a number of stresses during their service life, mostly resulting from traffic and climate conditions. In consideration of continuously increasing traffic and high exposure to freeze-thaw and deicing salt, maintenance and repair works of concrete pavements become of crucial importance. Thereby, the main aim is to find the most economically efficient construction method through minimizing life cycle costs without affecting the functionality and quality of concrete pavements.

Usually, maintenance of concrete pavements is conducted through the renewal of the damaged concrete road. An alternative method to the conventional replacement of the entire concrete structure is the renewal of the upper part of the existing pavement, which is impaired. Herein, the milled old concrete provides the base course layer. The crucial factor here is the bonding behaviour between the lower layer (old concrete) and the thin layer of new concrete. First pilot projects have been carried out in Austria on the motorway A14 Rheintal and the A1 between Salzburg Center and Salzburg West. So far, no damages could be detected.

The objective of this study at the Institute for Building Materials (Ruhr-University Bochum) is to investigate the influence of cyclic loading and freeze-thaw-cycles with deicing salts on the bonding properties of such concrete layers. Herein, a number of parameters are varied including the moisture condition of the old concrete, the application of adhesion primer, the consistency of the new concrete etc. On the basis of these investigations the pre-treatment of the structural joint and special measures during the application of thin layers of new concrete can be defined in order to ensure a durable bonding between these concrete layers.

The results acquired within the project so far have shown that the moisture condition of the old concrete does not play a significant role. Furthermore, the application of adhesive primer is only useful for concrete with a stiff consistency since the adhesive strength increased only slightly using concrete with a soft consistency.

Concrete Repair, Rehabilitation and Retrofitting IV – Dehn et al. (Eds)
© 2016 Taylor & Francis Group, London, ISBN 978-1-138-02843-2

Important factors in the performance, durability and repair of concrete façade elements

A.N. van Grieken
van Grieken & Associates, Melbourne, Australia

INTRODUCTION

Many existing and new facades feature prominent insitu and precast concrete elements. Whilst many elements have stood the test of time, others have not.

The disparity in the durability of concrete façade elements is linked to a number of key factors, which include amongst others, exposure conditions, concrete composition, and the surface geometry and finish. In addition, movements are a common cause of defects and are often not adequately anticipated and allowed for in the design and construction. These include shrinkage, creep, reversal thermal movements and deflections.

Where defects or deficiencies have developed, it is imperative that these are diagnosed and analysed to identify the cause or causes. In parallel, it is important to have an understanding of the construction and detailing. The remedial solutions and repairs need to address all of the causes. Frequently, the diagnosis is easier than specifying the cure!

In addition to technical requirements, commercial and site factors are important in successful repair projects, including repair quantities, minimising noise and disruption and access costs.

This paper examines important durability and performance factors related to concrete façade elements and the most important corresponding repair and protective works considerations. These are based on more than 30 years of involvement with concrete facades investigations and repair projects across Australia.

UNIQUE DURABILITY FACTORS

Surface finish

It appears that smooth polished exposed aggregate surfaces and surfaces with densely spaced exposed quartz and other aggregates, carbonate at a relatively lower rate.

Surface geometry

Ledges, rebates in the faces (photo 1) and water 'drips' at the base of beams and slabs (photo 2), are typically highly vulnerable areas.

Photo 1.

Photo 2.

Movements

The term 'movements' is used generically in this paper, as it covers many forms of movement with possible corresponding defects. Amongst others;

- *Deflections*
- *Shrinkage and Creep* in the form of vertical and horizontal shortening. This includes post-tensioned elements which can cause structural but also significant non-structural damage.
- *Differential thermal movements* are a particular concern where PCE elements are, or become, rigidly locked in place.

In general concrete elements move. Amongst others, buildings shorten over time, predominantly due to self-weight, beams deflect, and post-tensioned floors shorten due to creep. The latter can continue for many years and well beyond some expectations.

Other

A. Spalling has occurred at locations of concentrated water run-off.
B. Hotter facades dry out faster; predominant rain directions create more frequent and more severe wetting.
C. Differences have been noted at for example facade corners where negative local wind pressures create drier, vacuum-like conditions.

In summary, the diagnosis of defects and conditions needs to take all of the above into consideration. Rarely are defects contributable to a single cause.

Photo 3.

Photo 4.

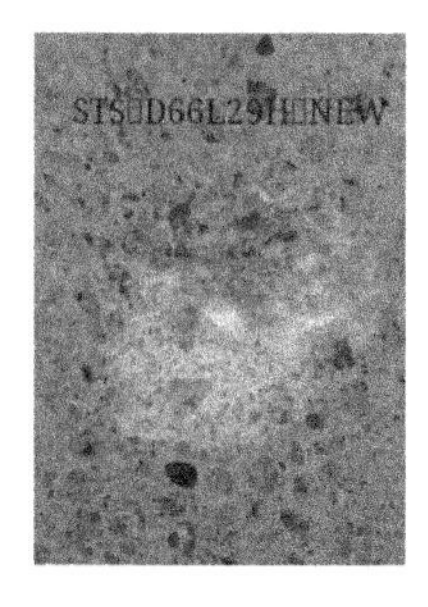

Photo 5.

Photo 6.

IMPORTANT FACTORS IN SUCCESSFUL REPAIRS AND REPAIR PROJECTS

Technical

In addition to generic repairs, specific targeted repairs may be required. For example, additional movement accommodation capacity may need to be introduced and the local geometric configuration may need to be altered. Protective treatments are required where the development of future defects is likely. Following are some examples:

A. Cracks and spalls may develop where PCE elements are rigidly connected to the façade structure (photos 3, 4).

B. Glass etching due to alkaline contaminated run-off water is a common problem. Impregnating silane treatment of the concrete surfaces above the windows was a very successful and very durable 'repair' at one highrise building.

C. Close examination of face spalls on a high rise façade, revealed that these were generally not due to corroding reinforcement, and the presence of small white particles (photos 5 and 6). As a result, petrographic examinations were carried out on thin sections Scanning Electron Microscopy (SEM) and Energy-Dispersive X-ray (EDX) analysis revealed that the white aggregates at the centre of the spalls are composed of Periclase (MgO) and Brucite [$Mg(OH)_2$)], the latter a hydration product of Periclase which is expansive. Clearly, conventional repairs are not applicable in this case, and as the spalling to date may be the 'tip of the iceberg', preventative remedial works are required.

D. Straightforward repairs to concrete walls can become more complicated at windows. Repairs may need to include replacement of connections.

E. The inclusion of galvanic anodes can prevent the formation of insipient corrosion and has no visual impact. For facades this is the preferred approach, rather than cathodic protection.

Commercial

A. *Quantities* are the most contentious and challenging aspects of façade repair projects.

Photo 7.

Photo 8.

B. *Noise and disruption* due to impact equipment is another major challenge.

C. *Communication.* Building occupants expect and need to be regularly informed.

D. *Access.* Inevitably, access and access equipment requires consent from apartment owners and building tenants.

Quality assurance

Quality Assurance starts from day one on-site and continues completion. This needs to include a range of components.

- Trials allow the client to see and approve the proposed outcomes, and hold points enable the designer/specifier to inspect and study the critical stages of the works (photo 7).
- Pull-off tests provide a quantitative evaluation of the quality and the bonding of the repairs (photo 8).
- Humidity, surface temperature and moisture content tests before coating are required by coating manufacturers.
- Monitoring quantities at agreed 'hold points'

CONCLUSION

Concrete façade repair projects are complex and involve many aspects. An adequate understanding of the cause of the defects, solutions which eliminate or accommodate these causes, experienced specialist contractors and a bit of luck are required for successful repair projects.

Concrete Repair, Rehabilitation and Retrofitting IV – Dehn et al. (Eds)
© 2016 Taylor & Francis Group, London, ISBN 978-1-138-02843-2

Evaluation of shear bond test methods of concrete repair

V.D. Tran, K. Uji, A. Ueno, K. Ohno & B. Wang
Tokyo Metropolitan University, Tokyo, Japan

ABSTRACT

A good adhesion between old concrete and repair material plays an important role in the success of repair work for aged concrete structures. However, the prediction of this bond strength is very difficult because concrete is a highly complex heterogeneous material. Up till now, although there are many test methods to estimate the shear bond strength between old and new concrete, their results are relatively different. In this paper, an experimental and theoretical study were performed to evaluate the shear bond strength of the interface between concrete substrate and repair material in 5 types of test method, namely, Direct Shear Test (DST), Push-Off Test (POT), Bi-Surfaces Shear test (BSST), Slant Shear Test (SST) and Punching Shear Test (POT) (Fig. 1).

A total of 63 specimens of all types were created. There are 3 kinds of surface roughness level (Smooth, Medium, High) divided equally in each method (Figure 2). During testing, failure processes were monitored by the Acoustic Emission (AE) method and analyzed by SiGMA (Simplified Green's functions for Moment tensor Analysis) procedure. The obtained results of all methods were compared with each other and evaluated in the correlation with roughness index. As can be seen in the Figure 3, SST always gives the highest value of the shear bond strength. The stress state of POT, PST is a combination of shear and compressive stress. Therefore, their shear bond strengths are higher than in DST and BSST which have only purely shear stress at the interface. Simultaneously, the important role of roughness level of the interface in the increasing of shear bond strength was confirmed.

In the remained part of this paper, a saw-tooth model is used to exemplify the macro texture of the interface between concrete substrate and repair material. It is assumed that in every saw-tooth always exist two bonding components on the surface to restrain slip between 2 parts, namely adhesion and friction. If external loading exceeds total of restrained forces, the shear failure will occur like a slip phenomenon between two materials. Based on stress state at the interface, these test methods can be divided to 3 groups; group 1 includes DST&BSST; group 2 has POT and PST; the third one has only SST. In each type of test, the normal stress appears and varies resulting from external force or the change of bond plane's angle are the main reasons make the results of each tests is significantly different. However, the theoretical calculations, which based on the shear-friction theory,

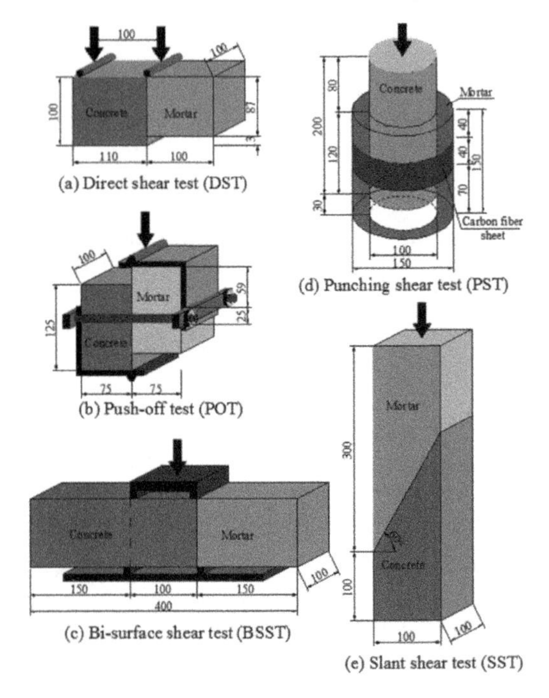

Figure 1. Shear bond test methods.

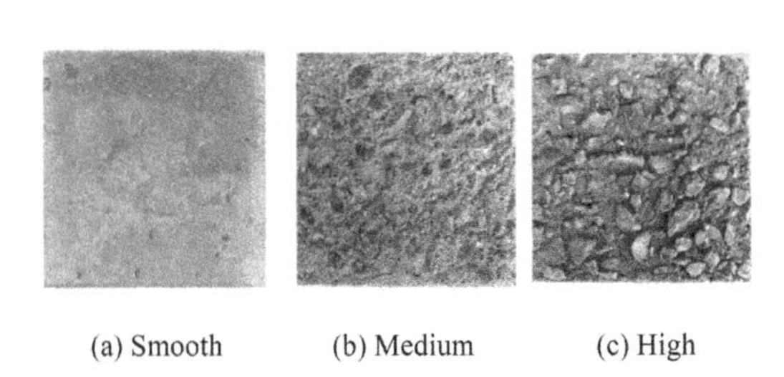

Figure 2. Three types of surface texture.

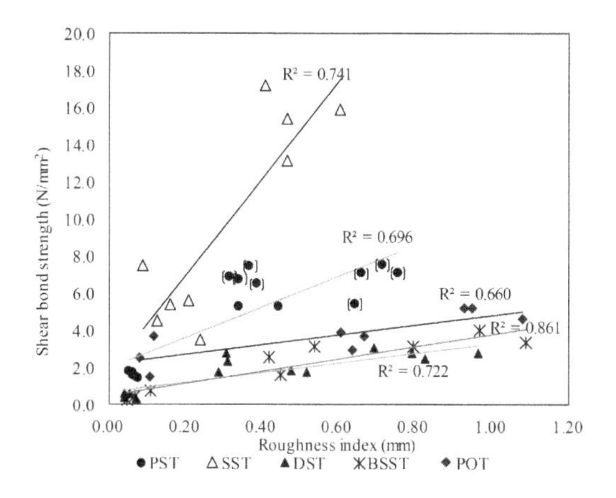

Figure 3. Relation between the roughness index and the shear bond strength from test results.

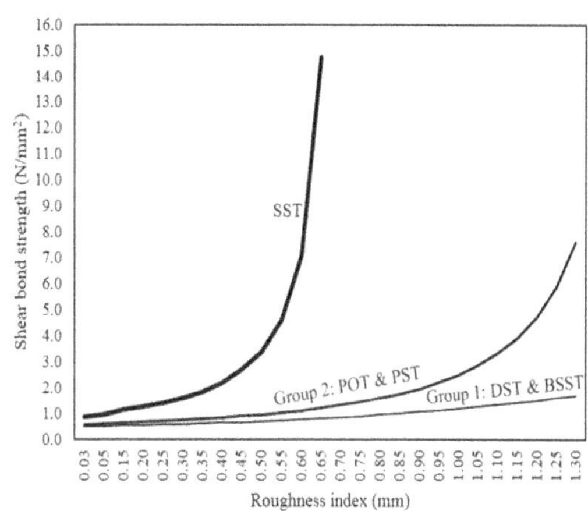

Figure 4. Relation between the roughness index and the shear bond strength from theoretical calculations.

were presented to some degree make clear about the alteration in the results of these tests (Figure 4).

In conclusion, the shear bond between old and new concrete is significantly vary depend on test method. In addition, it is obvious to say that roughness level has a great influence on shear bond strength. However, from a certain limit of roughness, increment of roughness would become meaningless. The theoretical calculations are presented somewhat make clear why the compress-shear stress state tests always give higher results than others and SST presents the highest. In the purely shear stress state test, BSST is the most suitable one.

Concrete Repair, Rehabilitation and Retrofitting IV – Dehn et al. (Eds)
© 2016 Taylor & Francis Group, London, ISBN 978-1-138-02843-2

Structural repair approach for reinforcement corrosion in concrete building structure: An application case

G. Gallina, M. Graziosi & A. Imbrenda
Proges Engineering S.r.l., Rome, Italy

ABSTRACT

The reinforcement corrosion in concrete building structures is one of great problem in preservation, durability and life services extension of existing building. The knowledge of structural safety level of building damaged by corrosion effects is necessary to estimate the residual service life.

In Italy, most of the reinforced concrete buildings were built in a period from the end of Second World War to the early 70s. Reinforced concrete structures of the buildings of this period have often shown problems of degradation due to corrosion of the reinforcement bars, which prompted the subsequent studies and research, which introduced the durability requirements among the criteria for a project. In this context, those who manage real estate assets will often have to decide whether to take action to repair the structures or defer interventions.

Structural deterioration due to corrosion of the bars depends on, as seen, many variables. Some of these are related to the structure itself (quality of concrete, water cement ratio, porosity, concrete cover thickness, etc.), others are completely independent from it (temperature, humidity, etc.). In addition to this we have to consider that the probability of collapse is significantly influenced by the occurrence of loads and in particular of that seismic (load not considered in the design phase for most of existing buildings). Derivatively the probability of failure will be conditioned by the depth of carbonation and the subsequent oxidation level of the bars that affects sections resistance.

The description above makes clear that the probability of failure comes from interrelation of various stochastic variables some of which are completely independent and others are conditioned one from each other or dependent.

The application of Bayesian Network seems to be the most appropriate instrument to take correct account of these interconnections and achieve, easily, the determination of a failure probability.

The study carried out show the application on a seven stories office building 56 years old. Each floor is 2000 smq. Columns-beams frame forms a façade structural system. High level of reinforcement corrosion damage was evident on many structural elements. The experimental tests on reinforced concrete elements was carried out to know the carbonatation level. The structural repair is carried through rust remove, process of rebars passivation and FRP application. The service life extension is calculated for repaired structures. The study shows

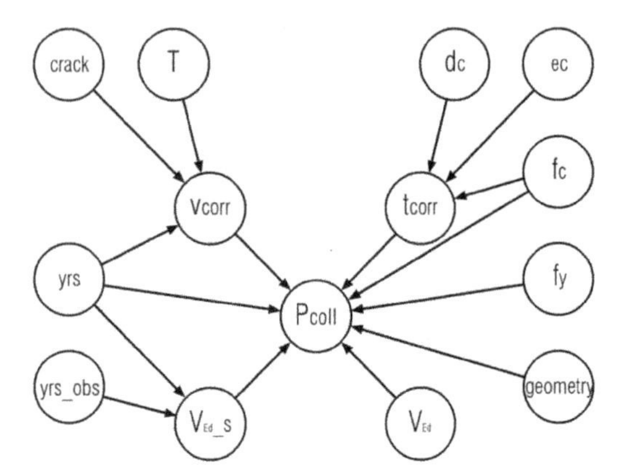

Figure 1. The BBN for probability of failure.

Table 1. Nomenclature.

Pcoll Probability of failure
V_{Ed} Shear caused by vertical loads
V_{Ed_s} Shear caused by seismic loads
tcorr Time from the beginning of the corrosion
vcorr Corrosion rate
crack Parameter opening of cracks
yrs Time from the construction of the structure
yrs_obs Time of observation
geometry Geometric dates of the structure
dc Concrete cover depth
fc Compressive strength of concrete
fy Yield strength
T Temperature
ec Exposure class

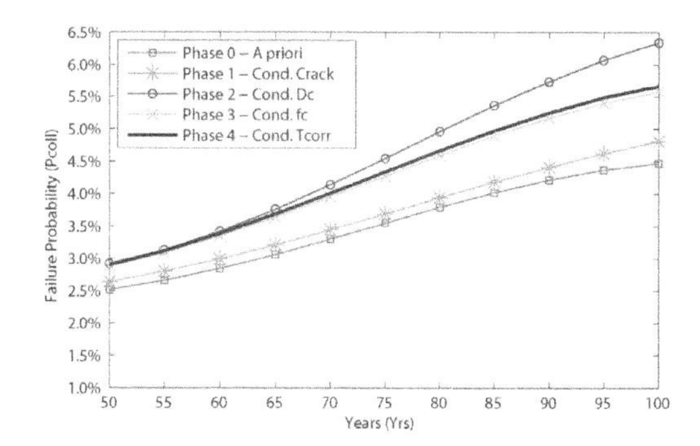

Figure 2. Failure Probability curves with increasing level of knowledge.

the potentiality of the method (the opportunity of an integral approach in this field) and limits of the approach (evaluation of confidence range in experimental data).far more complicated than previously considered and has important implications for upland water quality.

The structural repair is carried through rust remove, process of rebars passivation and FRP application. The FRP wraps allow the refurbishment of lost stirrups section and a reactivation of the bond between concrete and steel rebar. Before of the application of FRP, a migratory corrosion inhibitor was applied to protecting and passivate the rebars.

The paper shows that the bayesian net are an easy method to evaluate the probability of collapse in application case too.

The bayesian net are a very good method to synthesize a great number of stochastic variability and build an adequate model of physical phenomena.

In the application case, the opportunity to build an a priori evaluation jointed to data tests produce a valid collapse probability.

Next steps on this research pattern are the use of bayesian net to define a focused experimental tests.

Concrete Repair, Rehabilitation and Retrofitting IV – Dehn et al. (Eds)
© 2016 Taylor & Francis Group, London, ISBN 978-1-138-02843-2

Electro Active Repair of concrete for improved durability of conventional repair

R.B. Polder
TNO Technical Sciences, Delft, The Netherlands

M.R. Geiker
NTNU, Department of Structural Engineering, Trondheim, Norway

ABSTRACT

Chloride induced corrosion of reinforcement is a growing problem due to aging of concrete structures exposed to deicing and marine salts. Increased numbers of structures need repair of damage and protection against chloride induced corrosion [Polder et al. 2012]. The majority of repairs follow the conventional method: removal of chloride contaminated concrete, cleaning of the steel to bright metal and application of new concrete [EN 1504-10]. In practice, conventional repairs have poor durability and their life appears limited: according to a European survey, about 50% of repairs fail within ten years [Tilly & Jacobs, 2007; Tilly, 2011]; similar observations were made in a Dutch study [Visser & Zon 2012]. Failure of conventional repairs is due to various mechanisms. Corrosion related factors of repair failures are: insufficient removal of contaminated concrete; insufficient cleaning of affected reinforcing steel; and electrochemical effects (incipient anode or "patching" effects) between repaired and surrounding non-repaired places where chlorides are present. In such cases, the chlorides left behind in concrete, corrosion products, and corrosion pits reactivate the corrosion process. Consequently, the loss of steel cross section continues and new cracking results, necessitating new repairs.

Removing chlorides and cleaning of steel is not critical for electrochemical methods for corrosion protection [Bertolini et al. 2013]. Cathodic Protection (CP) works by passing a low direct current to the reinforcing steel, suppressing corrosion. CP is applied on a wide scale and long working lives can be obtained [Polder et al. 2014]. However, continuous current flow must be secured and monitoring must be carried out over the remaining life of the structure. In particular the obligation to monitor is a disadvantage for many owners. Electrochemical chloride extraction (desalination) works by passing a high current for a period of up to several months, removing chloride ions from the concrete (and from pits) [Polder & Hondel 2002]. The main disadvantages are the long period of execution and the uncertainty of when the process can be stopped without compromising the effect and without causing major changes of the concrete microstructure [Castellote et al. 1999].

The authors propose to improve the conventional method of concrete repair by adding a step in which a direct current is applied for a short period of time. This additional step is called Electro Active Repair (EAR). It causes electrochemical removal of chlorides from the steel surface, corrosion products and corrosion pits. The overall repair process then becomes as follows. After removal of spalled concrete, a temporary material is placed in the excavation, a temporary anode is installed and a cathodic current is applied to the steel for a short time (24 hours) as illustrated in Figure 1. The temporary material can be a cementitious material or an ion conductive gel or paste. Following steps are removal of temporary materials, removal of contaminated concrete and conventional repair as usual. A European patent was granted in 2014 and a worldwide patent applied for. The method has the advantage that it fits in the conventional repair process. Monitoring is not necessary. The additional cost is expected to be moderate. Applying the invented pre-treatment to conventional repair is expected to increase the life of repairs and consequently reduce the life cycle cost of corrosion affected structures.

Pitting of reinforcing steel is caused by accumulation of chloride ions at the steel surface, exceeding the (local) threshold [Angst et al. 2011]. When the pit develops, more chloride ions are attracted that auto-catalyse the iron dissolution reaction. Iron chloride ($FeCl_2$) and iron (hydr)oxides are the main constituents of the pits. Because the iron (hydr)oxides are poorly soluble, the liquid contains a concentrated solution of $FeCl_2$, with a low pH; pH values were suggested of about 2 [Pacheco et al. 2011, Polder et al. 2011], which was attributed to HCl. However, the pit solution also contains a rather high concentration of $FeCl_2$. After intentionally contaminating concrete near steel bars with chloride, chloride contents of corrosion products have been observed of 3.5 to 7% of

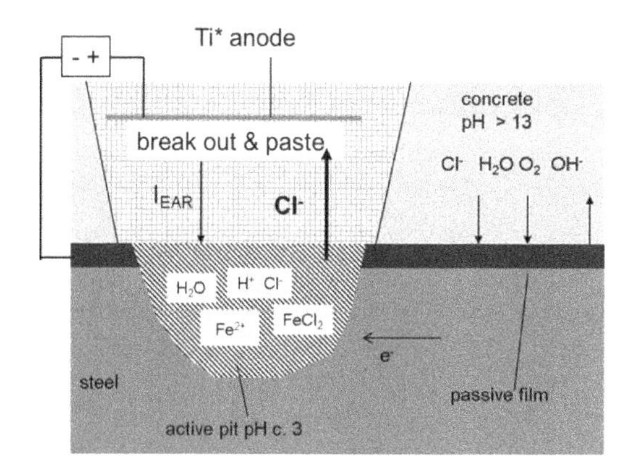

Figure 1. Principle of EAR, showing break out with temporary material ('paste'), temporary anode (Ti* anode) and current application (I_{EAR}), which removes chloride from corrosion pit.

chloride ion by mass of corrosion product [Polder & Hondel, 2002]. High levels of 3% chloride by interaction volume have been found by SEM/EDS, corroborating such high chloride contents [Wong et al. 2010], see also [Scheiner & Hellmich 2007]. This can be interpreted as corresponding to a solution of approx. 2.5 mole/l of $FeCl_2$ or 5 mole/l of Cl.

Summarising, active corrosion pits contain a solution with a pH between 2 and 4. A low pH is buffered by $FeCl_2$ at about pH 3 to 4. Consequently, for effective repassivation by obtaining a high pH (preferably above 10), the majority of chlorides need to be removed from the pit solution. The target level of reduction of the chloride concentration for successful treatment was set at 90%. This is what the present invention aims at and what the experiments have documented to be achievable within the parameters set.

The principle of the proposed method was confirmed in laboratory experiments. They will be reported briefly here and in more detail elsewhere [Geiker & Polder 2015]. A cementitious material was placed between NaOH and $FeCl_2$ solutions as anolyte and catholyte, respectively, using activated titanium electrodes for applying a voltage. More than 90% of the chloride was removed from the catholyte ($FeCl_2$) in 20 hours at 8 A/m^2 and the pH increased from 3 to 12. Such a reduction of chloride concentration and increase of pH will suppress re-activation of corrosion and consequently improve the durability of the repair.

It is proposed to apply a direct current to rebars in repair areas for a short period of time, which will remove chlorides from corrosion products including pits and from concrete adjacent to the rebar. This treatment, called Electro Active Repair (EAR), is an additional step to conventional repair. The authors expect that adding this treatment will increase the durability of repairs. For two hypothetical cases, differing in orientation and amount of damage, it has been described how EAR can be applied in practice. Arguments have been give why other options are not attractive and EAR provides a better solution.

REFERENCES

Angst, U., Elsener, B., Larsen, C.K., Vennesland, Ø., 2011, Chloride induced reinforcement corrosion: Rate limiting step of early pitting corrosion, Electrochimica Acta, 56, 5877–5889.

Bertolini, L., Elsener, B., Pedeferri, P., Polder, R.B., 2004, Corrosion of Steel in Concrete: Prevention, Diagnosis, Repair, Wiley-VCH Verlag GmbH & Co. KGaA, Weinheim, ISBN 3-527-30800-8, 392.

Castellote, M., Andrade, C., Alonso, C., 1999, Changes in pore size distribution due to electrochemical chloride migration trials, ACI Materials Journal, 96 (3), 314–319.

Geiker, M.R., Polder, R.B., 2015, Experimental support for new Electro Active Repair method, in preparation.

Pacheco, J., Polder, R.B., Fraaij, A.L.A., Mol, J.M.C., 2011, Short-term benefits of cathodic protection of steel in concrete, Proc. Concrete Solutions, Dresden, Eds. Grantham, Mechtcherine, Schneck, Taylor & Francis, London, 147–156.

Polder, R.B., Hondel, A.W.M. van den, 2002, Laboratory investigation of electrochemical chloride extraction of concrete with penetrated chloride, HERON, 47 (3) 211–220.

Polder, R.B., Peelen, W., Stoop, B., Neeft, E., 2011, Early stage beneficial effects of cathodic protection in concrete structures, Materials and Corrosion, 62 (2), 105–110.

Polder, R.B., Leegwater, G., Worm, D., Courage, W., 2014, Service life and life cycle cost modelling of Cathodic Protection systems for Concrete Structures, Cement and Concrete Composites, 47, 69–74.

Scheiner, S. Hellmich, C., 2007, Stable pitting corrosion of stainless steel as diffusion-controlled dissolution process with a sharp moving electrode boundary, Corrosion Science, 49, 319–346.

Tilly, G.P., Jacobs, J., 2007, Concrete repairs – performance in service and current practice, IHS BRE Press, Bracknell.

Tilly, G.P., 2011, Durability of concrete repairs, in Concrete Repair, a practical guide, ed. Grantham, Taylor & Francis, 231–247.

Visser, J.H.M, Zon, Q. van, 2012, Performance and service life of repairs of concrete structures in The Netherlands, ICCRRR III, Cape Town, Alexander et al. (eds.), Taylor & Francis.

Wong, H.S., Zhao, Y.X., Karimi, A.R., Buenfeld, N.R., Jin, W.L., 2010, On the penetration of corrosion products from reinforcing steel into concrete due to chloride induced corrosion, Corrosion Science, 52, 2469–2480.

Concrete Repair, Rehabilitation and Retrofitting IV – Dehn et al. (Eds)
© 2016 Taylor & Francis Group, London, ISBN 978-1-138-02843-2

A quantitative approach to the concept of concrete repair compatibility

B. Bissonnette
CRIB, Department of Civil Engineering, Laval University, Quebec City, QC, Canada

F. Modjabi-Sangnier
SNC-Lavalin, Montreal (QC), Canada

L. Courard
GeMMe Building Materials, ArGEnCo Department, University of Liège, Liège, Belgium

A. Garbacz
Warsaw University of Technology, Warsaw, Poland

A.M. Vaysburd
Vaycon Consulting, Baltimore, MD, USA

ABSTRACT

The work reported in this paper is part of a wider research program intended to provide the repair industry with improved fundamental knowledge to implement rational design methods and rules for repairs. The importance of dimensional compatibility between existing concrete and repair materials has been addressed conceptually by various authors. One of the main challenges to be faced now lies in evaluating quantitatively compatibility and determining what it requires under given circumstances.

This paper summarizes the work achieved to relate quantitatively the individual dimensional compatibility-related properties (notably elastic properties, creep, drying shrinkage) to the corresponding stress and strain values recorded in an annular restrained shrinkage test, commonly referred to as the *ring* test. In the first part of *the program, classical formulas* derived for thick cylindrical specimens were used to analyze the tensile stress buildup in restrained shrinkage test specimens. A quantitative approach for the evaluation of concrete repair with a single dimensional compatibility parameter, referred to as the *compatibility index,* was then developed. *Compatibility index* evolution curves were finally calculated for a range of repair concrete mixtures in order to validate the approach and highlight material behavior relating to composition parameters and temperature.

The dimensional balance inside a repair is governed by the volume changes and mechanical properties of the repair material, together with the level of restraint provided by the existing structure. Thus, in order to evaluate the restrained shrinkage cracking sensitivity of materials, the analysis must take into account the combined effect of these properties and phenomena. This complex task can be simplified by addressing the evolution of the stress state in a simple restrained-shrinkage element such as that used in ring-type tests. On the basis of the theory of elasticity and a given geometry of the annular test device, it is possible to calculate the evolution of the average shrinkage-induced stress ($\sigma_{c\ avg.}$) over time with equation 1, where ε_{fs} is the concrete free shrinkage, E_c and E_s are the elastic moduli of concrete and steel, ϕ_c is the creep coefficient of concrete, ν_c and ν_s are the *Poisson's* ratios of concrete and steel, and a, b, and c, are the internal, interfacial and external radii of the composite steel-concrete ring specimen.

$$\sigma_{c\ avg.}(t) = \frac{b(b+c)}{c^2-b^2} \times \frac{\varepsilon_{fs}(t)}{\left[\frac{1}{E_s}\left(\frac{b^2+a^2}{b^2-a^2}+\nu_s\right)+\frac{1+\phi_c(t)}{E_c(t)}\left(\frac{b^2+c^2}{c^2-b^2}+\nu_c\right)\right]} \tag{1}$$

By comparing the ring test results with the calculated tensile stresses for a range of repair concrete mixtures, the validity and accuracy of the theoretical approach could be appraised. Overall, a good correlation was found between the ring test results and the tensile stress values calculated based on individual concrete properties/phenomena.

The aforementioned mathematical formulas were further used to define a quantitative parameter for the evaluation of the performance of concrete repair in terms of dimensional compatibility.

Derived from the basic strain balance approach (ratio between the total deformability in tension and the drying shrinkage deformation), the dimensional *Compatibility Index* (CI) was thus introduced in order to analyze the evolution of dimensional compatibility as a function of time for a given concrete mixture, taking into account the actual degree of restraint in the test specimen. The dimensional *compatibility index* can be expressed in the following general form, where f_t is the concrete tensile strength, α_r is the instantaneous elastic restraint, α'_r is the creep-dependent restraint, and C_c, C_g are geometrical parameters:

$$C.I.(t) = \frac{\left[\frac{f_t(t)}{E_c(t)} + \phi_c(t)\varepsilon_{fs}(t)\alpha_r(t)\cdot\frac{C_c}{C_g}\alpha'_r(t)\right]}{\varepsilon_{fs}(t)\alpha_r(t)} \quad (2)$$

As it requires the evaluation of individual properties that for most are readily available, the *compatibility index* carries much potential as a relatively simple and convenient analytical tool for assessing the cracking potential of concrete repair materials.

The CI evolution allows to assessing whether the concrete can withstand shrinkage-induced cracking over time, taking into account the various phenomena involved and their complex interaction. Based upon comprehensive experimental work, it can be concluded that the composition parameters and the thermal conditions during the setting and hardening period may influence the susceptibility to cracking of a cementitious material undergoing restrained shrinkage and thereby need to be taken into account in the identification of performance criteria for dimensional compatibility. The diagram presented in Figure 1 shows an example of the evolution of the *compatibility index* for different repair concrete mixtures.

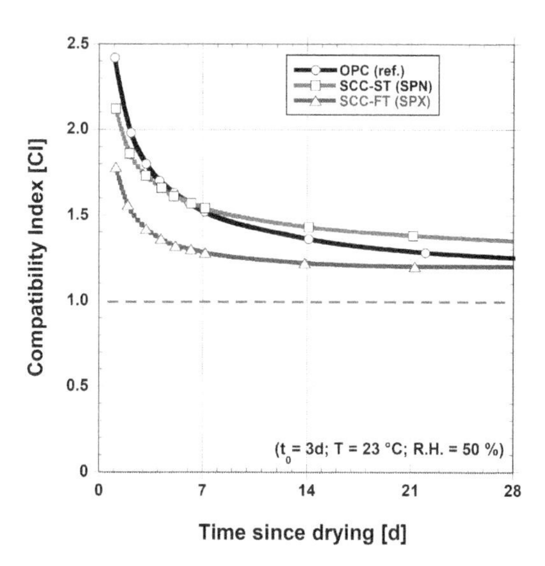

Figure 1. Evolution of the *Compatibility Index* (CI value) with time for a reference repair concrete (OPC) and two self-compacting repair concrete mixtures prepared with a slag-based ternary cement (SCC-ST) and a fly ash-based ternary binder (SCC-FT), exposed to drying at 50% R.H. at the age of 3 days.

The proposed *compatibility index* appears as a quite promising analytical tool for predicting the performance of repair materials in terms of shrinkage-cracking resistance. The CI parameter provides a sound basis for the identification of dimensional compatibility criteria. Such performance criteria are much awaited in the repair industry, to assist both the development of crack-resistant materials and the issuance of improved materials specifications.

Concrete Repair, Rehabilitation and Retrofitting IV – Dehn et al. (Eds)
© 2016 Taylor & Francis Group, London, ISBN 978-1-138-02843-2

The compatibility in concrete repair—random thoughts and wishful thinking

A.M. Vaysburd
Vaycon Consulting, Baltimore, MD, USA

B. Bissonnette
CRIB, Department of Civil Engineering, Laval University, Quebec City, QC, Canada

K.F. von Fay
U.S. Bureau of Reclamation, Denver, CO, USA

R. Morin
City of Montreal, Montreal, QC, Canada

ABSTRACT

The durability of repaired concrete structures and their useful service life depend on many factors. One of the critical factors is the achievement of compatibility in such a composite system. As a prerequisite for meeting the repair service life objectives, the compatibility/incompatibility issues should be analyzed and adequately addresses at the design stage of the project. Despite the number of publications on durability and service life of repaired concrete structures, fundamental discussion and guidance on compatibility issues are lacking. The opinions and recommendations issued on the subject are in some cases confusing, misleading or incorrect, regrettably leading to fallacies about compatibility in the practicing community.

This paper is an attempt to raise and discuss some critical issues concerning the compatibility of concrete repair composite systems. Ambiguities and misconceptions in the current understanding of fundamental compatibility considerations in concrete repairs (dimensional compatibility, electrochemical compatibility and permeability compatibility) are addressed. The authors share their random thoughts on what we know, what we do not know, and what we need to know.

Compatibility in repair systems is defined as the balance of physical, chemical and electrochemical properties and volume changes between the repair and existing substrate, which ensures that the composite repair system withstands the stresses induced by all loads, chemical effects, electrochemical effects and restrained volume changes without distress and deterioration over a designed period of time.

The essence of dimensional compatibility can be formulated as follows:

- Drying shrinkage of the repair material relative to the substrate;
- Thermal expansion or contraction differences between repair and substrate materials;
- Differences in modulus of elasticity, which may cause unequal load sharing and strains resulting in interface stresses;
- Differences in creep;
- Relative fatigue performance of the phases in composite repair system.

During service life, incompatibilities in the form of dissimilar strength and moduli of elasticity between repair and substrate concrete can create difficulties, while drying shrinkage of repair materials may reduce longer-term structural efficiency by either inducing high tensile strains in the repair, or due to cracking that may reach the repair/substrate interface. Also, creep of the repair material under sustained stress may render the load-sharing capacity of the repair less effective with time.

It is widely accepted that the driving force for the phenomenon of corrosion in repair systems is generally attributed to the electrochemical incompatibility between the repair and substrate. The electrochemical incompatibility is defined as the imbalance in electrochemical potential between different locations of the reinforcing steel because of their dissimilar environments caused by a repair. The dissimilar environments can be due to the differences in physical properties, chemistry and internal environments.

Unless the repair project provides for "global" cathodic protection, the risk of corrosion is omnipresent. Therefore, performing a repair that will last successfully without problems for 10 to 15 years is not an easy task to achieve. The design engineer must be realistic and honest when setting out project objectives.

Permeability of repair materials is one of the properties of importance in achieving compatibility and durability in repair projects. The majority of repair publications strongly recommend using low permeability materials in repair situations. If the use of low permeability concrete in new construction is a key to achieving durability, this rule does not necessarily apply to concrete repairs, where the situation is more complex. Unfortunately, in the case of repairs there is no "rule of thumb". There are "horses for courses", each situation is different and requires to being accurately analyzed and addressed.

The use of low permeability repair materials regardless of repair specifics can lead to unsuitable choices, incompatibility problems and eventual repair failures. Durability of the repair can be negatively affected in many situations when repair and substrate have different, incompatible permeability.

Most likely, there is no single recommendation as to whether materials with very low permeability or materials having compatible permeability with the existing concrete, are more effective. It depends, in the authors' view, on particular transport mechanism in the repair system. The effects of such variables as location of the repair in the structure, chemical environment in the composite repair system, amount and distribution of cracks in both materials (substrate and repair), temperature, moisture and stresses need to be considered.

Better understanding of compatibility/incompatibility issues in concrete repair composite systems will allow the repair professionals to "silt the grain from the husk" in the design of concrete repair projects.

Concrete Repair, Rehabilitation and Retrofitting IV – Dehn et al. (Eds)
© 2016 Taylor & Francis Group, London, ISBN 978-1-138-02843-2

Influence of the interface mechanisms on the behavior of strengthened with reinforced concrete or steel existing RC slabs

Marina Traykova
Department of Reinforced Concrete Structures, University of Architecture, Civil Engineering and Geodesy, Sofia, Bulgaria

Raina Boiadjieva
Department of Structural Mechanics, University of Architecture, Civil Engineering and Geodesy, Sofia, Bulgaria

ABSTRACT

Extending the life of the existing structures is the base of the sustainable design. The reinforced concrete slabs are usually subjected directly to different actions and overloading and very often are seriously damaged.

The reinforced concrete and the steel jackets are one of the most popular techniques in the retrofitting practice for reinforced concrete slabs. The quality of the load transfer and the behavior of the strengthened slabs are related to the type of the interface mechanisms: mechanical interlocking and adhesive bonding, friction, dowel action of reinforcement and/or connectors crossing the interface. The choice of interaction mechanisms depends on the type of the member and of the type of loading.

According Model Code 2010 the main parameters for the load bearing capacity in the case of concrete-to-concrete load transfer are: interface roughness, cleanliness of surface, concrete strength and concrete quality, inclination of the shear forces, bond conditions, ratio of reinforcement, crossing the interface. Concrete to steel interfaces Model Code 2010 play a governing role with regard to the design of strengthening with steel sheeting and profiles for different reinforced concrete elements. Interaction between concrete and steel components can be classified as follows: adhesion between two materials, frictional interlock, mechanical interlock and dowel action provided by anchor devices and systems.

The case study in the numerical example is a real design situation and represents a reinforced concrete one—way spanning continuous beam slab. The thickness of the slab is 10 cm. New imposed load is provided with value 10 kN/m^2. Two design solutions for strengthening are presented: with an additional concrete layer and with steel sheeting.

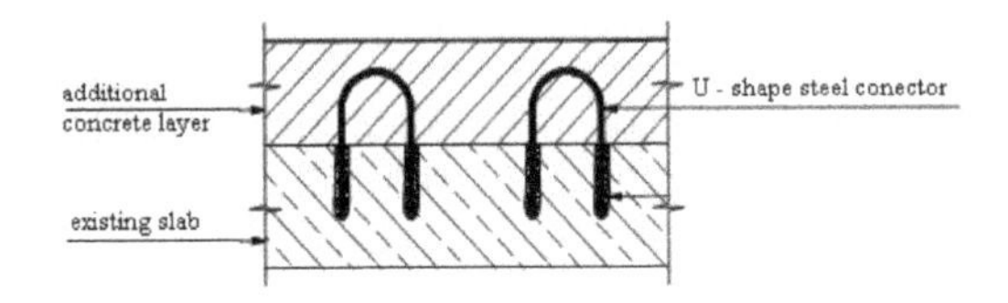

Figure 1. Solution with additional concrete layer.

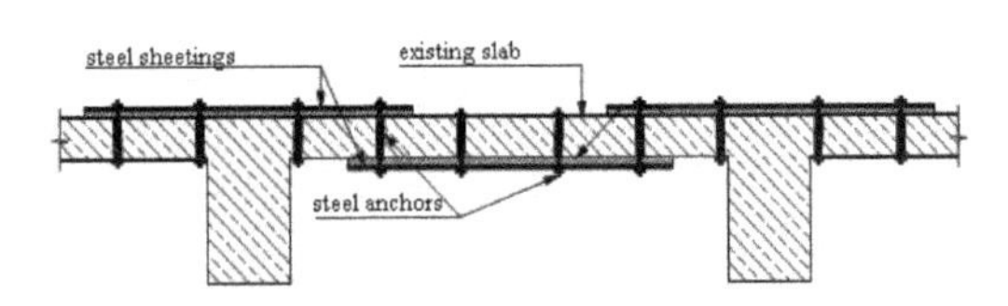

Figure 2. Solution with steel sheeting.

The investigation is addressed to the following parameters:

- in case of design solution with additional concrete layer—interface roughness characteristics, arrangement of connectors and the reinforcement ratio crossing the interface;
- in case of design solution with steel sheeting—number (spacing) and type of anchors, interface between the existing reinforced concrete surface and the new steel sheeting.

The numerical analysis is performed using the program SAP2000 based on the Finite Element Method. All models are developed in linear-elastic formulation of the problem.

In the case of design solution with additional concrete layer the slab is modeled by multi-layer shell elements. The beams are modeled by linear (frame) finite elements positioned in their geometric center.

Figure 3. Solution with additional concrete layer—3D model.

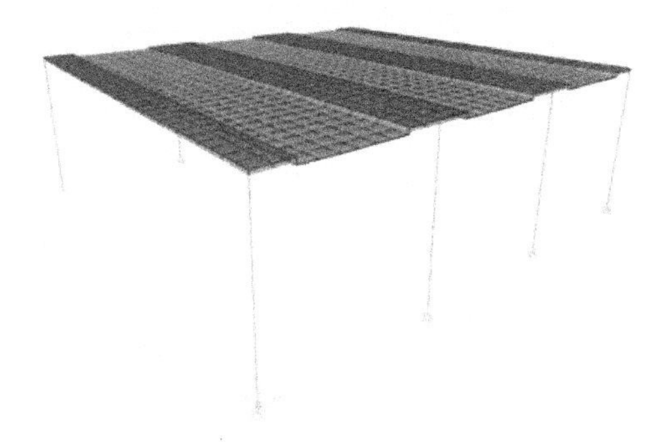

Figure 4. Solution with steel sheeting—3D model.

Several numerical models are developed to obtain the most reliable identification of actual behavior and bearing capacity of the strengthened slab. They are focused on the joint work of the different concrete layers and the most adequate reflection concrete-to-concrete connection.

In the case of design solution with the steel sheeting the slab is modeled by shell (area) elements. The beams are modeled by linear (frame) finite elements positioned in their geometric center. The steel sheeting is modelled by shell elements positioned in geometric level of this strengthening. The connection concrete-to-steel is with anchors modeled in the example with stiff elements.

The analysis of the final results can be summarized in the following main conclusions:

- The amount of reinforcement bars that crosses the interface as well as the roughness of the interface are the two parameters of main importance of the shearing mechanism. The shear strength increases accordingly to the increase of the steel ratio and the roughness.
- During the reinforcement design it is important to evaluate the evolution of the building construction process and the load actions.
- When strengthening with additional concrete layer, it's very important to carry out a global solution for the slab and the beams. Increasing only the stiffness of the slab has unfavorable effect on the deflections.
- The variations of the thickness of the interface layer, in the case of additional concrete layer, show an insignificant influence on the numerical results.
- Increasing the stiffness of the interface layer between the existing and the new concrete layer, leads to a serious reduction of the deflections in the strengthened slab.
- Strengthening with steel sheeting (regardless of the changes in the different parameters) has no special contribution to the deflection of the slab and the final deflections are very similar to the deflections before the strengthening. These results are expected in view of the fact that the proposed model is not enough accurate for the real process of deformation.
- The solution with steel sheeting is more sensitive for the bending moments.
- The position and the number of anchors for the steel sheeting have significant influence on bearing capacity for the bending moments of the strengthened slab.
- Finally the proposed results and conclusions require a validation with real experimental results. The experiments will give an additional opportunity to find more precise values of the different parameters.

Concrete Repair, Rehabilitation and Retrofitting IV – Dehn et al. (Eds)
© 2016 Taylor & Francis Group, London, ISBN 978-1-138-02843-2

Fundamental approach for the concept of concrete repair compatibility

L. Courard
University of Liège, Liège, Belgium

B. Bissonnette
Laval University, Québec, Canada

A. Garbacz
Warsaw University of Technology, Warsaw, Poland

ABSTRACT

Before being translated in terms of physical, mechanical and chemical of materials, the initial step for evaluating compatibility is the interface creation. The thermodynamic properties of the materials as well as transport mechanisms—diffusion, capillary succion—at the interface and roughness of the concrete substrate are acting from the beginning and influencing the durability of the bond strength.

The concept of adhesion has firstly to be clearly defined because of the "duality" of the term: *"on one hand, adhesion is understood as a process through which two bodies are brought together and attached—bonded—to each other, in such a way that external force or thermal motion is required to break the bond. On the other hand, we can examine the process of breaking a bond between bodies that are already in contact. In this case, as a quantitative measure of the intensity of adhesion, we can take the force or the energy necessary to separate the two bodies".*

Adhesion has therefore two different aspects, according to whether our interest is mainly (1) in the conditions and the kinetics of contact or (2) in the separation process. The intensity of adhesion will depend not only on the energy that is used to create the contact, but also on the interaction existing in the interface zone. Generally speaking, mechanism of adhesion has to be considered from two origins: specific adhesion and mechanical interlocking.

When the materials are in contact, the effective area, that means the surface where contact really exists, will be a fundamental parameter to be taken into account to explain the adhesion process. This is the result of the wetting procedure of the solid body by the liquid phase. The wetting procedure can be explained as follows: the surface energies of the solid and the liquid interact each other and a change of the energy conditions occurs due to surface decrease of liquid/vapour and solid/vapour interfaces while a new interface (liquid/solid) is created. At this point of view, contact angle is an interesting representation of this phenomenon: the lower is the contact angle, the better is the spreading on the surface and the more effective will be the inter-molecular interactions at the interface.

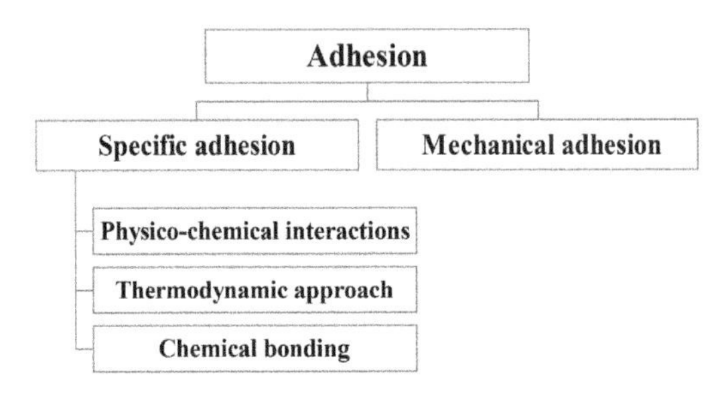

Figure 1. Principles of the theory of adhesion.

The investigation concerning the behaviour of the interface between repair systems and concrete substrate have shown that quality of concrete substrate is important factor affecting adhesion in repair system and has to be evaluated prior to repair. Knowledge about synergetic effects of parameters characterizing surface quality (surface roughness, microcracking, wettability) is fundamental for concrete repair compatibility.

Concrete Repair, Rehabilitation and Retrofitting IV – Dehn et al. (Eds)
© 2016 Taylor & Francis Group, London, ISBN 978-1-138-02843-2

Strengthening/retrofitting of coupling beams using advanced cement based materials

M. Muhaxheri, A. Spini, L. Ferrara & M. di Prisco
Department of Civil and Environmental Engineering, Politecnico di Milano, Milano, Italy

M.G.L. Lamperti
European Commission, Institute for the Protection and Security of the Citizen, Ispra, Italy

ABSTRACT

In this paper the effectiveness has been assessed of an upgrading/retrofitting system for coupling beams (span-to-depth ratio equal to 1.5) not designed to resist earthquake actions. The proposed retrofitting technology employs either High Performance Fiber Reinforced Cementitious Composites (HPFRCCs) or Textile Reinforced Cementitious Composites (TRCCs). 14 mock-ups of coupling beam units, scale 1:2, have been casted and subjected to either monotonic or cyclic loading tests. The different resisting mechanisms in conventional plain and reinforced concrete coupling beams have been first of all investigated (tensile strength of concrete in plain concrete coupling beam; strut-and-tie mechanisms in coupling beams reinforced with only longitudinal bars; enhancement of the aforementioned mechanism due to stirrups). The influence of the boundary conditions due to the shear wall shafts on the coupling beam spanning between, has been investigated. Longitudinal and transverse reinforcement have been designed to comply only with EC2 minimum requirements for non-seismic design situations. In a second stage the effectiveness of the upgrading/retrofitting techniques has been checked testing non- and pre-damaged coupling beams. In the latter case two different drift levels have been selected for pre-damage, namely 1% and 2%, respectively meant as representative of ductility demand for SLS and ULS. Results highlight the effectiveness of the proposed upgrading/retrofitting techniques, which, also in view of their ease of execution and reduced invasiveness, can stand as a reliable alternative to other more commonly employed ones.

INTRODUCTION

In this paper the use of High Performance Fiber Reinforced Cementitious Composite (HPFRCC) and Textile Reinforced Cementitious Composite (TRCC) as detailed hereafter has been explored as an upgrading/retrofitting solution for poorly designed or damaged coupling beams. 14 mock-ups of coupling beam units, scale 1:2 featuring minimum longitudinal and transverse reinforcement according to Eurocode 2 prescriptions, were casted and subjected to either monotonic or cyclic loading tests. The different resisting mechanisms in conventional plain and reinforced concrete coupling beams were first of all investigated (tensile strength of concrete in plain concrete coupling beam; strut-and-tie mechanisms in coupling beams reinforced with only longitudinal bars; enhancement of the aforementioned mechanism due to stirrups). The effectiveness was then investigated of the upgrading/retrofitting techniques has been checked by testing non- and pre-damaged coupling beams. Two different drift levels were selected for pre-damage, namely 1% and 2%, respectively meant as representative of ductility demand for SLS and ULS.

Results, analyzed in terms of load bearing and energy dissipation capacity, highlight the effectiveness of the proposed upgrading/retrofitting techniques, which, also in view of their ease of execution and reduced invasiveness, can stand as a reliable alternative to other more commonly employed ones.

The application to the case of a five storey r/c shear wall was finally addressed through numerical analyses also to study the effectiveness of tailored retrofitting/upgrading alternatives.

EXPERIMENTAL PROGRAMME

Experimental mock-ups have been designed making reference to a shear wall containing a typical door opening, 900 mm wide (equal to the length of the coupling beam) and 2.1 m high. Assuming an inter-storey height equal to 2.7 m, this has resulted in a depth of the coupling beam equal to 600 mm and hence in a span to depth ratio of 1.5. The coupling beam specimen was hypothesized to be "poorly designed". To comply with maximum load capacity of the testing equipment, a scale ratio 1:2 was adopted. The specimen was $450 \times 300 \times 100$ mm^3 ($l \times h \times b$).

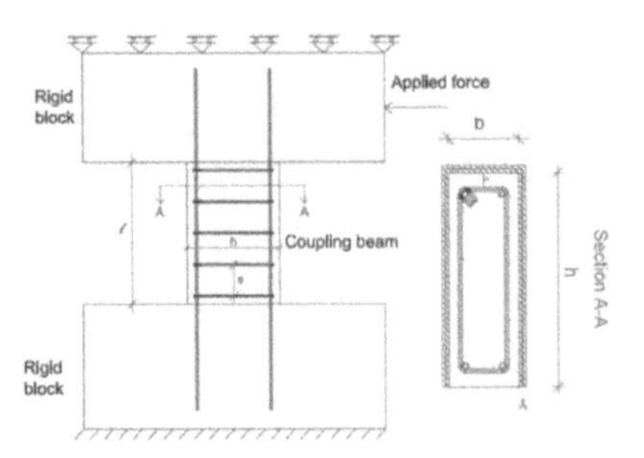

Figure 1. Schematic representation of the test mock-up with intended boundary conditions.

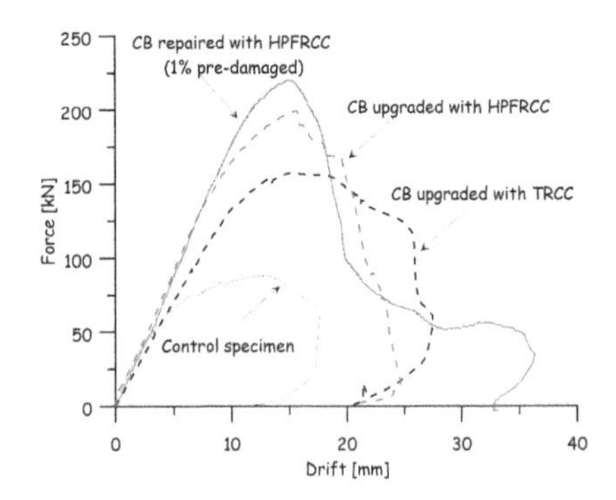

Figure 2. Experimental results under monotonic loading considering different upgrading/retrofitting techniques.

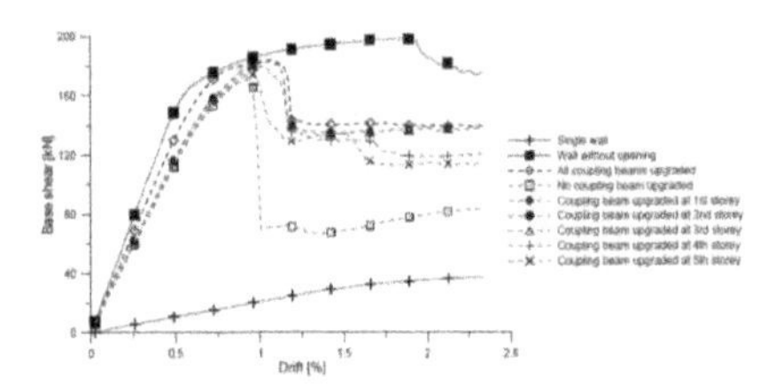

Figure 3. Push-over curves for different retrofitting options for the numerically analysed shear wall, compared to uncoupled two-shaft wall and monolithic wall with no opening.

The experimental campaign performed involved tests on "non-retrofitted" coupling beam specimens, through which the influence of the different resisting contributions as well as on retrofitted ones was studied. As for the first part, the programme consisted of tests on two plain concrete coupling beams, two beams reinforced only with longitudinal bars and further two with both longitudinal and transverse reinforcement detailed as above. Only monotonic tests were performed on plain and longitudinally-only reinforced mock-ups whereas one monotonic and one cyclic test were performed on the mock-ups with both longitudinally and transverse reinforcement.

As for the tests on retrofitted/upgraded specimens, two "repairing" materials were employed, namely a HPFRCC and a TRCC. With for the latter case only two "upgraded" specimens were tested (the TRCC was applied to undamaged mock-ups), for the former both cases of upgraded and retrofitted specimens were considered. Besides two undamaged (upgraded) specimens, also four retrofitted ones were tested, two of which were pre-damaged at 1% drift level and two at 2% drift level. The pre-damage drift levels were selected as representative of SLS and ULS behaviour. For each couple of nominally identical retrofitted and/or upgraded mock-ups one monotonic and one cyclic test was performed.

As a first retrofitting option, a 20 mm thick layer was applied of a HPFRCC was applied. As a second option, a 6 mm thick layer of TRCC, selected from a previous investigation.

In order to reproduce realistic boundary conditions the testing set-up in Figure 1 was designed.

EXPERIMENTAL RESULTS: ANALYSIS AND DISCUSSION

With reference to the tests performed on retrofitted beams, employing either the HPFRCC or the TRCC technique, a synopsis of their efficiency can be glimpsed from the force-drift curves obtained from monotonic tests on retrofitted mock-ups and compared to the reference one (Figure 2).

The efficacy of the retrofitting techniques has been finally applied, in a numerical example to a true structural application dealing with a 5 storey shear wall. The results (Figure 3) highlight the efficacy of the employed retrofitting technique in getting back, even if not completely, to the behavior of a monolithic shear wall. On the other hand, the significance also appears of a tailored retrofitting, applied only to coupling beams at "selected" storeys.

CONCLUSIONS

A series of experimental tests on coupling beams, either non retrofitted or upgraded/retrofitted with advanced cement based materials was performed. The use of HPFRCC/TRCC as an upgrading/repair solution for poorly designed coupling beams leads to enhanced performance under both monotonic and cyclic loadings.

The efficacy of the proposed retrofitting techniques was assessed, via a simple numerical example with reference to the structural behaviour of a 5-storey shear wall and the engineering significance of a selected retrofitting was demonstrated.

Concrete Repair, Rehabilitation and Retrofitting IV – Dehn et al. (Eds)
© 2016 Taylor & Francis Group, London, ISBN 978-1-138-02843-2

Cyclic behaviour of R.C. column with corroded reinforcement repaired with HPFRC jacket

S. Mostosi
Tecnochem Italiana S.p.A., Italy

A. Meda & Z. Rinaldi
"Tor Vergata" University, Rome, Italy

P. Riva
University of Bergamo, Italy

ABSTRACT

The reduction in the useful service-life of reinforced concrete structures, mainly due to reinforcement corrosion, is a cause of concern for several RC buildings. The corrosion damage, often, makes necessary to define strengthening techniques allowing for recovering the initial capacity of the structure. Furthermore, the reduction of strength and ductility, due to corrosion, can jeopardize the behaviour for horizontal loads. In this research, the technique based on the use of High Performance Fiber Reinforced Concrete (HPFRC) jackets with thickness of 30–40 mm, is considered and applied on RC corroded columns. Three full-scale experimental cyclic tests were performed. Besides the reference specimen (UC), one column was tested after the rebars artificial corrosion (c), and the last one, after the reinforcement corrosion, was strengthening with 40 mm thick HPFRC jacket (CR). The tests were performed up to failure by applying cyclic horizontal loads of increasing amplitude for verifying the performances of the proposed solution.

The analysis of the obtained results allows drawing the following remarks:

- The corrosion phenomenon can strongly affect the global behaviour of columns subjected to cyclic loads. In the tested case (characterized by rebar corrosion of about 20% in mass loss) a reduction of about 30% of the ultimate force and mainly a reduction of ultimate displacement of about 50% was detected, together with a significant stiffness decrease in the last cycles. Therefore, the corrosion phenomenon can strongly affect the local and global behaviour of structures in seismic zones and can significantly change their failure modes;
- With the application of a high performance jacket it is possible to increase the bearing capacity of the column with corroded rebars reaching a maximum strength greater than the one of the undamaged element. This technique is suitable for strengthening existing RC structures characterized by low concrete strength, low reinforcement ratio, concrete damage and corrosion of rebars.

Finally, the proposed technique can be easily used in structural applications as it allows strengthening R/C elements by means of a thin jacket (40 mm). Furthermore, a curing at ambient temperature and humidity is sufficient to grant the development of the full strength and a simple sandblasting ensures a good bond between substrate and HPFRC.

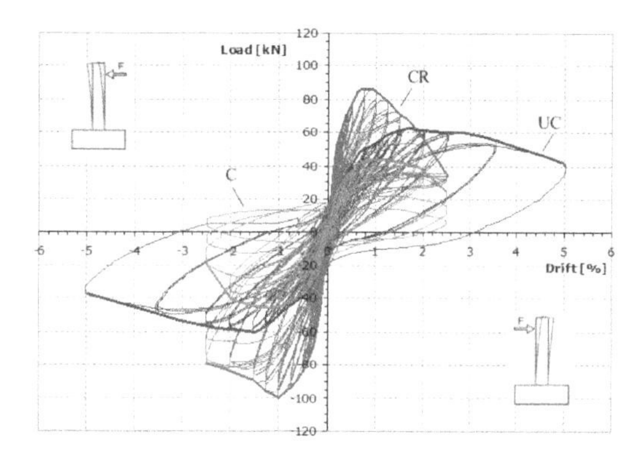

Figure 1. Load-drift relationships: comparisons.

Concrete Repair, Rehabilitation and Retrofitting IV – Dehn et al. (Eds)
© 2016 Taylor & Francis Group, London, ISBN 978-1-138-02843-2

Concrete hinge bearing replacement: A case study of concrete hinge collapse and broader implications for concrete hinge bearings under seismic effects

R.K. Dickson
SMEC South Africa Consulting Engineers, South Africa

E.J. Kruger
The South African National Roads Agency SOC Limited, South Africa

ABSTRACT

In February 2012, several concrete hinge bearings collapsed under one of the spans of the Kakamas Bridge on National Route 14 over the Orange River in the Northern Cape, South Africa. The bridge was immediately closed to traffic after a span settled by approximately 100 mm. Four out of six concrete hinge bearings had collapsed at one pier, the two remaining bearings were split and severely damaged.

This paper provides a case study of the collapse of concrete hinge bearings and the broader implications for concrete hinge bearings under seismic load effects. The paper describes a method of bearing replacement as a retrofit to existing bridges with concrete hinge bearings.

The bridge consists of fifteen simply-supported spans of 24 m, comprised of post tensioned I-girders, constructed in 1962. It is a strategic link of the N14 over the Orange River, and is a critical link during harvest, as well as carrying high volumes of local pedestrian traffic.

During the investigation for the repairs, it was found that the area had experienced an earthquake in December 2011 of magnitude 5.0 on the Richter scale. This earthquake ranks in the top six largest earthquakes in South African history. The area also experienced an "earthquake swarm" in which 600 earthquakes were recorded in the magnitude range 0.5 to 5.0, from 2008 to 2012. It was found that the remainder of the concrete hinge bearings had been severely damaged under earthquake loads and were vulnerable to collapse in seismic events.

Damage to collapsed bearings at Pier 9 is illustrated in Figure 2 (and in the full paper).

After the collapse, the Client immediately initiated a make-safe response to the collapse and the road was ultimately re-opened to traffic 2 ½ weeks after the collapse. In this repair contract, the bridge deck was jacked up and supported on temporary supports, then opened to traffic whilst bearings were removed and replaced with elastomeric bearings on new plinths in the following weeks.

Following the initial collapse, a complete inspection was done of all the remaining 174 concrete hinge bearings. It was found that several of the remaining concrete hinge bearings had been

Figure 1. Elevation of Orange River Bridge with collapsed hinge bearings.

Figure 2. Splitting collapse of concrete hinge bearing at Pier 9.

displaced by up to 40 mm. This resulted in rotations about the hinges of up to 1/11 (0.089 radians), which is substantially in excess of the recommended 1/20 limits. Numerous concrete hinges exhibited cracks of up to 2 mm in the throats of the hinges. In addition, splitting cracks up to 3 mm wide and spalls were noted. The compressive working stress in the throats was determined to be approximately 30 MPa under full traffic. However, due to displacement and cracking through the throats, it is estimated that compressive stresses of up to 95 MPa have developed in these hinges.

It was clear from the cracking and visible displacement that the remaining hinges had been overstressed and there was a risk of further bearing collapses. Therefore, all of the remaining 174 bearings were replaced in a second six-month contract. This was carried out in a contract to retrofit modified bearing plinths and elastomeric bearings to the entire bridge. In this second contract, the contractor opted to use simplified temporary supports, which allowed traffic to be more freely accommodated on the bridge during construction. The complete bearing replacement retrofit contract was successfully carried out on programme and within budget.

In 1962 (the year of construction of the Kakamas Bridge), bridges in South Africa were conventionally not designed for earthquake load effects. Had earthquake loading with Peak Ground Accelerations of up to 0.1 g been a consideration, such concrete hinge bearings would not have been used.

Whilst it is debatable what level of vibration could cause the failure of concrete hinges, they are not robust when subject to large longitudinal displacement and do not accommodate any lateral displacement whatsoever. A major consideration is the physical height of the concrete hinge required to create double rockers able to accommodate longitudinal displacement. For the Kakamas Bridge, the double rockers hinges were 800 mm high. In the event of a bearing splitting failure, this 800 mm height represents a large potential energy which can suddenly be released in an earthquake.

The client has found from inspection of the Kakamas Bridge bearings and others at the Great Fish River, that this form of concrete hinge bearing (without a transverse taper at the throat) has suffered cracking and spalling to the transverse side faces. In addition, cracking in the throat section of the hinges has been observed. This is consistent with the mode of failure and collapse at the Kakamas Bridge.

As a risk mitigation measure, the Client will carry out close-up inspections of all bridges with concrete hinge bearings in the national bridge network in South Africa. In the case of the Great Fish River Bridge on National Route 2, the existing concrete hinge bearings will be replaced by retro-fitting with elastomeric bearings as part of a bridge widening.

CONCLUSIONS

Following the collapse of a series of concrete hinge bearings at the Kakamas Bridge over the Orange River, it was found that these bearings had suffered damage after an earthquake "swarm" in the area, which included a 5.0 magnitude earthquake. It is therefore, a likely scenario that the bearings were severely compromised by earthquake and then collapsed at Pier 9 under the passage of heavy vehicles. After the collapse, the remaining bearings were found to be cracked and severely damaged after seismic events and normal traffic loading.

The primary mechanisms of deterioration of these concrete hinge bearings were cracking and spalling of the faces of the bearings. When subjected to longitudinal displacements, the bearings had also cracked through the throat of the hinges and split under applied vertical loads. In the remaining bearings which had not collapsed, cracks of up to 2 mm were measured in the throats of the hinges and splitting cracks of up to 3 mm were measured on the transverse side faces of the bearings.

In this rehabilitation project, existing damaged concrete hinge bearings were successfully replaced after 50 years' service with elastomeric bearings and seismic restraint blocks to comply with modern standards. This rehabilitation project demonstrated that bearings can be replaced in an emergency in extremely short time frames, given effective collaboration with an experienced Client, Engineer and Contractor.

ACKNOWLEDGEMENTS

The authors acknowledge the collaborative efforts which helped make this work possible:

- Client: South African National Roads Agency SOC Limited (SANRAL);
- Consulting Engineer: SMEC South Africa (previously t.a. Vela VKE);
- Contractor for initial emergency repair contract: Botes & Kennedy Manyano
- Contractor for balance of bridge repair contract: Haw & Inglis Civil Engineering

Concrete Repair, Rehabilitation and Retrofitting IV – Dehn et al. (Eds)
© 2016 Taylor & Francis Group, London, ISBN 978-1-138-02843-2

Seismic reinforcement of the URM by FRP system

O. Simakov
CJSC "Prepreg-SKM" Moscow, Russia

G. Tonkikh
FCSHT RSRICDE, Moscow, Russia

O. Kabancev
Moscow State University of Civil Engineering, Moscow, Russia

A. Granovsky
Central Research Institute of Building Constructions V.A. Kucherenko, Moscow, Russia

ABSTRACT

At the moment, there is a considerable experimental basis for reinforcement of masonry (especially for buildings located in seismic areas) with external materials based on carbon fibers. In addition, sufficient experience and results of monitoring is collected in system of external reinforcement of objects. However, there are still significant theoretical gaps in calculation bases of this reinforcement method. One of these outstanding issues is reinforcement of the piers of masonry using external reinforcement based on carbon fibers. In order to manage this issue there have been experimental studies on natural samples, reinforced with the external reinforcement system FibArm (products of CJSC "HC "Composite", Russia).

The studies have been conducted on the following samples:

Series I (S-I)—check samples of brick masonry without reinforcement, in amount of 3 pcs.;

Series II (S-II)—experimental samples of brick masonry, reinforced by a carbon one-directional tape FibARM Tape 230/300. The number of samples in the series is 3 pcs.;

Series III (S-III)—experimental samples of brick masonry, reinforced by a carbon one-directional tape FibARM Tape 530/300. The number of samples in the series is 3 pcs.;

Series IV (S-IV)—experimental samples of brick masonry, reinforced by three carbon one-directional tapes FibARM Tape 230/300. The distance between the tapes is 200 mm. The number of samples in the series is 1 pce;

Series V (S-V)—experimental samples of brick masonry, reinforced by a carbon two-directional cloth FibARM Twill 240. The number of samples in the series is 1 pce.

Series VI (S-VI)—experimental samples of brick masonry, reinforced on both sides by three carbon one-directional tapes FibARM Tape 230/300. The distance between the tapes is 200 mm. The number of samples in the series is 2 pcs.;

Series VII (S-VII)—experimental samples of brick masonry, reinforced on both sides by a carbon two-directional cloth FibARM Twill 240. The number of samples in the series is 2 pcs.

Test configuration is shown in Fig. 1.
Results of experiments are presented in Table1.

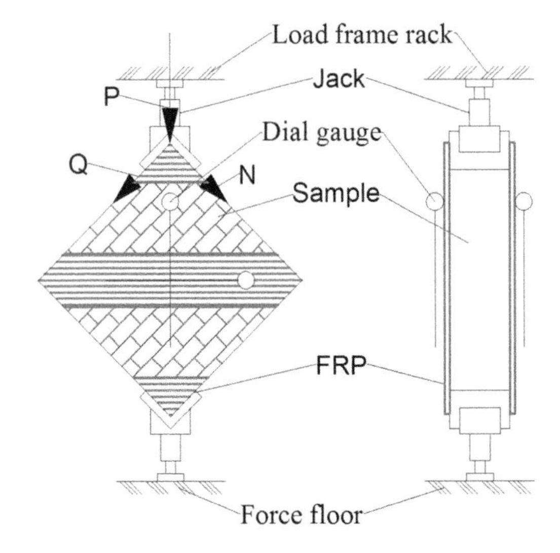

Figure 1. Test configuration.

Table 1. Evaluation of shearing forces of the samples.

Series identification number	P, kgf	Q, kgf	Increase in shearing force	
			Q, kgf	%
S-I	50392	35288	–	–
S-II	66120	46303	11015	31.21
S-III	71322	49945	14657	41.53
S-IV	76506	53576	18288	51.82
S-V	84570	59223	23935	67.82
S-VI	99400	69608	34320	97.25
S-VII	102471	71758	36470	103.35

As a result of obtained experiments data processing and the results of numerical simulation the method has been developed calculating masonry load bearing capacity increase under the action of shearing forces. This method is described with a formula:

$$Q = \left(0.35 \cdot ln(k+80) - 0.53\right) \cdot Q_0, \tag{1}$$

where Q_0, the load bearing capacity of masonry without increase, is taken to be equal to the load bearing capacity with the action of the principal tensile stresses;

k – coefficient of external reinforcement

$$k = \frac{A_{a_i}}{A_k} \cdot b_{fib} \cdot \delta_{fib} \cdot R_{fib} \cdot n \tag{2}$$

A_k is surface area of the masonry, m^2:
b_{fib} is width of carbon fiber, m:
δ_{fib} is thickness of the one layer tape, m:

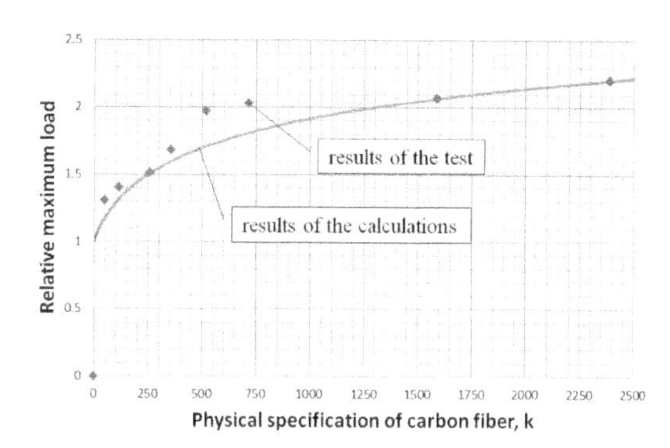

Figure 2. Convergence of the developed method with the experiment data.

R_{fib} is average value of carbon fiber tensile breaking stress, Pa:
n is number of carbon fiber layers.

The values calculated by this method with sufficient accuracy converge with the experimental results (Fig. 2).

CONCLUSION

Conducted studies have shown efficiency of masonry external reinforcement. Increase of load bearing capability of this work amounted to 103%, and reinforcement is calculated analytically up to 120%.

Also as a result of experimental studies the method has been developed calculating sufficient accuracy that describes the work of the masonry sample reinforced by external reinforcement.

Concrete Repair, Rehabilitation and Retrofitting IV – Dehn et al. (Eds)
 ISBN 978-1-138-02843-2

Influence of surface concrete preparation on adhesion properties of repair materials

M. Skazlic
Faculty of Civil Engineering, University of Zagreb, Zagreb, Croatia

K. Mavar
Institute IGH, Zagreb, Croatia

A. Baricevic
Faculty of Civil Engineering, University of Zagreb, Zagreb, Croatia

INTRODUCTION

Durability under conditions of aggressive environment is nowadays among one of the major problems of concrete structures. Concrete structures rapidly deteriorate under such conditions due to poor design requirements, mistakes made during the performance of works, and lack of structure maintenance. This is why the repair procedure is necessary and unavoidable in order to preserve the bearing capacity and usability of a concrete structure.

One of the most common forms of concrete structure repairs pertains to the procedure of removing damaged or contaminated concrete, as well as to its reshaping with repair mortars or concretes. The reshaping procedure usually takes place in the zone of the reinforcement protective layer, and consists of a number of stages as follows:

- Surface preparation (using well-known technologies for removing concrete layers and for preparing for the application of new layers),
- Application of the bonding coat,
- Application of repair material in accordance with the conditions of placing a new reshaped layer by using repair mortars and concretes.

EXPERIMENTAL PROGRAMME

The aim of the investigation is to determine optimal technology for the concrete surface preparation and the type of repair concrete which meet the new adhesion criteria ($f_a \geq 2.0$ N/mm^2), instead of the former criteria for tensile adhesion strength ($f_a \geq 1.5$ N/mm^2).

The investigation programme is designed to present the performance of repair on horizontal surfaces through trial works on models, such as for example on bridge concrete slabs which have been damaged and in which case the contaminated protective layer needs to be removed and the pavement slab reshaped by applying the repair system procedure.

Figure 1. Placement of repair concrete on prepared surface of concrete substrate (left) and overview of model slabs (right).

The following parameters have been varied in the investigative trial works:

- Concrete quality of the surface (concretes with the compressive strength class C35/45 (A) and C25/30 (B)),
- Surface concrete treatment technology (hydrodemolition—HD, sandblasting—PJ, manual pneumatic removal of concrete—PH),
- Type of material for repair (concrete without additives and binder, concrete with silica fumes without the binder, concrete with latex without the binder, concrete with latex with the binder).

In this way six different slabs (marks: AHD, BHD, APH, BPH, APJ, BPJ) were prepared with three different technologies of concrete removal (HD, PH, PJ) for each existing concrete quality (A and B), Figure 1.

To asses properties of underlying concrete and repair material properties following methods were used: compressive strength, static modulus of elasticity and optical microscope analysis. Investigation methods applied for evaluating the quality of the surface concrete preparation and the quality of the performed repair by reshaping with repair concretes was performed by testing adhesion with pull off method and determining the roughness by measuring with a calliper.

CONCLUSIONS

Based on the results of conducted experimental research following conclusion are made regarding adhesion properties of repair material:

- The quality of concrete substrate has minor influence on the adhesion of repair system. Its influence is pronounced only in case when surface is prepared using mechanical pneumatic technology, due to the appearance of microcracks in underlying layer.
- The roughness of substrate has no direct effect on the accomplished adhesion of repair system, but presence of micro cracks in existing concrete found to be important. Surface preparation using mechanical pneumatic technology is therefore considered inappropriate.
- Adhesion of repair system is increased with its age.
- At the age of 28 day, adhesion requirement (f_a >2.0 N/mm^2) is achieved by following repair systems: concrete with addition of silica fume without bonding layer and concrete with addition of latex when bonding coat is applied.

Concrete Repair, Rehabilitation and Retrofitting IV – Dehn et al. (Eds)
 ISBN 978-1-138-02843-2

Management of the M4 Elevated Section substructures

C.R. Hendy & C.T. Brock
Atkins, Epsom, UK

A.D.J. Nicholls
Connect Plus (M25) Ltd., UK

S. El-Belbol
Highways Agency, UK

ABSTRACT

The M4 Elevated Section in West London is a 1.9 km concrete viaduct structure providing a major arterial route in to London. An intervention model bringing together structural assessment and forecast deterioration, corrosion modeling and cracking has been developed to prioritise structural rehabilitation. To evaluate residual strength an initial assessment of crossheads was undertaken, which identified a deficiency in tensile capacity at the ends of the crosshead cantilevers. Further assessment has been undertaken including three dimensional strut and tie, non-linear finite element analysis and plastic analysis to justify continued trafficking of the structure, confirm public safety and to determine the need for strengthening.

Additionally extensive monitoring of the crossheads has been implemented including mapping of all cracks, and remote crack monitoring to safeguard the substructures and indentify early signs of structural distress. The long term maintenance strategy brings together strengthening and cathodic protection concrete preservation methods with removal and repair of concrete delamination.

Figure 1. General view of the M4 Elevated Substructures.

This paper discusses the development of the prioritisation process including deterioration modelling, together with the extended structural analysis to formulate a strengthening programme for these substructures.

The use of BIM to aid design and capture all condition information and intervention plans in a 3D environment is also discussed.

Concrete Repair, Rehabilitation and Retrofitting IV – Dehn et al. (Eds)
© 2016 Taylor & Francis Group, London, ISBN 978-1-138-02843-2

Modern technique and concrete technology used to conserve a unique marine heritage breakwater

S. Hold
Civil and Structural Engineering C.Eng.

ABSTRACT

The history of the breakwater dates back to the 1850s when a large harbour was being planned to encompass an area of water from the La Collette rocks to the east to a new breakwater from Elizabeth Castle to the west.

Work began from both areas of rock outcrop in the 1870s with the sections from Elizabeth Castle to Hermitage rock completed successfully. The breakwater at the La Collette rocks was never really begun in earnest in such an aggressive wave environment, particularly from the south-west prevailing wind direction which made construction conditions extremely hazardous. The method of construction chosen for the breakwaters was to form large blocks of mass concrete with large rock aggregate cast in ground level pits. The pre-cast concrete block method was used as an alternative to the large mass granite rock blocks previously used for the construction of other historical breakwaters in Jersey. All of these structures have outer walls of dressed stone with an inner core of 'secondary' rock fill. The hermitage breakwater was also constructed using this method but with cross-walls or diaphragms, creating independent cells filled with loose core material.

In 1876 the original construction method was also abandoned after only 100 m of breakwater was completed and a newer concept of 'slice blocks' used to construct a 'solid' breakwater that required no secondary fill. The change of construction to the 'slice-blocks' method enabled an 'end over end' placement of the blocks at a steep incline that were notched, grooved and tessellated together for the progressive construction out to sea.

This method too had to be abandoned before the breakwater was completed approximately 100 m short of the original plate rock objective that was the intended full length of the breakwater. The abandoned breakwater has remained in place since 1880 and has protected the port of St. Helier from the south-westerly swells for the intervening 130 years.

A regular monitoring of the structure revealed that damage had been sustained on the sea side of the pre-1876 breakwater in recent years that required repair and protection. Site investigations and concrete analysis showed that despite severe wear damage in some areas only small 'patch' repair replacement was necessary in comparison to the more radical concrete armour 'overlay' method of protection used elsewhere in Jersey which was the preservation method originally selected. This was as a result of the petrographic analysis of concrete cores taken from the concrete blocks which exhibited Delayed Etringite Formation (D.E.F.) which as the outer layers of the blocks are eroded away has the ability to 'self-heal' as the newly exposed concrete has the D.E.F. released.

Concrete Repair, Rehabilitation and Retrofitting IV – Dehn et al. (Eds)
© 2016 Taylor & Francis Group, London, ISBN 978-1-138-02843-2

Some conclusions of durability and behavior of structural rehabilitation solutions applied to deteriorated reinforced concrete elements after ten years of intervention

G. Croitoru
Telekom RMC, Bucharest, Romania

A. Popaescu
Transilvania University Brasov, Romania

ABSTRACT

This study analyzed the durability and the behavior in service in difficult conditions of reinforced concrete structural elements rehabilitated 10 years ago.

Structure analyzed is "Turbines Hall" and is part of a great objective with function of electric and thermal power stations from Romania.

For structure "Turbines Hall" (type industrial hall) construction started in 1967 and is currently in operation (almost 50 years in service).

In time, the service conditions were generally very difficult: water leakage from technological pipes, the emission of technological hot steam and major values of humidity in some areas of the building.

It is very important to mention the inefficiency of ventilation system inside "Turbines Hall" during service.

The main types of deterioration in reinforced concrete structural elements of "Turbines Hall" construction was: deterioration from durability (Major) and faulty from execution and deterioration from operating in service.

Major contribution to deterioration from durability was confirmed by the conclusions of the technical expertise.

Deterioration from durability were manifested as follows: intense corrosion of reinforcement until detachment of corrosion products and complete disintegration and cracks and dislocations of concrete cover due to corrosion of reinforcement.

The main solutions for structural rehabilitation (applied in 2004) were: strenghtening with reinforced concrete, strenghtening with laminated profiles and repairs with special mortars (e.g. epoxy type).

Following analysis of behavior in service of structural rehabilitation solutions (bad conditions in service in some areas of the building continued in the 2004–2014), the conclusions were: new deterioration from durability have been observed.

Even though have been used some high quality materials for rehabilitation of structural elements, conditions in service has led in many cases to restarting of the corrosion of reinforcement.

There are situations where restarting of the corrosion of reinforcement led to the emergence of new cracks in strengthened beams.

Analysis of durability of structural rehabilitation solutions applied to reinforced concrete elements for 10 years in service (2004–2014) concludes that ensuring of good conditions in service is very important even for strengthened structural elements.

Concrete Repair, Rehabilitation and Retrofitting IV – Dehn et al. (Eds)
© 2016 Taylor & Francis Group, London, ISBN 978-1-138-02843-2

Case study for the repair of Berths 4 & 6 in Guernsey

John Drewett
Concrete Repairs Ltd., London, UK

Kevin Davies
CorroCiv Limited, Manchester, UK

Paul Segers
CH2M Hill, Birmingham, UK

INTRODUCTION

In 2007 Guernsey Harbours commissioned a review of Berths 4, 5 & 6 which included a structural condition survey of the facilities. The report recommended that the existing rail-mounted cranes be replaced with two modern mobile, wheeled, harbour cranes and associated refurbishment and strengthening work to quay structures be carried out.

The commission was extended to produce a detailed defect assessment specification and tender package which identified the methods and extent of the structural investigation and testing required. A detailed assessment of the reinforced concrete and metallic structures was undertaken.

The findings of the site investigation confirmed the feasibility of a 50 year life extension and the original rehabilitation methodology proposed.

THE REHABILITATION SPECIFICATION

The rehabilitation contract was awarded to Concrete Repairs Ltd (CRL) in 2012 with a scheduled start date on site in 2013.

CONCRETE REPAIR DESIGN

There were significant areas of cracked and spalling concrete identified during the detailed investigation that required repair prior to the ICCP installation. Primarily these repairs were to the primary and secondary downstand beams.

The beam repairs were undertaken by using hydro-demolition techniques to remove defective areas of concrete and reinstating using a prebagged flowable high performance repair concrete compatible with the ICCP system.

Following concrete removal, localised welding repairs were carried out to the reinforcement at areas exceeding the allowable section loss tolerance.

The repair areas of the beams were shuttered and recast. The soffits were repaired using a dry sprayed concrete mortar. In total some 80 m^3 of concrete repairs were undertaken.

CATHODIC PROTECTION DESIGN

The system was designed in accordance with BS EN ISO 12696:2012, BS EN ISO 13174:2012 and the Works Information.

Berth 4—RC components above + 2.00 M OD (ICCP and GACP)

The anode zones were split into a combination of two primary beams and the interconnecting soffits and secondary beams. The anode system was a MMO/Ti mesh fixed to the concrete surface and overlaid with a sprayed cementitious overlay nominally 20 mm thick.

For the columns which were subject to tidal influence MMO/Ti discrete ribbon anodes installed into cut chases within the cover concrete were used.

Berth 4—RC components below +2.00 m OD (GACP)

The reinforced concrete columns and existing crane foundations below +2.0 mOD were protected using conventional seawater galvanic aluminum alloy anodes.

Berth 4—Metallic fenders and Knuckle—sheet steel piling +0.24 mOD (GACP)

The existing fender and sheet steel piles were protected with a combination of conventional seawater galvanic aluminium alloy anodes which provided galvanic cathodic protection below mean tide level +0.24 mOD and a high performance glass flake epoxy protective coating system for protection of tidal and atmospherically exposed areas above mean tide level.

Berth 6—RC deck slabs above +2.50 m OD (ICCP)

The selected anode system was a MMO/Ti mesh fixed to the concrete surface and overlaid with a sprayed cementitious overlay 20 mm thick.

Berth 6—Concrete encased steel I beams below +2.50 m OD (GACP)

A distributed standard aluminium anode system was used to protect the 8 concrete encased steel columns.

Cathodic protection monitoring system

The installed CP systems are monitored using monitoring clusters consisting of one manganese manganese dioxide (Mn/MnO_2) reference electrode, one silver silver chloride 0.5 M potassium chloride (Ag/AgCl 0.5 M) reference electrode and one monitoring connection to the steel.

Cabling and cable management

All the cabling back to the junction boxes was embedded within the concrete or anode overlay. All cable connections were made within the junctions boxes, which were IP68 rated for submersion in water to a depth of one metre. From the junction boxes multicore cables run on 316 L stainless steel cable tray, mounted under the berth.

The DC positive feeders to the uncoated titanium conductor bars feeding the MMO/Ti mesh run in ring circuits from the junction boxes.

The DC negative connectors are made off site and required only welding of the connector to the exposed steel with sufficient redundancy to ensure durability.

Power and control units

With 34 impressed current anode zones and 372 reference electrode inputs this is a large and quite complex ICCP installation which required a high performance power and control unit.

CRL selected Rectifier Technologies to manufacture and supply the control equipment. The reference cells were buffered by differential multiplexers with high impedance (>100 MΩ) and inherent surge protection.

THE SITE WORKS

With the ICCP design agreed the work on site commenced in March 2013 with a 104 week contract programme for completion in early 2015.

To facilitate access under Berth 4 when the berth was in use, a temporary access hole was cut through the deck.

COMMISSIONING

Commissioning comprised; detailed testing of each core of the multicore cables interlinking the below deck junction boxes with the control units in the electrical switch room to prove correct connection, electrical continuity of the DC negative and DC positive cabling, electrical separation of the anode and cathode in each zone and stability of the steel potentials measured against the embedded reference electrodes.

Energisation involved; reconnecting the switches that physically disconnect the DC positive output feeders at the terminal rails and energising the DC outputs to each zone in turn. Initial output limits were set at 2 A and 5 VDC. In all cases the circuit resistances were such that the 2 A output current was reached first.

The steel potentials were recorded as the ICCP was applied (polarisation) and left on for an hour or so. The DC outputs were then switched off for another hour or so to allow the steel potentials to recover (depolarise) whilst recordings were made on a more frequent basis. In all cases the steel potentials became decidedly more negative against the two reference electrodes in the clusters when the ICCP system was applied (polarisation) and less negative (depolarisation) as it was removed. This is as expected and proves that even after a short period of time the ICCP system achieved its requirements.

SUMMARY

This has been one of the largest marine repair and cathodic protection systems ever installed in Europe. The technical and logistical details have been challenging for all the project team but the scheme has been successful to date

The impressed current and galvanic cathodic protection systems are functioning correctly in accordance with BS EN ISO 12696 and 13174. During the next 12 months the impressed system will be adjusted to its optimum operating performance. With such a large and quite complex system there will be some minor maintenance required, but providing this is undertaken and the system regularly monitored we are confident that the 25 year design life will be achieved.

Concrete Repair, Rehabilitation and Retrofitting IV – Dehn et al. (Eds)
© 2016 Taylor & Francis Group, London, ISBN 978-1-138-02843-2

Southern Europe pipeline: New life of a 1960s pipeline

C. Chanonier & C. Raulet
Diadès, Vitrolles, France

F. Martin & C. Carde
LERM, Arles Cedex, France

J. Resplendino
Setec TPI, Vitrolles, France

ABSTRACT

The Society of South European Pipeline (SPSE) ensures the supply of petrochemical refineries via the Fos-Lyon-Karlsruhe (769 km) pipeline. Given the environmental issues, an inspection program and rehabilitation of the equipment of the 34 inch pipeline were launched, in order to once again exploit it.

As part of this restoration, SPSE asked DIADES, in collaboration with LERM, to carry out a structural diagnosis of an aerial crossing of the Durance. This analysis was completed by the study and evaluation of the conformity of the structure in two respects: static and seismic resistance.

Investigations (structure, materials and geotechnical) and static and dynamic recalculations as part of the diagnosis have led to a reinforcement and rehabilitation of the structure through the use of external prestressing.

The diagnosis was made on the basis of a program of structural investigations, to design and manage the repair and reinforcement project of the structure:

- Recalculation,
- Crossbow test
- "Corrosimetry",
- Concrete lab analysis
- Soil investigations.

The dynamic reinforcement of the structure was planned from the start, with a concern for the sustainability of the project. Indeed, it seemed appropriate to take advantage of dynamic reinforcement to sustain the structure.

The rehabilitation of an existing structure, by means of external prestressing, stresses the need

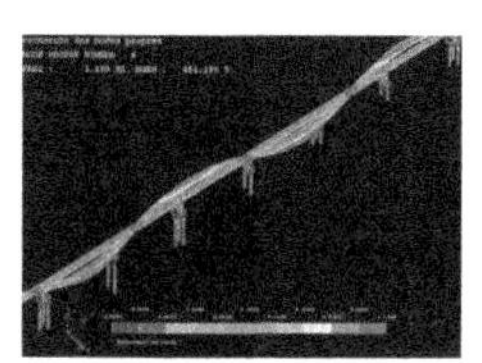

for reliable assumptions that include the determination of actual residual stresses, not estimated on the basis of the fixed losses. These should form the basis for the definition and the "true conception" of the reinforcement required. This example also highlights the value of prestressing and composites in reinforcement and preventive maintenance of assets, in particular if seismic compliance settings are required.

This type of rehabilitation requires a thorough structural diagnosis and mastery of the technical prestressing processes. The technical development has enabled to achieve a "modular" and "removable" solution to adapt as well as possible to the structural needs of the structure without changing the exploitation of the structure and by ensuring ease of monitoring and overcoming potential defects in internal prestressing concrete.

Concrete Repair, Rehabilitation and Retrofitting IV – Dehn et al. (Eds)
© 2016 Taylor & Francis Group, London, ISBN 978-1-138-02843-2

Innovative subsequently applied shear strengthening techniques for RC members

N. Randl & P. Harsányi
Carinthia University of Applied Sciences, Austria

INTRODUCTION

Research significance

In a research project in cooperation with the HILTI Corporation and SIKA Austria new strengthening methods for shear deficient bridges are being developed and tested. Although there exists a variety of techniques for subsequent strengthening of RC structures, there is still lack of knowledge on the efficiency of available shear repair methods. Most existing techniques require also construction activities on the top side of the structure which in turn leads to restrictions of the traffic flow. The avoidance of such restrictions was one of the main targets in this project. Other decisive criteria were the efficiency of the applied strengthening method, economic aspects, simplicity of installation and the avoidance of significant adjustments of the member's cross section.

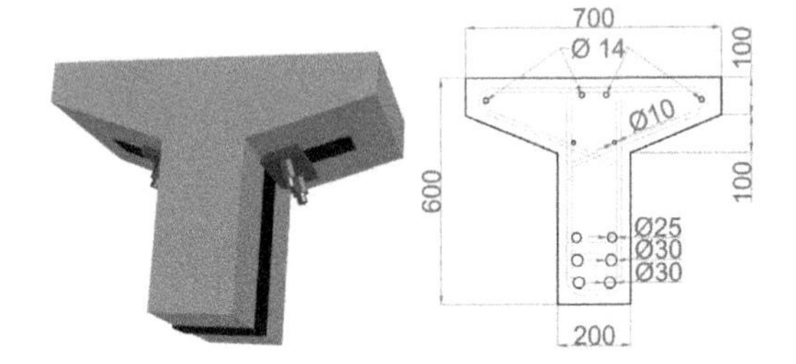

Figure 1. CFRP strengthening and layout of T-shaped cross section.

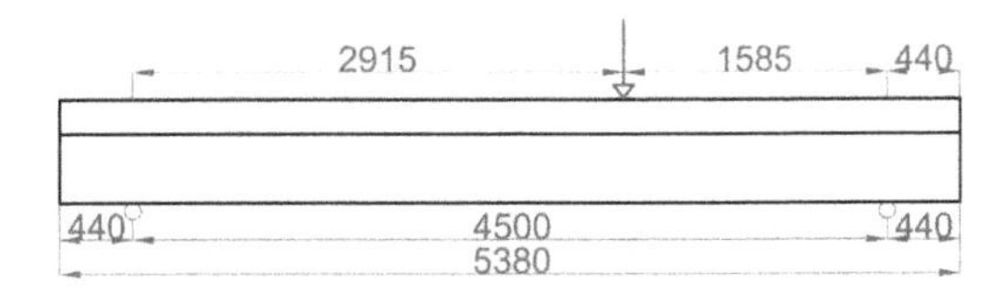

Figure 2. Test setup and side view of T-beam.

DESCRIPTION OF APPLIED STRENGTHENING TECHNIQUES

CFRP sheets with end anchorage

For practical reasons and with respect to typical cross sectional shapes. CFRP sheets are usually bonded only to the sides or to sides and bottom of RC members in a U-shaped constellation. The usually observed failure mode will then be lateral delamination, i.e. de-bonding, of the CFRP ends from the concrete (Belarbi et al. 2012, Ozden et al. 2014).

In order to rectify the mentioned deficits, a new and very simple end-anchorage system has been designed which allows a direct load introduction into the compression zone: A steel plate 300 × 150 × 30 mm is fixed to the upper flange of the T-section by means of rapidly curing bonded anchors (Figure 1).

Post-installed undercut anchors

The basic technique of strengthening shear deficient beams by applying post-installed reinforcement has been investigated successfully by Randl & Kunz (2009). In the present study, in cooperation with the Hilti Corporation, a new anchorage system especially designed for this type of application was developed. The anchor is set vertically into the pre-drilled hole and the load is then introduced at the very end inside the structure by means of a self-undercutting expansion sleeve. A tubular sleeve around the whole length of the inner threaded rod enables subsequent prestressing of the rod.

EXPERIMENTAL CAMPAIGN

Test setup

The test specimen was a single span beam made of concrete C30/37, with the load introduction at a distance of 1.585 m from the support (Figure 2). Taking into account the span of 4.5 m, the shear span-to-depth ratio a/d was 3.2 (with an effective depth to the bottom reinforcement d = 494 mm).

To avoid premature flexural failure, an over-reinforced T-shaped cross section was designed,

Table 1. Applied strengthening systems.

Test no.	Type	Strengthening system	Spacing/number [mm]/-
1	Reference	None	-/-
2	CFRP_4-2*	CFRP, 2 layers	190/4
3	CFRP_4-2	CFRP, 2 layers	190/4
4	PUAnc_7	Post-inst. anchor	190/7
5	CFRP_4-1	CFRP, 1 layer	190/4
6	PUAnc_4	Post-inst. anchor	380/4
7	PUAnc_7	Post-inst. anchor	190/7
8	CFRP_4-2	CFRP, 2 layers	190/4
9	CFRP_3-1	CFRP, 1 layer	570/3
10	PUAnc_4	Post-inst. anchor	380/4
11	CFRP_4-1	CFRP, 1 layer	190/4
12	PUAnc_4	Post-inst. anchor	380/4

* Alternative shape of end anchorage.

Table 2. Test results.

Test no.	Type	F_{cm} [N/mm²]	F_{max}[kN]	V_{max} [kN]	$F_{max}/F_{max,1}$
1	Reference	44,5	370,9	285,3	1,00
2	CFRP_4-2	45,7	813,8	612,8	2,15
3	CFRP_4-2	46,4	957,4	719,0	2,52
4	PUAnc_7	42,6	681,7	515,1	1,81
5	CFRP_4-1	41,1	831,3	625,8	2,19
6	PUAnc_4	43,3	612,0	463,6	1,62
7	PUAnc_7	41,2	714,4	539,3	1,89
8	CFRP_4-2	43,1	961,4	722,0	2,53
9	CFRP_3-1	40,5	752,1	567,2	1,99
10	PUAnc_4	44,5	515,0	391,9	1,37
11	CFRP_4-1	37,9	853,5	642,2	2,25
12	PUAnc_4	51,1	500,4	381,1	1,34

using high grade steel S670/800 as longitudinal tensile reinforcement at the bottom of the beam (Figure 1). No stirrup reinforcement was provided in the beam sections between load introduction and closest support.

Application and type of strengthening

After a cyclic preloading phase, all specimens except for the reference test no. 1 were provided with an adequate strengthening. An overview of applied strengthening methods and tested constellations is given in the following Table 1. All strengthening elements were installed at a constant spacing of either 190 or 380 mm, in one case 570 mm.

TEST RESULTS

All specimens exhibited typical shear failure modes, flexural failure was never decisive (see example of beam strengthened with 4 post-installed anchors in Figure 3). The reference beam reached an ultimate load of 370.9 kN, corresponding to a back-calculated maximum shear force $V_{max,1}$ = 286 kN. All strengthened beam specimens exhibited a significant load increase due to the applied strengthening elements (Table 2).

Figure 3. Critical shear crack at beam no. 6.

CONCLUSIONS

Both tested strengthening techniques exhibited a satisfying load enhancement in shear. With the CFRP sheets the load gain was, depending on the number of sheets and layers, between 100% and 150%. The end anchorage of the CFRP sheets proved very effective so that either rupture of the CFRP or concrete failure in compression was observed.

The installation of post-installed anchors, on the other hand, yielded a 40% load increase with four and 85% with seven anchor rods.

ACKNOWLEDGEMENTS

The authors express their sincere gratitude to the Austrian Research Promotion Agency (FFG), the Austrian Railway Company ÖBB and the Austrian motorway operator ASFINAG for the funding of this research. Likewise the support by the Hilti Corporation and SIKA Austria and the fruitful exchange with Jakob Kunz (Hilti), Guenther Grass and Rainer Planer (both SIKA) is gratefully acknowledged.

Concrete Repair, Rehabilitation and Retrofitting IV – Dehn et al. (Eds)
© 2016 Taylor & Francis Group, London, ISBN 978-1-138-02843-2

Development of a multi-disciplinary graduate course on rehabilitation of structures

T. El-Maaddawy
United Arab Emirates University, Al Ain, United Arab Emirates

ABSTRACT

This paper describes the development of a multi-disciplinary graduate course on rehabilitation of structures. The course is offered for graduate students by the Civil and Environmental Engineering department at the United Arab Emirates University. The course offers practical tips, design details, and case studies collected by the author upon comprehensive research and outreach activities. The course comprises two modules, four topics each. Course modules and topics are summarized in Table 1. Module I is about condition assessment and conventional repairs whereas Module II is about innovative strengthening with composites. Active-learning and problem-solving strategies are adopted in teaching the course. Cooperative learning and collaboration are prompted throughout students' group work. Self-learning, and inquiry-based learning are inspired throughout a research paper assignment and laboratory project. Students are exposed to real-life problems and hands-on training. They conduct research, analysis, and comparisons. Students are asked to present and critique recent research on assessment and rehabilitation of structures, report findings and submit technical reports.

Course outcomes that are observable, measurable, and capable of being understood by students, faculty, external agencies, and stakeholders are developed and mapped to the graduate program outcomes in compliance with the QF*Emirates* (2012). The statement of each course outcome begins with an action verb describing knowledge, skills, or competencies. The outcome-based education concept has been implemented in several engineering programs worldwide (Lindholm et al. 2009, Osman et al. 2012). Quantitative (direct) and qualitative (indirect) assessment tools are developed to evaluate the level of attainment of course outcomes. The quantitative assessment tools used to assess the level of attainment of course outcomes include homework assignments, exams, laboratory project, research paper, and presentation. The qualitative assessment tools include students and faculty self-perceptions surveys.

Table 1. Course modules and topics.

Module	Topic
Condition assessment and conventional repairs	Defects and Deterioration Repair Procedure and Materials Structural Condition Assessment Structural Strengthening and Stabilization
Innovative Strengthening with Composites	Principles of Strengthening with Composites Flexural Strengthening with Composites Shear Strengthening with Composites Column Strengthening with Composites

In the laboratory project, students conduct the following tests:

- Covermeter survey.
- Half-cell potential.
- Linear polarization.
- Concrete resistance.
- Schmidt hammer.
- Ultrasonic pulse velocity.
- Compression tests on columns with and without carbon fiber-reinforced polymer wrapping.

In the research paper assignment, students present and critique recent research on assessment and rehabilitation of structures. They report findings and communicate effectively with pees and clients.

REFERENCES

Lindholm, J. A. (2009). Guidelines for developing and assessing student learning outcomes for undergraduate majors, 1st Edition.

Osman, S., Jaafar, O., Badaruzzaman, W., and Rahmat, R. (2012). The course outcomes (COs) evaluation for civil engineering design II course. Procedia—Social and Behavioral Sciences, 60: 103–111.

QF *Emirates* (2012). Qualifications Framework Emirates Handbook, National Qualifications Authority (NQA). Abu Dhabi. UAE.

Concrete Repair, Rehabilitation and Retrofitting IV – Dehn et al. (Eds)
© 2016 Taylor & Francis Group, London, ISBN 978-1-138-02843-2

Design of externally bonded FRP systems for strengthening of concrete structures

T.A. Mukhamediev & V.R. Falikman
Scientific Research Center "Construction", Moscow, Russia

ABSTRACT

For the recent 10–15 years in Russia and abroad the amount of work relating to rehabilitation of various purpose buildings with the aim to extend their life cycle has been essentially increased. Composite materials on the basis of glass, basalt and carbon fibers (FRP) are widely used to strengthen various structures.

Just few studies and insufficient documented experience in the field of using composites to strengthen of reinforced concrete structures, as well as unavailability of the regulatory framework complicate the problem facing design engineers and relating to evaluation of the adopted design solution reliability referring to structural strengthening with composite materials and their wide use in construction practice.

The Code of Practice "Strengthening of reinforced concrete structures using composites. Design rules" (hereinafter—the Code) was developed in Russia. It was enacted by the Ministry of Construction of the Russian Federation on 1 September 2014.

The Code establishes the requirements to analysis and design of reinforced concrete structures rehabilitated or strengthened with external reinforcement in the form of fabrics, strips and laminates made from composite materials using carbon (Carbon Fiber-Reinforced Plastics, CFRP), aramid (organoplastics, AFRP) or glass (glass fiber plastics, GFRP) fibers (externally bonded FRP reinforcement). The Code scope is limited by reinforced concrete structures from ordinary and fine-grain concrete that falls under the guidelines of the Russian Code 63.13330.2012 [1].

The new Code comprises guidelines on regulation of composite material strength and strain characteristics, on the analysis of composite-strengthened structures for the strength of their sections, being normal and oblique to a structure longitudinal axis, for the normal crack opening width and deflections. The Code formulates structural requirements and short guidelines on the strengthening process technology and quality control also.

The Code embraces reinforced concrete structures strengthened along normal sections by means of external reinforcing in the direction of the longitudinal axis or by means of external reinforcing of a casing in the lateral direction, as well as the structures strengthened with stirrups along oblique sections.

Standard design values of composite material strength and strain characteristics are defined according to their specifications with account of experimentally established features of strengthened structure failures. According to the test results the bearing capacity of flexural structures failed due to strengthening element breakaway depends on not only strength and strain characteristics of composite materials and concrete in the strengthened structure, but also on the total thickness of the strengthening element. The relationship for coefficient γ_{f2}, accounting for these features, is taken from the ACI 440.2R-08 [2].

One of the features of composite material properties is reduction of their tensile strength with increase of the period under stress. With the first group limit state design of strengthened structures under action of only sustained dead loads this feature is taken into account by adding the decreasing coefficient of load duration to the composite material tensile strength. This coefficient is accounted for only in the analysis of sustained dead load increment after structural strengthening and is introduced instead of partial safety for materials.

According to few experimental studies including those of composite polymer reinforcement the ratio of long-time and short-time strengths is: for GFRP—0.29…0.55, for AFRP—0.47…0.66 and for CFRP—0.79…0.93. Given that by the moment of strengthening a part of sustained load has been already applied to the structure, the Code prior to accumulation of experimental data assumed the lowest values of the decreasing coefficient in the aforementioned intervals.

The methods of strengthened structure design assumed in the Code are formulated with account of the initial stress and strain state of a structure prior to its strengthening.

Ultimate limit state design procedures for strengthened structures are developed basing on design models given in the Code 63.13330.2012. As in the Code 63.13330.2012, the basic design method relating to normal section strength of a reinforced concrete structure strengthened with external reinforcement is assumed as the deformation model design method, while for special cases of structural cross-section shapes and force impacts it is allowed to use the breaking stress method.

The design relationships to consider behaviors of composite strengthened structures are assumed on the basis of the test result analysis using foreign code recommendations. The accumulated experimental base (about 1000 test results from foreign and domestic studies) was used for model verifications and evaluations of calculated dependences adopted in Code of Practice. The paper presents the results of statistical processing and the accuracy coefficients of calculations, taking into account above-mentioned base.

In Code principles are normalized and formulas for calculation are delivered taking into account ultimate limit state design procedures (the limit states of the first group): the normal section strength design of flexural and eccentrically compressed structures by the breaking stress method; the normal section strength design basing on the nonlinear deformation model; the strength design of casing-strengthened compressed structures; the strength design of sections being oblique to the element longitudinal axis and strengthened with external composite reinforcement in the form of bilateral, trilateral or closed stirrups.

In conformance with Code 63.13330.2012 requirements to the second group limit state designs presented in the Code comprise the crack formation design, the crack opening design and the strain design.

Because of the fact that experimental studies performed in Russia and abroad are apparently insufficient, the Code recommendations on the second group limit state design of strengthened structures proceed from the guidelines for steel-reinforced structures and direct consideration of strengthening element stiffness.

In the simplified method of deflection design of strengthened elements the geometric characteristics of element sections in the design relationships are defined with account of external composite reinforcement.

The crack opening width design, when cracks are normal to the element longitudinal axis, is recommended to be performed in conformance with the guidelines given in Code 63.13330.2012, while assuming the element section geometric characteristics in the design relationships with account of external composite reinforcement.

In the deformation model design of deflections the values of stiffness coefficients in the system of physical correlations are determined in the Code with account of strengthening elements, while the relationship between axial stresses and relative strains in external reinforcing elements are assumed in the form of:

$$\sigma_{fk} = \frac{E_f \cdot \varepsilon_{fk}}{\psi_{fk}},$$

where

$$\psi_{fk} = 1 - \frac{1}{1 + 0{,}8\dfrac{\varepsilon_{fk,crc}}{\varepsilon_{fk}}}$$

where $\varepsilon_{fk,crc}$—relative strain of the external reinforcing element in the section with a crack immediately after normal crack formation;

ε_{fk}—averaged relative strain of the external reinforcing element.

Implementation and application of the present Code will permit to find a sound approach to revision of a number of structure strengthening and repair designs in cast-in-situ and precast reinforced concrete construction.

REFERENCES

[1] Code of Practice 63.13330.2012. "Concrete and Reinforced Concrete Structures. General Provisions". Updated version of SNiP 52-01-2003. Ministry of Regional Development of the Russian Federation, 2012.

[2] ACI 440.2R-08 "Guide for the Design and Construction of Externally Bonded FRP Systems for Strengthening of Concrete Structures," American Concrete Institute, 2008.

Concrete Repair, Rehabilitation and Retrofitting IV – Dehn et al. (Eds)
© 2016 Taylor & Francis Group, London, ISBN 978-1-138-02843-2

Conservation and restoration of exposed cement concrete structures in habitable buildings: A case study of historic cement concrete surfaces at Chandigarh

J.S. Ghuman & Janbade Prafulla Tarachand
Chandigarh College of Architecture, Chandigarh, India

ABSTRACT

Early concrete buildings in India and especially in Chandigarh are threatened by deterioration. Effective protection or maintenance is the key to sustain good health & durability of exposed cement concrete structures. This calls for creating awareness, consistent attention for repair or replacement of the sensitive deteriorated façade, taking effective continues preventive treatment of CC exposed surface in habitable buildings. This also calls for creating a trained work force.This paper tries to identify those parameters of the cement concrete properties and analyze them for conservation aspects. Though the analysis is qualitative and an attempt has been made to quantify it and thus making open for empirical understanding. The interpretations drawn could be the first step towards the goal of having explicitly described strategy for future course of suitable actions in use for conservation project. Chandigarh-City Beautiful is a unique expression of urbanism in the machine age civilization, use of natural building material, exposed cement concrete construction and the citizen's faith to sustain its Architectural Heritage. Chandigarh Administration College of Architecture, Research Cell Study, focuses on grading, maintenance, and effective treatment of the exposed cement concrete *surface texture* of buildings in the City Landscape. It also focuses on norms-standards and construction code for sustainable conservation of exposed cement concrete *surface textures*, to ensure, a healthy work and living environment in habitable buildings. Part-II of this study "Grading and Restoration of Corbusier's Concrete" covers the demonstration exercise undertaken for conservation of exposed concrete structures and surface textures in CCA Building.

KEYWORDS

Historic cement concrete Conservation, Historical monument Conservation, Urban conservation, heritage management, Architectural conservation, and Material characterization.

Concrete Repair, Rehabilitation and Retrofitting IV – Dehn et al. (Eds)
© 2016 Taylor & Francis Group, London, ISBN 978-1-138-02843-2

Investigations into the cause and consequence of incipient anodes in repaired reinforced concrete structures

C. Christodoulou
AECOM Ltd., Birmingham, UK

C.I. Goodier
School of Civil and Building Engineering, Loughborough University, Loughborough, UK

G.K. Glass
Concrete Preservation Technologies, University of Nottingham Innovation Lab, Nottingham, UK

ABSTRACT

The incipient anode (or halo) effect often occurs on repaired reinforced concrete structures. The diagnosis of this problem is widely reported to be macrocell activity. It is deemed that the cause of incipient anodes is the loss of the natural cathodic protection provided by the corroding steel to the steel in the parent concrete adjacent to the patch repair. This diagnosis is based on very limited data. Indeed potential measurements on field structures repaired with proprietary materials have provided data that suggest that macrocell activity is not a cause of incipient anode formation but it is a consequence. Alternative mechanisms that may cause incipient anode activity include repair/parent material interface effects, residual chloride contamination within the parent concrete, and/or vibration damage to the steel/parent concrete interface during repair area preparation. The aim of the work presented here was to assess the impact of macrocell activity on the formation of incipient anodes around the perimeter of repairs in patch-repaired reinforced concrete structures. This was examined based on a major multi-storey car park and a bridge structure both located in the UK. The analysis challenges the view that macrocell activity is a cause of incipient anode formation. Indeed this work shows that the data supporting the existing diagnosis is not convincing and suggests that macrocell activity is primarily a consequence of incipient anode formation and the cause probably, results from other factors.

INTRODUCTION

Corrosion of steel reinforcement affects many concrete structures. Patching is a common repair technique that involves the removal of physically deteriorated concrete (by hydro-demolition or jack hammer), cleaning the steel reinforcement within the patch and finally restoring the concrete profile with a proprietary repair mortar by hand, trowel, spray or others. This process aims to protect the repaired area and makes the steel within the repair area passive by removing the previously corroding anodic area. However, in many cases further corrosion induced deterioration has been observed in the parent concrete in the immediate area around the patch repairs, sometimes within a few months following completion of the repair process. This phenomenon is known as incipient or ring anode formation, or the halo effect.

The aim the current work was to assess the impact of macrocell activity on the formation of incipient anodes around the perimeter of repairs in patch-repaired reinforced concrete structures. A multi-storey car park and a bridge, both constructed

Material	Structure type	Repair location	Chemical base & characteristics
A	MSCP	Deck	Shrinkage compensated, pourable, polymer modified concrete, trowel finished
B	Bridge	Soffits and vertical faces	Shrinkage compensated, dry sprayed, polymer modified micro-concrete, trowel finished

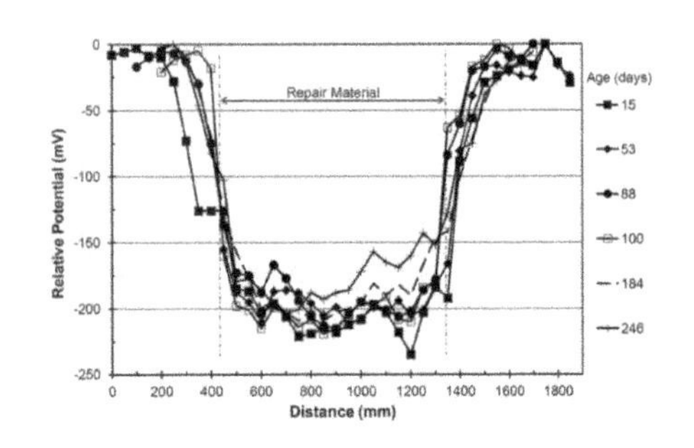

Figure 1. Evolution of average potential readings through the patch repair with material A as a function of time.

of reinforced concrete provided first-hand data following concrete repairs. Details of the repair materials are included in Table 1, alongside a description of their chemical base and characteristics.

ANALYSIS

Multi-storey car park

Figure 1 illustrates the potential monitoring of a repair using Material A over a period of 246 days. The early age results (15 days) show that the steel potentials within the patch repair were depressed to very negative values as a result of the fresh alkalinity provided by the repair mortar. The steel potentials shifted to less negative values as the age of the patch repair increased, however, at no point within the 246 days of monitoring did the potentials of the steel within the patch rise above the steel potentials in the parent concrete.

Similar behavior was observed for all the patch repairs monitored as part of this work, confirming consistency on the results obtained.

Bridge structure

Similar behavior was observed for all the patch repairs from the bridge structure too and a random sample of the monitoring data is reported here.

Figure 2 illustrates the potential monitoring on a bridge repair using material type B. The potentials of the steel within the patch repair remained more negative than the potentials of the steel in the adjacent parent concrete over a period of 83 days.

DISCUSSION

The textbook understanding of the cause of incipient anode formation is that steel within the repair passivates as a result of the alkalinity of the fresh repair material, the absence of chlorides and the abundance of dissolved oxygen in the pore solution of the freshly mixed concrete or repair mortar. The steel potential in the repair rises above the passive steel potential in the parent concrete resulting in a macrocell that induces passive film breakdown and causes an incipient anode to form adjacent to the repair.

However, the results of the present study suggest the steel in the patch repair had a more negative potential than the steel in the parent concrete. Although potentials shifted to more positive values with time, they were always more negative than the potential of the steel in the parent concrete.

Some possible reasons for this behavior include the build-up of the oxide film or that a membrane or streaming potential exists between the parent and the patch concrete.

The pH of the environment can also have a strong impact on equilibrium potentials with a higher pH

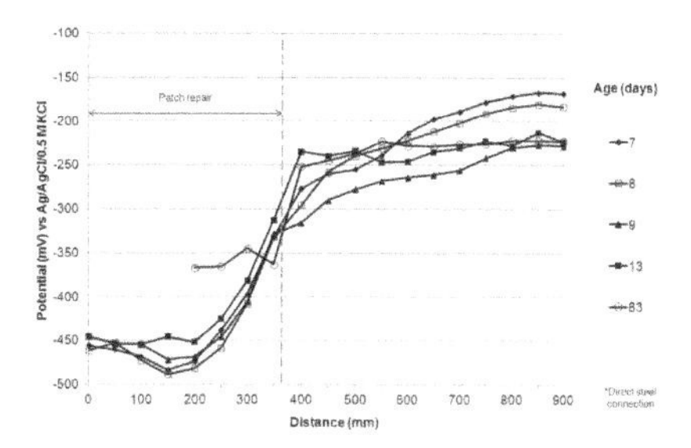

Figure 2. Potential mapping monitoring on a bridge repair with material type B.

resulting in more negative equilibrium potentials. Fresh repair concrete may well have a higher pH than aged concrete because the reaction between hydroxide and silica within the concrete or between hydroxide and carbon dioxide in the air will tend to reduce the pore solution pH to that of one of the more abundant buffering solid phases present.

The change in pH between the repair material and the parent concrete can also give rise to a membrane potential. This results due to diffusion of hydroxide ions from the patch repair to the parent concrete and a resulting build-up of charge on the walls of the pore system at the repair interface.

Cracks may also occur at the interface between the parent concrete and the repair material following patch repair of concrete structures. These may provide a path for chlorides to penetrate preferentially into the substrate. The extent of this effect will be dependent on surface preparation, application techniques, curing, material properties and compatibility with the parent concrete.

All of the above suggest that, on balance, macrocell activity is a consequence, not a cause, of incipient anode formation.

CONCLUSIONS

From potential measurement data obtained over a period of up to 250 days, no evidence was found in this work to support the hypothesis that macrocell activity is a cause of incipient anode formation. In fact, the use of proprietary repair materials may permanently depress steel potentials within the repair area probably due to their low permeability and high pH.

Cracks developing at the repair/substrate interface may provide an easier path for chlorides to penetrate into the substrate. Chlorides may also enter the concrete through the interface between the parent and repair material, parent concrete adjacent to the repair area may have an above average level of residual chloride contamination that is sufficient on its own to cause corrosion, and/or preparation of a repair area may result in vibration damage at the steel interface with the adjacent parent concrete.

Performance and health monitoring

Concrete Repair, Rehabilitation and Retrofitting IV – Dehn et al. (Eds)
© 2016 Taylor & Francis Group, London, ISBN 978-1-138-02843-2

Experimental in-situ investigation of the shear bearing capacity of pre-stressed hollow core slabs

G. Schacht
MarxKrontal GmbH, Hannover, Germany

G. Bolle
Hochschule Wismar, Wismar, Germany

St. Marx
Insitut für Massivbau, Leibniz Universität Hannover, Germany

ABSTRACT

As a result of bad quality sealing, moisture and de-icing agents penetrated into pre-stressed hollow core slabs of the top floor in a freely exposed parking garage. Caused by the chlorides the slabs showed especially heavy damage in the support regions near the joints. The theoretical shear bearing capacity according to the technical approval of these slabs was strongly questionable because of the damage. With the help of an experimental loading test in-situ the influence of the damage on the shear bearing capacity of the slabs with different damage-levels could be determined.

Because of the great amount of damaged slabs it was impossible to experimentally investigate all of the questionable slabs to determine the shear bearing safety. Therefor a limited number of slabs were cho-sen which represented a characteristic sample of the whole population of the damaged slabs. In total 5 characteristic slabs (two with heavy and three with medium damages) were tested and the results of these tests used as the basis for an overall evaluation of the shear bearing safety of all slabs of the City-Center. The experimental results allowed a secure evaluation on how many slabs had to be replaced or could be repaired.

For the experimental investigation of the shear bearing capacity of the hollow core slabs a combination of photogrammetry, Acoustic Emission Analysis and section-wise curvature measurement was used to determine the beginning shear damage on a very low level during the test. The comparison of the results of the different measuring techniques allowed a clear identification of the damage processes and increased the information quality about the bearing condition of the slabs during the experimental investigation significantly.

The loading tests proved a sufficient shear bearing safety of the medium damaged slabs. The results of the tests on five slabs could be transferred to the whole population of slabs because sufficient margins of safety were considered in the calculation of the aim test load. With the results of the loading tests it was concluded how many slabs had to be replaced and which could further be used.

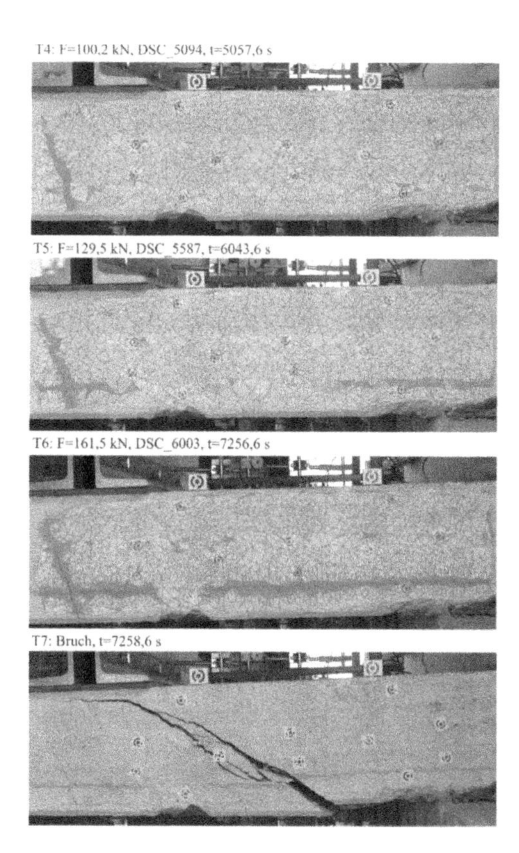

Figure 1. Photogrammetric evaluation of the shear failure process.

Concrete Repair, Rehabilitation and Retrofitting IV – Dehn et al. (Eds)
 ISBN 978-1-138-02843-2

Monitoring chloride concentrations in concrete by means of Ag/AgCl ion-selective electrodes

Yurena Seguí Femenias, Ueli M. Angst & Bernhard Elsener
Institute for Building Materials (IfB), ETH Zürich, Zurich, Switzerland

ABSTRACT

This work investigates the applicability of Ag/AgCl Ion-Selective Electrodes (ISEs) for the non-destructive measurement of chloride concentrations in concrete.

The Ag/AgCl ISEs exhibit a Nernstian electrochemical potential as a function of the chloride ion activity in solution (Koryta, 1972, Janata, 1989, Bard et al., 2002); thus, for concrete, the Ag/AgCl ISE allows measuring the free chloride content in the pore solution.

However, ISEs are also sensitive to other species that can form compounds normally of lower solubility with the constituent ion (Ag^+ for the Ag/AgCl ISE). The sensitivity in environments containing interfering species and the stability at the high pH characteristic of the concrete pore solution are thus important requirements that need to be fulfilled.

To illustrate the effect of the interfering species on the ISE response, the Ag/AgCl ISE potential-logarithm of the interfering specie activity plot can be schematically divided in three different regions, as shown in figure 1.

In zone A, the Ag/AgCl ISE behaves as an ideal chloride sensor; it exhibits a stable potential independent of the interfering specie (Angst et al., 2009, Koryta, 1972, Rhodes and Buck, 1980). It is in this range where the Ag/AgCl ISE is suitable for field measurements without interference.

When the concentration of the interfering specie increases (zone B in figure 1), the response of the ISE is altered and it shows a potential determined by the simultaneous action of the primary (chloride) and interfering ion. This interference is due to the replacement of the chloride by the interfering specie on the surface of the ISE (Rhodes and Buck, 1980, Koryta, 1972).

At sufficiently high concentrations of interfering specie, the ISE surface becomes totally covered by the salt formed between the silver and the interfering specie and acts as an ISE sensitive to this specie (Rhodes and Buck, 1980, Hulanicki and Lewenstam, 1977) (zone C in figure 1). The ISEs virtually exhibit Nernstian as a function of the activity of the interfering specie.

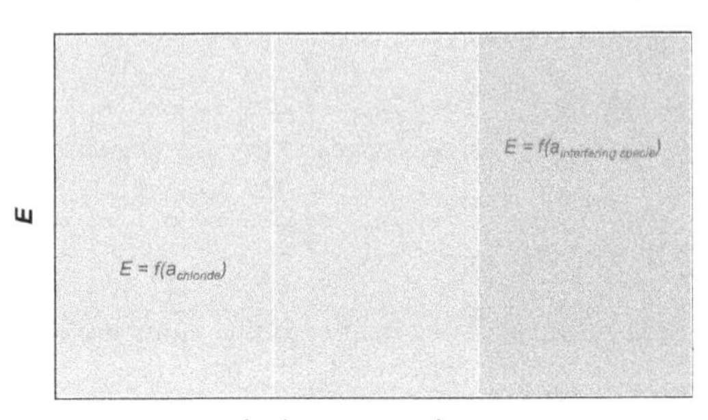

Figure 1. Schematic Ag/AgCl ISE potential as a function of the activity of the interfering specie (at room temperature).

The response of the Ag/AgCl ISE will be on one of the three zones of figure 1 depending on how severe is the interference and the experimental conditions (Koryta, 1972, Lindner and Umezawa, 2008, Hulanicki and Lewenstam, 1977, Rhodes and Buck, 1980).

The hydroxide interference has been treated in detail in this study. The results indicated that the Ag/AgCl ISEs potential is overall plateau (zone A in the graph) for a large range of hydroxide concentrations.

When it comes to chloride-induced corrosion, a concentration ratio chloride to hydroxide $c_{Cl^-}/c_{OH^-} = 0.6$ is sometimes considered as threshold value for corrosion initiation (Hausmann, 1967). From the results obtained, no interference is expected even for lower ratios.

Nonetheless, the instability of the Ag/AgCl ISE at high pH with no or low amount of chlorides has also been questioned (Elsener et al., 2003, Angst et al., 2009, Švegl et al., 2006). In chloride-free alkaline solutions, the ISEs were unstable over time – probably due to transformation reactions with the environment. Upon addition of chloride, however, the sensors responded again according the Nernst's law.

The effect of the possible interference arising from bromide, sulfate, and fluoride was also investigated, as these are the main species present in seawater that could cause interference. The results indicated negligible interference for fluoride and sulfate (zone A in figure 1) but relatively severe for bromide. Nevertheless, due the high chloride/bromide concentration ratio in seawater, the interference of bromide is considered negligible (zone A in figure 1) for applications related to seawater exposure.

The ions present in pore solution of hydrating slag cement are principally the same as those in Portland cement, with a remarkable amount of sulfide ion in the pore solution (Chen, 2006, Gruskovnjak et al., 2008, Lothenbach et al., 2012). The sulfide interference is thus also studied and discussed.

The results showed that sulfide interference is particularly severe; relatively small amounts of sulfide cause high potential shifts. Therefore, the applicability of the Ag/AgCl ISE in concretes with mid-high amounts of slag could be impaired. For the concretes containing mid-low amounts of blast furnace slag (CEM III/A), it could be possible that the amount of sulfide in solution remained low enough. However, the hydration of cement containing slag is not well understood (Gollop and Taylor, 1996, Taylor, 1997).

Based on the current experimental observations, it was concluded that the studied Ag/AgCl ISEs are feasible for practical monitoring of the chloride concentration in concrete exposed to chloride-containing environments. An exception is concrete with high contents of blast furnace slag, where the presence of sulfide could strongly disturb the measurements. For the case of low to mid content of blast furnace slag, further research regarding the kinetics and the hydration is needed to evaluate the severity of the sulfide interference.

Concrete Repair, Rehabilitation and Retrofitting IV – Dehn et al. (Eds)
© 2016 Taylor & Francis Group, London, ISBN 978-1-138-02843-2

Structural health monitoring of the Scherkondetalbrücke: A semi integral concrete railway bridge

S. Marx
Institute of Concrete Constructions, Leibniz Universität Hannover, Hannover, Germany

M. Wenner
Marx Krontal GmbH, Hannover, Germany

ABSTRACT

Semi integral bridges are characterized by the monolithic connection between the superstructure and the piers. Thereby every element of the structure participate to the transfer of forces due to traffic loads. Simultaneously, high constraining forces due to thermal variation, creep, shrinkage or settlements appears.

On the High-Speed Railway-Line between Erfurt and Leipzig (Germany), long railway viaducts have been designed and executed as semi integral structures for the first time in Germany. In order to verify the calculation assumptions and to collect experience about the long-term behavior of this bridge type, the authorities required an extensive monitoring of the bridges. In this paper, a few results of the measurements on the Scherkondetalbrücke (Figure 1) will be presented.

The Scherkondetalbrücke is a 576.5 m long viaduct. Nearly all the piers are monolithically connected to the superstructure. The developed monitoring concept follows two aims: (1) the measurement of the longitudinal deformations of the superstructure and (2) the measurement of the section curvature on the pier head.

The results of the last five years allow to draw first conclusions about the long-term behavior of the bridge.

The measurements of the bridge temperature and of the joint displacements permit to describe the thermal behavior of the superstructure. A thermal coefficient of $8.7 \cdot 10^{-6}$ $1/K$ has been determined. Measurement on several other concrete bridges shows that the thermal characteristics of concrete fluctuate between $8.0 \cdot 10^{-6}$ and $13.0 \cdot 10^{-6}$ $1/K$. These observations demonstrate the influence of the used aggregates and of the receipt of the concrete mixture on the material characteristics.

Figure 1. Scherkondetalbrücke, photograph: Ludolf Krontal.

The appearing time-delayed concrete deformations have been determined. The results show that the material models in the code provide a good approximation of the long-term development of creep and shrinkage strains. The seasonal development of these strains is obviously very sensitive to variation of environmental conditions as temperature and relative humidity. Comparative investigations showed that the material models in the norm are currently not able to explain this seasonal behavior.

The measured curvatures in the monolithic fixed piers permit to establish the long-term solicitation of the piers. The results confirm that the chosen geometry is optimal for the site topography. A comparison with the design assumptions indicates that the structure reacts "softer" than expected. This behavior is favorable for the structure in terms of constraint forces due to temperature, creep and shrinkage.

The performed monitoring over the last 5 years on the Scherkondetalbrücke delivered important data for the understanding of the behavior of long semi integral bridges and confirms that semi integral bridges are reliable structures. The experiences derived from the results allow to identify actual inaccuracies and to specify the design basics for future projects.

Concrete Repair, Rehabilitation and Retrofitting IV – Dehn et al. (Eds)
 ISBN 978-1-138-02843-2

Maintaining and monitoring durable cathodic protection systems applied on 30 concrete bridges with prestressing steel

R.N. ter Maten
Vogel Cathodic Protection, Zwijndrecht, The Netherlands

A.W.M. van den Hondel
Cathodic Protection Advice, Capelle aan den IJssel, The Netherlands

ABSTRACT

Cathodic Protection (CP) is a successful technique to guaranty structural integrity by which corrosion of reinforcement is inhibited to a negligible rate by lowering its potential. In a project named 'Liggerkoppen' (which translates to 'beam heads') 30 concrete bridges in the Netherlands are provided with a Impressed Current Cathodic Protection (ICCP) system. The project was executed in 2012 and 2013 in commissioned by the Dutch Highway Administration ('Rijkswaterstaat'; DHA).

While the long-term processes surrounding the maintenance and monitoring of the CP systems had to be secured, an innovative solution was established which included 20 years of maintenance and monitoring, which will be performed according to NEN-EN-ISO 12696. As a result of the multi-year contract the structures will (safely) be maintained by the contractor in collaboration with the client in the coming decades.

Due to chloride initiated corrosion, caused by leakage through the joints in combination with insufficient concrete cover, numerous beams were damaged and it was urgent to find a solution to stop further corrosion at short notice in order to maintain structural integrity. Distributed over the 30 bridges, a total of over 1.500 prestressed beam heads are provided with a conductive, graphite filled aluminosilicate, coating as anode system.

Each beam head is equipped with (2 or 3) reference electrodes for monitoring purposes, which results in a total of more than 3.000 reference electrodes. Near the prestressing steel a true reference electrode (silver chloride electrode) with a long lifetime expectancy was placed. Special care could therefore be taken for sufficient protection of the steel without overprotecting any part of the prestressing steel in order to avoid hydrogen embrittlement.

Due to the large number of beams per bridge, beam heads were grouped to zones with decentralized power supplies to limit the cabling. The CP systems are powered with solar power with a battery back-up with a up-time over 90% on a yearly basis. The CP-systems are, as tender documents specified, remotely controlled.

During a preliminary assessment of the possible risks and their effect on the function of the CP systems, key factors of possible failure where determined. These threats are either technical, which focus on CP-systems, or procedural, due to the organizational challenges for maintaining anything technically complex for a period of 20 years.

The procedural challenges are dealt with for special care was taken to avoid the loss of information on the side of the DHA in the transition from 'project' (execution phase) to 'process' (20 year maintenance). Evaluating the execution and transition phase, the assessed risks and threats proved to be valid. Without the attention demanded for the maintenance period following on the execution, this project would easily have become too complex to handle in the 'low attention' phase of monitoring. At this point, it has been rewarding to see a vital installation being taken care of to function as intended for decades to come.

The technical threads were dealt with in specifying minimum demands for anode durability, technical provisions for control, and integrated quality control procedures during all phases of the execution

Figure 1. As-built CP system with conductive coating and decentralized power supply and monitoring units.

Figure 2. Central measuring and control unit and solar panels.

and maintenance period. Specific demands where made on the long term performance of the CP system and the monitoring therefore involves at least two depolarisation measurements per year and an annual visual inspection, but with special emphasis on large scale evaluations in year 4, 9 en 14 of the total project.

The status of the zones are frequently monitored, confirming the good operation of the hardware, sufficient protection of the reinforcing steel and the absence of overprotection at the prestressing steel. This results in large sets of data (48 million during the total maintenance period) which have to be processed and evaluated.

The acquisition and processing of the data is challenging and demands automated procedures. Based on automatic processes on a yearly basis all general results are published and statistically treated. Specific measurements that could indicate problems, i.e. failure, insufficient protection or over-protection, are reported without delay on an individual basis. Acceptable outcomes (within bandwidths) are specified in the software to reduce the workload without risking neglect or ignorance of important results.

The results of the monitoring during the first two years are promising. All bridges and beam heads show sufficient protection, while no risk of over-protection was observed for this first period. Typically the initial current density at start up of the zone is very high but drops rapidly to normal values. At the same time this current output requires a steady increase of applied cell voltage (cathode to anode) from an initial 2 volts to values in the range of 3–7 volts. It can be concluded that the apparent system resistance increases.

Initial measurements of depolarisation, to prove sufficient protection, show high depolarisations which are well over the minimum requirement of 100 mV with little variation over beams with active

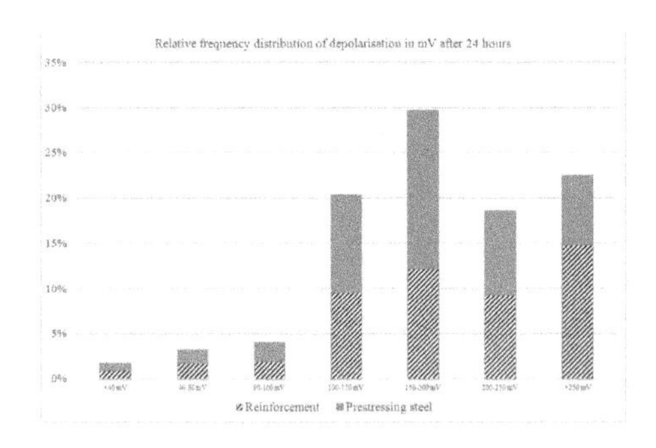

Figure 3. Overall data for frequencies (for degree) of depolarization.

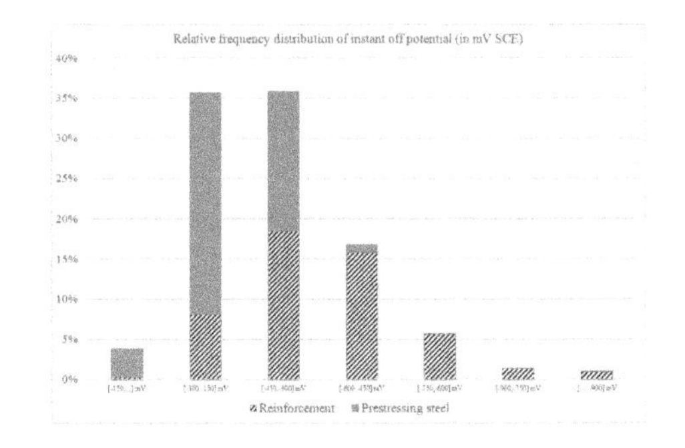

Figure 4. Overall data for frequencies exclusion of hydrogen embrittlement.

corroding steel and beams included in the project where the protection is more or less preventive. Over time this tends to reduce to values from 100–250 mV depolarisation, with no structural difference between protected and prevented sides. This might be due to the repassivation of the active corroding steel, resulting in comparable micro-environments in all beams. The difference between the monitored mild steel reinforcement and the monitored deep locations near the prestressing steel is present at start up, but tends to disappear over time.

During visual inspections several minor defects from the execution and the maintenance period where detected and dealt with. There are no signs of aging, as expected during this relatively short period, but some evaluated risks have proven to be valid. All observed problems proved to be self-explanatory, were not related to the CP technique itself, and were unproblematic to deal with.

With the application of the CP system the corrosion of the reinforcement is inhibited to a negligible rate by lowering the potential of the reinforcement. The result is that further degradation and the change of possible failure of the mild and prestressed steel is removed and the concrete repairs will endure much longer than without a CP system.

Concrete Repair, Rehabilitation and Retrofitting IV – Dehn et al. (Eds)
 ISBN 978-1-138-02843-2

Case studies in the practical application of pulse echo technology

D. Corbett
Proceq SA, Switzerland

ABSTRACT

One of the key areas of development in Non-Destructive Testing (NDT) of concrete in recent years has been in imaging using tomography methods. There is a real need to see the internal structure of the concrete element, be it the reinforcement structure, the location of post-tensioning cables, the presence and extent of delaminations or honeycombing. Ground penetrating radar, eddy current and impact echo instruments have all been used for this purpose. Each technology has its advantages and disadvantages, yet none of them, used independently can provide all of the information required. A combination of instrumentation is necessary to build up a more complete image of the internal structure of the concrete element. Pulse echo technology is an additional string to the bow of the NDT specialist.

Although it has been around for several years, the use of ultrasonic pulse echo technology for the assessment of concrete structures is not widely understood. This paper will demonstrate the practical application of this technology by means of a number of real case studies. It will show what information can be uniquely obtained from a structure with this method and will also show how it is complementary to other non-destructive measurement techniques. It will also demonstrate the limits of the technology as it is equally important to understand what it cannot do.

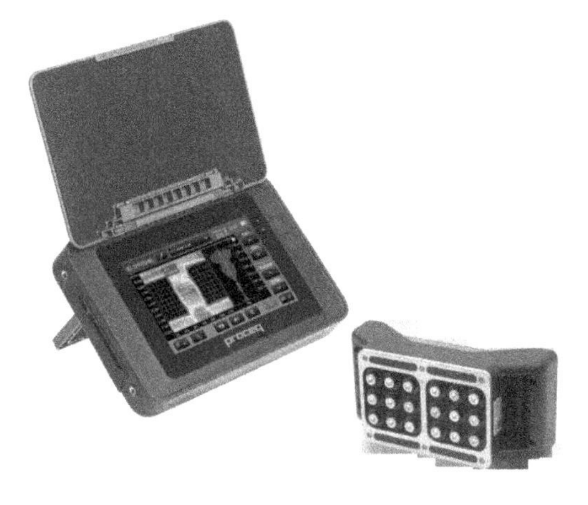

Figure 1. Pundit PL-200PE Pulse echo instrument from Proceq.

Concrete Repair, Rehabilitation and Retrofitting IV – Dehn et al. (Eds)
© 2016 Taylor & Francis Group, London, ISBN 978-1-138-02843-2

Corrosion rate measurements in concrete—a closer look at the linear polarization resistance method

U.M. Angst & M. Büchler
Swiss Society for Corrosion Protection (SGK), Zurich, Switzerland

ABSTRACT

The rate at which steel corrosion occurs in reinforced concrete is an essential parameter for assessing the residual service life of a structure and thus for maintenance planning. However, determining corrosion rates reliably, particularly on site, is not straightforward. In principle, there exist a variety of experimental methods to determine the corrosion rate such as electrochemical procedures, gravimetric (weight loss) measurements or other approaches (McCafferty, 2010). For reinforcement steel embedded in concrete, only electrochemical methods permit *non-destructive* measurements of instantaneous corrosion rates. These are typically based on applying an external polarization current to excite the system and recording the system response. The by far most common method is the so-called *Linear Polarization Resistance* (LPR) method.

This paper discusses the theoretical background of this method and focuses particularly on the severe discrepancy between theory and the conditions in practice when the method is applied to localized corrosion, which is the typical morphology in the case of chloride-induced reinforcement corrosion. Some of the considerations are highlighted with selected experimental results that were obtained in a recent research project performed at the Swiss Society for Corrosion Protection.

The linear polarization resistance method relies on the Stern-Geary equation (Stern and Geary, 1957) that states an inverse proportionality between the corrosion current, I_{corr}, and the so-called polarization resistance R_p. The Stern-Geary equation was derived on the basis of the *mixed potential theory*, which was postulated in the first half of the last century by Wagner and Traud (Wagner and Traud, 1938). One fundamental aspect of the mixed potential theory is that the corrosion morphology is uniform, i.e. it is assumed that cathodic and anodic partial reactions occur at microscopically small locations that are uniformly and randomly (with time) distributed across the metal surface. This assumption permitted to establish the concept of the *mixed potential*.

In many practical situations, the corrosion morphology is, however, not uniform. This is particularly true for chloride-induced reinforcement corrosion, where local anodes (corrosion pits) interact galvanically with typically much larger cathodic zones (Bertolini et al., 2004). Due to the finite conductivity of the electrolyte (the alkaline solution in the pore system of the concrete), the ohmic drop in the galvanic cell is not negligible and thus anodic and cathodic sites have *different potentials*. The direct consequence of this is that a well-defined, unique mixed potential in the sense of the theory by Wagner and Traud does for the case of macro-cell corrosion—or localized corrosion in general—not exist.

Upon polarization of a macro-cell from an external counter electrode (no matter where this counter electrode is located), any change in potential that can be recorded, ΔE, is dependent on the location of the reference electrode. This is apparent from the experimental measurement results shown in Fig. 1. A macro-cell formed between a local anode (1 cm^2) and a large cathodic reinforcement steel mesh (ca. 1 m^2 in size) was polarized with a galvanostatic current pulse (60 μA) injected from a small counter electrode located close to the anode,

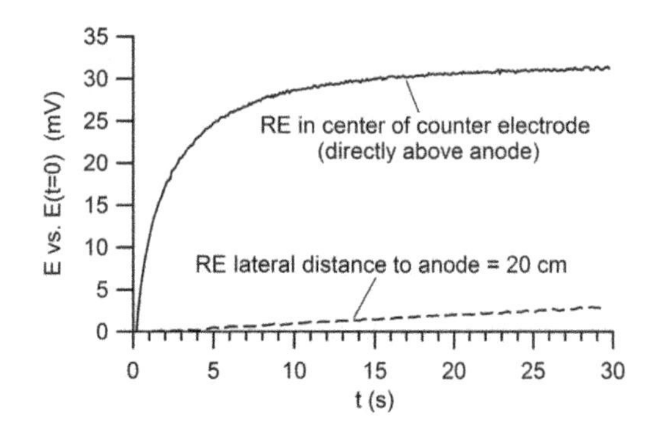

Figure 1. Potential transients during galvanostatic polarization of a macro-cell, recorded with Reference Electrodes (RE) posi-tioned at two different locations (initial IR jump removed). For more details about the geometry of the setup it is referred to the full paper.

and the potential was recorded over time with two reference electrodes. The observed transients are clearly different and different steady state polarization values ΔE are reached (Fig. 1). The essential question now is: which of the two ΔE is the correct one? Since $R_p = \Delta E/\Delta I$, also the polarization resistance is a function of the position of the reference electrode. The fundamental difficulty is that *there exists no correct position for the reference electrode* in such a situation and as a result, there is no correct ΔE to be used for computing R_p. In short, owing to the absence of a well-defined, unique mixed potential, the Stern-Geary relationship is in the case of macro-cell corrosion fundamentally not valid.

Nevertheless, several laboratory investigations have shown a certain agreement between the LPR method and other corrosion rate measurements. Whereas some of the studies led to the conclusion that the LPR method underestimates the true corrosion rate, by factors ranging from 2 to 10, other authors concluded that the LPR method overestimates the true corrosion rate, by factors ranging from 1.5 to 5 (Andrade and González, 1978) (Law et al., 2004) (Liu and Weyers, 2003) (Nygaard et al., 2009). Despite these over- and underestimations of the true corrosion rate, there seems to be some sort of literature agreement on the practical applicability of the LPR method also to localized corrosion. One may thus conclude that despite the fundamental difficulties with the method for the case of localized corrosion, an inversely proportional relationship between the corrosion rate and an apparent "polarization resistance" can still be observed. It is, however, important to note that due to the absence of theoretical grounds, this relationship is entirely empirical.

In a recent publication by the present authors, experimental measurements and theoretical considerations allowed to explain the reasons for this empirical agreement (Angst and Büchler, (in press)). In summary, although the Stern-Geary equation is in the case of macro-cell corrosion fundamentally not valid, two compensating errors cause the result to be close to correct. The extent to which this is possible, however, is strongly dependent on the actual conditions, such as the geometrical configuration, the kinetics of the partial reactions, etc.

In the full paper, also the issue of "current confinement" is discussed. Experimental results of the current distribution between cathodic and anodic parts of a macro-cell during external polarization indicated that the polarization current flows predominantly through the cathodic metal surfaces. It has in the past been attempted to counteract this by "current confinement" and "guard ring" approaches. However, it is in the full paper argued that the lateral current spread-out should not be considered as a problem that needs to be eliminated, but that it is in fact one of the reasons for the empirical "apparent applicability" of the LPR method to macro-cell corrosion.

Finally, further research work should concentrate on characterizing in more detail the conditions (e.g. geometrical configurations) under which the empirical relationship between corrosion rate and apparent polarization resistance still holds. Alternatively, successful research attempts to develop a measurement method for macro-cell corrosion rates that eliminates the shortcomings of the LPR method, e.g. by avoiding the need to rely on the Stern-Geary relationship, would present a major advance for science and engineering.

REFERENCES

Andrade, C. & González, J. A. 1978. Quantitative measurements of corrosion rate of reinforcing steels embedded in concrete using polarization resistance measurements. *Materials and Corrosion*, 29, 515–519.

Angst, U. & Büchler, M. (in press). On the applicability of the Stern-Geary relationship to determine instantaneous corrosion rates in macro-cell corrosion. *Materials and Corrosion*.

Bertolini, L., Elsener, B., Pedeferri, P. & Polder, R. 2004. *Corrosion of steel in concrete*, Weinheim, WILEY VCH.

Law, D. W., Cairns, J., Millard, S. G. & Bungey, J. H. 2004. Measurement of loss of steel from reinforcing bars in concrete using linear polarisation resistance measurements. *NDT&E International*, 37, 381–388.

Liu, Y. & Weyers, E. W. 2003. Comparison of guarded and unguarded linear polarization CCD devices with weight loss measurements. *Cement and Concrete Research*, 33, 1093–1101.

Mccafferty, E. 2010. *Introduction to Corrosion Science*, New York, Springer.

Nygaard, P. V., Geiker, M. R. & Elsener, B. 2009. Corrosion rate of steel in concrete: evaluation of confinement techniques for on-site corrosion rate measurements. *Materials and Structures*, 42, 1059–1076.

Stern, M. & Geary, A. L. 1957. Electrochemical polarization. I. A theoretical analysis of the shape of polarization curves. *Journal of the Electrochemical Society*, 104, 56–63.

Wagner, C. & Traud, W. E. 1938. Über die Deutung von Korrosionsvorgängen durch Überlagerung von elektrochemischen Teilvorgängen und über die Potentialbildung an Mischelektroden. *Z. Elektrochem. Angew. Phys. Chem.*, 44, 391–402.

Concrete Repair, Rehabilitation and Retrofitting IV – Dehn et al. (Eds)
© 2016 Taylor & Francis Group, London, ISBN 978-1-138-02843-2

Impact of chloride redistribution on the service life of repaired concrete structural elements

A. Rahimi, T. Reschke & A. Westendarp
Federal Waterways Engineering and Research Institute (BAW), Karlsruhe, Germany

C. Gehlen
Centre for Building Materials (cbm), Technische Universität München, Munich, Germany

ABSTRACT

This paper presents a simplified probabilistic approach to estimating the residual service life of concrete structural elements after repair. The repairs consist of replacing the concrete cover either partially or entirely with a cement-based repair material. In particular, the presence of a residual contamination in the remaining concrete layer and how the redistribution of the residual chlorides affects the determination of the service life are considered.

Partial removal of the concrete cover may be appropriate for economic, execution- or construction-related reasons or also as a preventive measure etc. In such cases, it must be asked whether the remaining concrete layer has to be free from chlorides or to what extent it may still be contaminated with chlorides and how the residual contamination will affect the residual service life of the concrete structural element after repair (i.e. after the concrete cover has been partially removed and replaced with a repair material).

As shown in Figure 1, the residual chloride gradient will be redistributed through both the new layer and the remaining layer over time. The chloride concentration at the rebar surface changes as a result and may at times exceed the initial value (even without the additional ingress of external chlorides). If the repaired structural element is again exposed to chlorides, the residual contamination will generally result in the chloride content at which corrosion is initiated being exceeded at the rebar surface at an earlier point in time.

An approach for estimating the residual service life of repaired structural elements with residual contamination was presented in *Rahimi et al. (2014)* and was based on the calculation of the service life of new structures. However, the chloride content at the rebar surface was the only element of the residual contamination that was taken into consideration. The chloride gradient and its redistribution were disregarded. In this paper, it is first discussed whether, and if so to what extent, disregarding the chloride redistribution affects the estimation of the residual service life of concrete structural elements.

To facilitate understanding of the paper, the methodology of the performance-based probabilistic service life design of new structures is briefly explained as it will be referred to later on in this paper when its application to existing structures is discussed.

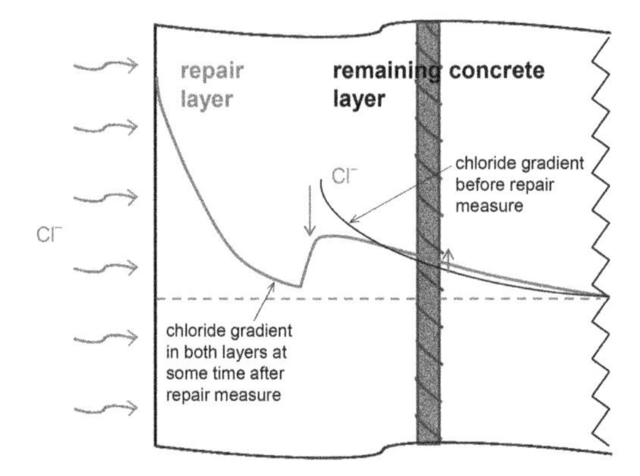

Figure 1. Chloride ingress and redistribution in a repaired concrete element (2-layer system) with residual chlorides.

REFERENCES

Rahimi A., Gehlen C., Reschke T., Westendarp A. 2014. Approaches for Modelling the Residual Service Life of Marine Concrete Structures. International Journal of Corrosion 1/2014, http://www.hindawi.com/journals/ijc/2014/432472/abs/.

Concrete Repair, Rehabilitation and Retrofitting IV – Dehn et al. (Eds)
© 2016 Taylor & Francis Group, London, ISBN 978-1-138-02843-2

Relationships between defects and inventory data of RC bridges and culverts in the Western Cape, South Africa

T.D. Mbanjwa & P. Moyo
Department of Civil Engineering, University of Cape Town, South Africa

ABSTRACT

Bridges and Culverts form critical links within transportation networks. They are expensive long term infrastructure investments and are therefore essential to a country's economic functioning and society's everyday lives (Tang, 2007). They are prone to deterioration due to unfavourable chemical, mechanical and physical consequences (Scharven, Hartmann & Dewulf, 2011). Nonetheless, they are expected to remain safe and functional for the duration of their design life. Thus bridge management systems have been fundamental to maintaining and preserving these (Nordengen, and Nell, 2005). However, the data they contain is solely used for structural prioritization. Hence, the purpose of the study was to utilize the Struman Bridge Management System database to investigate the relationships that exist between the inventory and inspection data of the RC bridges and RC culverts within the Western Cape Province. The study involved conducting data mining activities to simplify the data and extract useful information. These activities encompassed investigating structural defects, their predominance and spatial distributions (in terms of district municipalities) as well as their relationships with inventory data such as structure type, age and span length/width. STATA logit models were used to investigate whether the relationships were statistically significant and to determine odds ratios of several 'events'. In addition to this, the average CIs were also investigated to assess the condition of the structures relative to the inventory data. The results indicated that RC bridges and RC culverts had similar defects and consequently similar forms of deterioration. Several deterioration mechanisms and causes of defects were identified. These included chloride ingress, traffic loading, undermining and several others. There were no specific spatial distributions identified along the coast or inland as the prevalence of the predominant defects varied for the different district municipalities. Also, location (in terms of district municipalities) gave more meaningful results when considered with the mean age and the standard deviation of the age of the RC structures in that location. There were no statistically significant relationships identified between the predominant defects and the various structure types. However, findings from data mining suggested that the most dominant structure types had higher percentages of RC structures with defects and were in the worst condition. Moreover, there was a general increase in RC structures with predominant defects with increasing age categories as well as increasing span length/width. Also, the average structural condition showed a general decrease with age and span length/width. Nonetheless, the average CIs of the RC structures suggested that their average overall condition was good. Lastly, the study makes recommendations pertaining to the collection of BMS data, the analyses BMS software should be used to conduct, the parameters CI indicators should take into consideration and further research that would be a continuation of the study. more complicated than previously considered and has important implications for upland water quality.

Concrete Repair, Rehabilitation and Retrofitting IV – Dehn et al. (Eds)
 ISBN 978-1-138-02843-2

Safety assurance of problematic concrete bridges by automated SHM: Case studies

K. Islami & N. Meng
Mageba SA, Bülach, Switzerland

ABSTRACT

Structural Health Monitoring (SHM) is currently being more and more applied, all around the world, for the analysis of bridges and buildings. It greatly improves the knowledge available about structure's behavior and condition, both qualitatively and quantitatively, reducing uncertainties and increasing structural safety. This can have many uses, such as enabling inspection and maintenance programs to be optimized.

This paper shall present various recently designed and installed SHM systems. The diverse range of applications, designed in collaboration with structure owners and design engineers, includes concrete long-span box girder bridges and concrete viaducts. The systems are assigned to fulfil 4 major tasks such as monitoring of environmental loads (temperature, humidity), monitoring of the crack behavior, settlements, monitoring of strains, displacement (Fig. 3) and vibrations under rail and road traffic.

These case studies are excellent examples of ongoing monitoring activities, based on both static and dynamic approaches. As well as presenting the application of innovative monitoring techniques, the case studies demonstrate statistical models and data processing procedures that have been developed and applied in order to eliminate uncertainties and maximize the benefits gained from monitoring data. The recently installed automated SHM systems shall demonstrate their usefulness and ease of use, and the enormous gains in efficiency they offer over alternative manual monitoring methods.

Figure 1. Views of the monitored structures of the Hunter Expressway.

All of these techniques have been used in the present study, and more importantly, have been applied to important concrete bridges that were in danger of settlements (Fig. 1), deterioration (Fig. 2) or under modification of structural schemes.

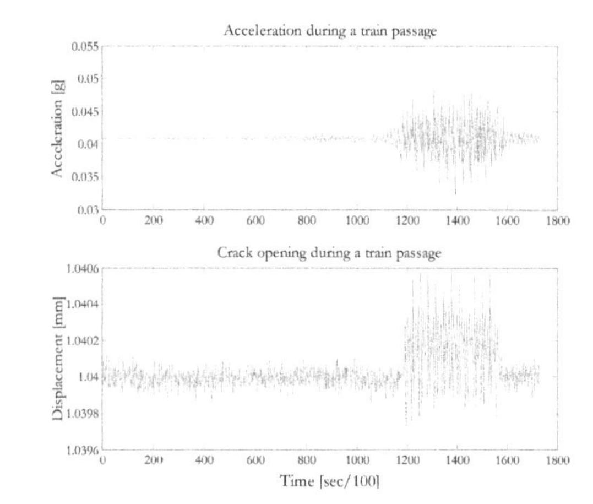

Figure 2. Train passage effects, in close view, on the accelerations and on the crack opening of the bridge.

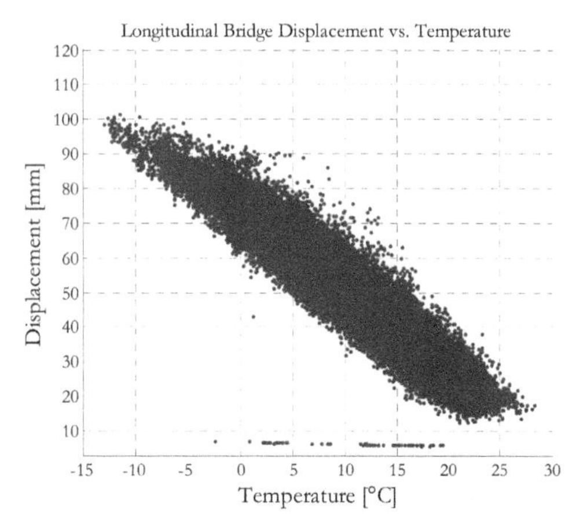

Figure 3. Relation between displacement and temperature measurements at Ponte Nanin.

Additionally, automated monitoring of structures does indeed have much to offer right throughout a bridge's life-cycle. They can provide continuous, real-time records of almost any variable in a bridge's condition, analyze the data gathered, present it in tabular or graphic format, and make it available to an authorized user anywhere in the world via the internet. Such systems can also be used, for example, to provide structural engineering or bridge usage data, or immediate notification of the reaching of predefined values of any measured variable. The examples of the use of such a system at various stages illustrate this—starting with a case study from the very start of a bridge's life.

Concrete Repair, Rehabilitation and Retrofitting IV – Dehn et al. (Eds)
 ISBN 978-1-138-02843-2

The Tannery bridge: A case study in structural health monitoring and rehabilitation of structures

L. Tassinari & Jérôme Sordet
Kung et Associés, Payerne, Switzerland

M. Viviani
HEIG-VD HES-SO, Yverdon-Les_Bains, Switzerland

ABSTRACT

Evaluating the structural health of a bridge is a complex activity. Small and medium bridges constructed in urban areas often undergo to many changes during their service life. Partial reconstruction, increased traffic loads, a new structural schema are some of the changes occurred in the Tannery Bridge (Switzerland) over time. These changes were only partially documented. Therefore it was necessary to understand the structure and its behavior prior to assessing its load carrying capacity and the pathologies, before proposing a refurbishment strategy. Furthermore, some important elements were documented but not constructed as stated in the plans. No transition slabs were ever planned and the ramps constructed to access the bridge were rapidly deforming due to the low quality of the foundation soil. Although the concrete deck is not in critical conditions, some degradation is visible and rapidly progressing. The curbs and barrier types obliged the engineers to plan the demolition and to change these important details. This article presents the analyses and the design process for the rehabilitation of the Tannery Bridge, an urban bridge that, in about 5 decades, was completely changed and resisted to unexpected loads such a number of heavy military tanks. Furthermore it will be shown how, sometimes, minor problems may become the real challenge, at least from the financial point of view

The "Tannery bridge" is a key infrastructure of the fast growing town of La Sarraz (about 42% population growth between 2006 and 2011). An ancient truss steel arch bridge built in 1887 was deeply modified in the 1976.

The 1976 bridge is a mixed structure with two simply supported steel beams, spanning 51.3 m. The concrete deck is 9.5 m wide (see fig. 1).

Some of the old features of the arc bridge remain visible: the abutments in stone masonry and the traces of the arch hinges. The tannery located under the bridge, is today an industrial laundry service facility. Two major events have brought this bridge to the attention of the authorities and of the engineers: the new Swiss code for roads, that today allows the transit of vehicles weighting up to 40 tons (previous limit was 28 tons) and the ownership of the bridge that passed from the state (canton) to the city of La Sarraz. In 2010, the need of reinforcement was evaluated: after calculations, the engineers stated that no adaptation was necessary and the bridge can stand the new loads as it is. However a complete analyses of the old project revealed some discrepancies on the thickness of the web of the steel beams. Although the bridge had undergone to severe loads, it is of capital importance that all details match. These details will be clarified when the work on the slab of the Ferreyres abutment will begin. If the real thickness will be even lower that the minimum found on the projects, this would lead to a new verification of the bearing capacity of the beams under the load cases provided by the norms.

Figure 1. The Tannery bridge today.

Concrete Repair, Rehabilitation and Retrofitting IV – Dehn et al. (Eds)
© 2016 Taylor & Francis Group, London, ISBN 978-1-138-02843-2

Evaluation of concrete structures durability under risk of carbonation and chloride corrosion

L. Czarnecki
Building Research Institute, Warsaw, Poland

P. Woyciechowski
Warsaw University of Technology, Warsaw, Poland

ABSTRACT

The mechanism of reinforced concrete structures degradation under the risk of carbonation and chloride corrosion and its relation to the life cycle of the structure are shown. The models of carbonation progress and chloride migration in concrete are analyzed. Preferences of hyperbolic model of carbonation and the special role of its asymptote as the limit defining the depth of carbonation [1] have been shown (fig. 1). Conditions for the initiation of corrosion in concrete elements with particular attention to the critical chloride concentration [2] are defined. Various methods of determining the distribution profile of chlorides in concrete (fig. 2) with particular emphasis on the diffusion coefficient of chloride are discussed.

Four possible procedures of determining apparent diffusion coefficient are shown On this basis an algorithm for predicting the durability of concrete structures under conditions of potential corrosion has been elaborated.

It is assumed that the durability of structures in corrosive environments is adequate to reaching the front of carbonation and chlorides the zone of reinforcement. From the moment of inhibition of rebar corrosion the further structure life time is determined by the steel corrosion rate. A concrete crack inhibition occurs due to the overgrowth of corrosion products of rebars; a consequences are the loss of concrete adhesion to reinforcement and the propagation of crack in the concrete cover. Subsequently the reinforcement cross-section area is reduced down to the loss of the bearing capacity. This assumption is also justified by practical considerations, as it often marks the end of the possibility of the repair action.

It is obvious in the light of presented study, that the moment of initiation of steel corrosion is simultaneously the start moment of the occurrence of structure duration threat. Two models of estimation of steel corrosion initial moments have been presented separately in case of carbonation and chloride corrosion risk. The structure residual life time and expected reduction of durability under the given conditions could be evaluated with use of author`s algorithm.

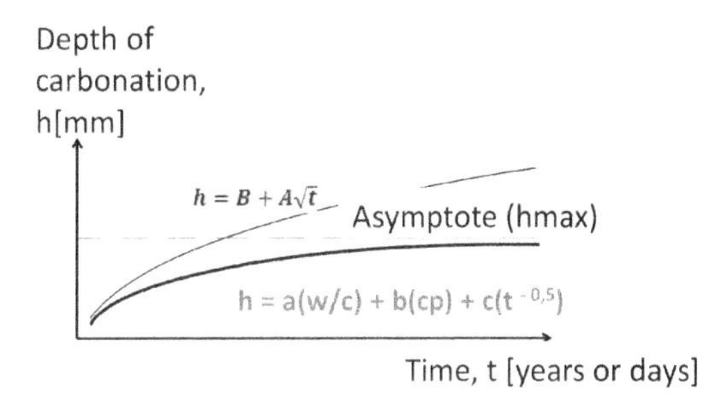

Figure 1. Sample of a figure caption.

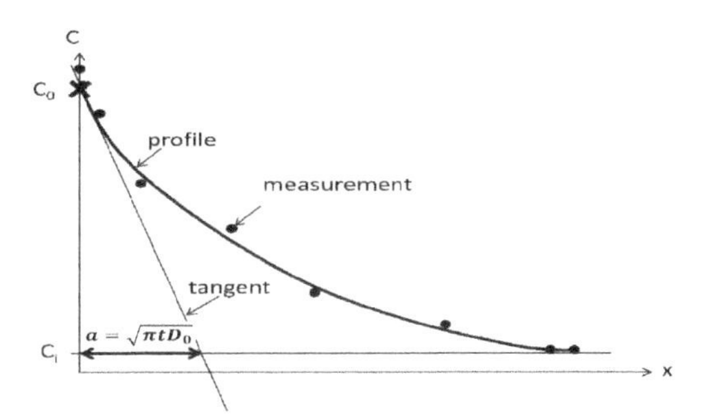

Figure 2. Determination of chloride diffusion coefficient.

REFERENCES

[1] Czarnecki L., Woyciechowski P., Concrete carbonation as a limited process and its relevance to CO_2 sequestration, *ACI Materials Journal* vol. 109, No 3, 2012, 275–282.

[2] Czarnecki L., Woyciechowski P., Application of self-terminating carbonation model and chloride diffusion model to predict reinforced concrete construction durability, *Bulletin of the Polish Academy of Science—Technical Sciences*, vol. 61, No 1, 2013, 173–181.

Concrete Repair, Rehabilitation and Retrofitting IV – Dehn et al. (Eds)
© 2016 Taylor & Francis Group, London, ISBN 978-1-138-02843-2

Using existing inspection data to probabilistically estimate the time-to-rehabilitation for concrete bridges exposed to deicing salts and humidity

Filippos Alogdianakis, Dimos C. Charmpis & Ioannis Balafas
Department of Civil and Environmental Engineering, University of Cyprus, Nicosia, Cyprus

ABSTRACT

During the last decades vast budgets have been spent to maintain aging infrastructures all over the world. The 20th century's increased construction activity has led, in particular, to the accumulation of bridges, forming stocks, which are deteriorating with age. Aging bridges need continuous interventions either in the form of maintenance or major rehabilitation.

In order to estimate future needs and optimally allocate available budgets, models to probabilistically predict the future bridge stock condition are essential. Various deterioration models exist and vary from simple linear regression models (Fitzpatrick et al. 1981) to much more involved Markov chain models (Madanat et al. 1995, Ng & Moses 1996), as well as newer approaches employing e.g. Artificial Neural Networks. These deterioration models use previous years' structural condition data to trace condition changes and estimate the probability of transition from a condition rating to another.

In the present work, a novel method is presented to probabilistically estimate the future structural condition of a bridge using the condition ratings of just one year. For this purpose, real data maintained by the Federal Highway Administration (FHWA) for USA bridges are exploited. The National Bridge Inventory (NBI) of FHWA, which is updated annually, contains a considerable amount of information and describes the structural condition of over 500,000 bridges (FHWA 1995). Bridge condition ratings are recorded on a scale 0–9, with 9 representing 'perfect' and 0 'failed' conditions. Condition coded with 5 ('fair' condition) can be regarded as a threshold for rehabilitation. Rehabilitating a bridge in that condition is important, before further deteriorating to a 'poor' (code 4), 'serious' (code 3), 'critical' (code 2), 'imminent failure' (code 1) or 'failed' (code 0) condition.

The novel method proposed herein utilizes the NBI bridge condition data of a single year to probabilistically estimate the time needed for a bridge to reach condition 5. Initially, the probability for a bridge of a certain age to be in a specific structural condition (0–9) is calculated. This is accomplished by categorizing bridges in age groups forming sub-stocks of the global bridge stock; thus, the condition probability of each sub-stock is independently calculated. To separate the bridge ages with acceptable variation ranges of calculated condition probabilities, a threshold age t_{cut} ('cutting age') is then determined to define the 'trusted' part of the overall bridge sample. Probabilities corresponding to bridge sub-stocks older than t_{cut} are considered to have unacceptably high variation ranges and are therefore 'untrusted'. Based on the 'trusted' data, the residual time to condition 5 is estimated by performing time-shifts to future years from condition distributions >5. While shifting, the time axis is scaled appropriately. Weibull distribution functions are finally fitted to the original and shifted/scaled data using non-linear regression to evaluate condition probabilities and estimate the time-to-rehabilitation.

A specific case study was considered to facilitate the description of the proposed method. The focus of this work is to study bridge deterioration with age, e.g. due to corrosion. Deicing salts are known to accelerate deterioration especially in combination with humidity (Alogdianakis et al. 2014). Hence, simply supported concrete bridges located at areas of USA using deicing salts have been identified. To ensure the presence of humidity, the sample was limited to bridges with water passing underneath the structure. After certain exclusions (bridges at high-seismicity areas, rehabilitated bridges, etc. have been ruled out), the final sample analysed counted 33,810 bridges.

The Weibull Cumulative Distribution Function (CDF) $F(t)$ for bridge condition ≤5 (Fig. 1) provides the sought probabilistic information for the time-to-rehabilitation of a concrete bridge exposed to deicing salts and humidity: by selecting a probability of reaching bridge condition ≤5, the respective time-to-rehabilitation for the bridge is determined. Table 1 gives the time-to-rehabilitation for selected probability-values.

The selection of a cutting age t_{cut} is a subjective decision, which specifies the trusted part of the available data ($t \le t_{cut}$), but also creates the need to predict the condition probabilities for $t > t_{cut}$. The effectiveness of the proposed method can be

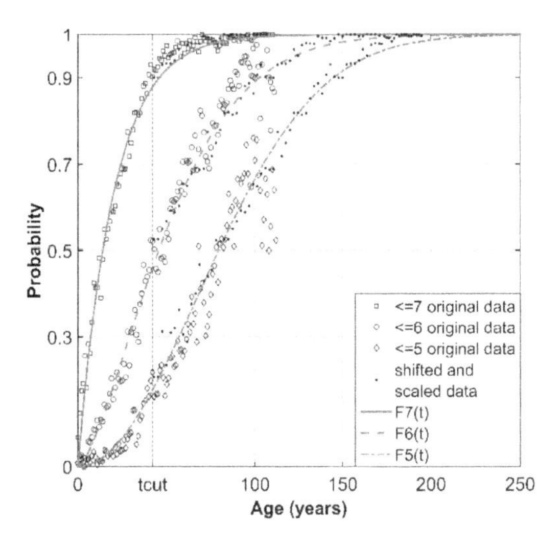

Figure 1. Weibull distributions fitted to the sample data assuming t_{cut} = 42 years. Original trusted and 'untrusted' data (before shifting/scaling), as well as shifted/scaled data, are shown for comparison.

Table 1. Time-to-rehabilitation for selected probabilities of a bridge to have reached condition ≤5.

Probability to reach condition ≤5	Time-to-rehabilitation (years)
30%	57
50%	79
70%	105
90%	145

statistically assessed with respect to the selected value of t_{cut}. Hence, Weibull CDFs are fitted to the same data using various t_{cut}-values. Then, the method's predictions for $t > t_{cut}$ are compared with the respective recorded data, which were excluded from the analysis.

Figure 2 presents time-to-rehabilitation results in the form of boxplots for various pre-selected $F(t)$-values (from 0.01% to 99.9%). Hence, for each $F(t)$-value, several t_{cut}-values are considered and the resulting variation of the bridge age to reach condition ≤5 is illustrated. It appears that a higher prediction variation is associated with a higher probability $F(t)$. Nevertheless, the selected t_{cut}-value does not seem to excessively influence the calculated age predictions.

In general, reasonable deviations between original data and data predicted with the new method

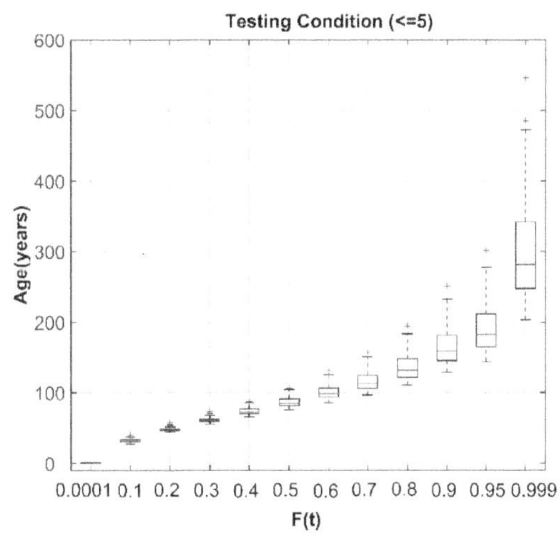

Figure 2. Boxplots for each $F(t)$-value and for all t_{cut}-values considered for bridge condition ≤5.

were observed for a large number of test configurations considered. The procedure followed herein can be applied to any type of bridges exposed to various environmental effects, provided that respective data are available. Enriching our knowledge on the deterioration rate of bridge stocks is a key aspect in cost-effectively handling and maintaining such important infrastructures.

REFERENCES

Alogdianakis, F., Balafas, I., Charmpis, D.C. 2014. Environ-mental effects on bridges—Statistical durability study based on existing inspection data. *2nd International Con-ference on Innovations on Bridges and Soil-Bridge Inter-action (IBSBI 2014)*, Athens, Greece, pp. 161–168.

Federal Highway Administration (FHWA). 1995. *Recording and coding guide for the structure inventory and appraisal of the nation's bridges*, Report No. FHWA-PD-96-001, Washington D.C.: U.S. Department of Transportation.

Fitzpatrick, M.W., Law, D.A., Dixon W.C. 1981. Deterioration of New York State highway structures. *60th Annual Meeting of the Transportation Research Board*, Washington D.C., USA, pp. 1–8.

Madanat, S., Mishalani, R., Ibrahim, W.H.W. 1995. Estimation of infrastructure transition probabilities from condition rating data. *ASCE Journal of Infrastructure Systems* 1(2):120–125.

Ng, S.K., Moses, F. 1996. Prediction of bridge service life using time-dependent reliability analysis. *3rd International Conference on Bridge Management*, University of Surrey, Guildford, UK, pp. 26–33.

Concrete Repair, Rehabilitation and Retrofitting IV – Dehn et al. (Eds)
© 2016 Taylor & Francis Group, London, ISBN 978-1-138-02843-2

Orphan sustainable sensor system for monitoring chlorides or CO_2 ingress in reinforced concrete

F. Barberon & P. Gegout
Pôle Ingénierie Matériaux, Bouygues Travaux Publics, France

ABSTRACT

Depassivation and corrosion of steel bars are the most severe threat of degradation for reinforced concrete elements. Actual available monitoring systems require contact or wired instrumentation, which can lead to delay in the measurement of the penetration front or create preferential path for the ingress of the deleterious agents.

The present system is designed to anticipate the corrosion of reinforcement and consists of isolated, wireless and embedded steel sheets which dimension are about 100 µm thick and few centimeters diameter. The sensors are placed into the freshly poured concrete and positioned at different depth between the surface and the first raw of steels. The sheets are oriented perpendicularly to the diffusion front. When in contact with chlorides or CO_2 the sheets are oxidized in the same manner as the reinforcement.

The principle is to use the thermal properties of iron to check its oxidation state. When the metallic iron is subjected to a variable magnetic induction, a current is created at the surfaces (Lenz's law) with a sudden and important raise of temperature of the iron sheet (Ohm's law).

The embedded sheets are warmed-up by induction from the surface of the concrete element. The thermal response is then measured with an infrared reader, from the surface of the concrete, and the signal analyzed to assess the state of oxidation. Indeed, dissimilar to a sound sheet, the oxidized one will provide with a different thermal response, which signal is of low amplitude and rapid decay. The position and the depth of the sheets being previously located in the structure, it is thus possible to determine, at any given time, the front of diffusion.

It is believed that the proposed non destructive method permits to anticipate the level of propagation of chlorides or CO_2 and promote mitigation measures prior the depassivation of the reinforcement steel and need of major repair.

From 2013, a hundred sensors were installed in a concrete structure exposed to sea spray to try to follow diffusion front during time in a real situation. Follow-up actions are planned in the coming years and will make a return of experience and refine the method if necessary.

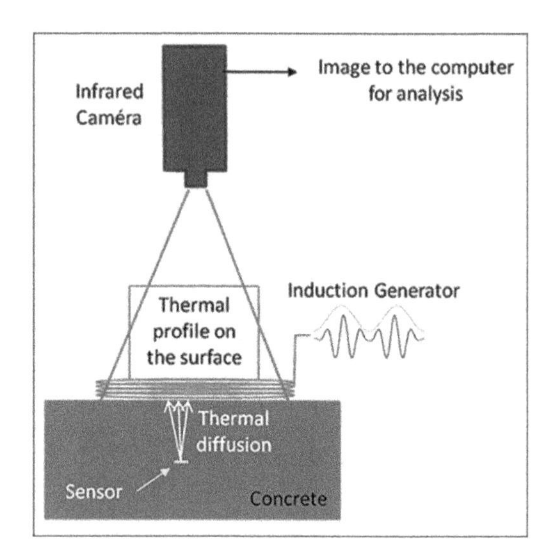

Figure 1. Principle of measurement of the state of corrosion of the sensor.

REFERENCES

Dollet A. et Taché G. 2010: *Méthodes de diagnostic du béton armé, in Anticorrosion et durabilité dans le bâtiment et le génie civil*, S. Audisio et G Béranger, pp. 659–668. Presses polyt. et univ. Romandes ed.

Garcia V., 2013: *Contribution du taux de chlorures pour l'amorçage de la corrosion des armatures du béton armé.* Thèse Univ. de Toulouse 3 Paul Sabatier, LMDC.

Mc Cartera W.J., Ø. Venneslandb, 2004: *Sensor systems for use in reinforced concrete structures* (Review), Construction and Building Materials, 18, 20, 351–358.

Raupach M., Schiessl P. 1997: Monitoring system for the penetration of chlorides, carbonation and the corrosion risk for the reinforcement. *Construction and Building Materials*, Vol. 11, 4, pp. 207–214.

Education, research and specifications

Concrete Repair, Rehabilitation and Retrofitting IV – Dehn et al. (Eds)
 ISBN 978-1-138-02843-2

Numerical modeling of basic creep of concrete under different types of load

N. Ranaivomanana, S. Multon, A. Turatsinze & A. Sellier
Université de Toulouse, UPS, INSA, LMDC, Toulouse Cedex 4, France

ABSTRACT

Despite a relative abundant literature on the topic, delayed mechanical behaviour of concrete remains a problematic of contemporary research. Underestimating concrete creep during the designing phase may lead to well-known consequences such as excessive strains, losses of prestress and even sometimes failure. This study aims to a better understanding of concrete's delayed mechanical behaviour by means of the modeling of basic creep of concrete under different types of load namely direct tension, compression and bending. For the purpose, an original model was developed; it takes into account physical phenomena, namely consolidation and damage induced by creep, to reproduce experimental findings. Such an approach allows a better understanding of the coupling between shrinkage, damage and creep in concrete.

INTRODUCTION

Concrete has a very complex delayed behaviour, including delayed strain occurring under loading (creep) and without loading (shrinkage). A common way to carry out measurements is to split delayed strains of concrete into different components including autogenous shrinkage, drying shrinkage, basic creep and drying creep, which assumes a decoupling of underlying phenomena. Actually, there is a strong coupling between these components, which has to be taken into account in the modeling. In this study, an original approach for modeling the basic creep of concrete is developed. The originality of the model relies on physical phenomena taken into account, namely consolidation (allowing creep potential depletion under hydro-mechanical loading to be reproduced) and damage induced by creep. Such an approach allows a better understanding of the coupling between shrinkage, damage and creep. Finally, a comparison of the model prediction with basic creep of concrete under different types of load is proposed.

DESCRIPTION OF THE MODEL

Classical poro-mechanics formulation allows different aspects of delayed behaviour of concrete such as shrinkage or basic creep to be explained. In this formulation, the total stress tensor σ can be expressed as follows:

$$\sigma = \sigma' - b_0 \pi \tag{1}$$

Where σ' represents the effective stress tensor corresponding to the stress part supported by the solid skeleton, b_0 the Biot tensor representing the part of pressure transmitted to the solid skeleton and π the equivalent pressure.

The classical poro-mechanics formulation does not however allow "Pickett Effect" to be reproduced. Drying creep phenomenon is usually interpreted as a structural effect: a loaded specimen will exhibit a more pronounced shrinkage that a non-loaded one. From a statistical point of view, the probability of existence of tensile zones decreases in the direction which was loaded in compression. This can be expressed by the following modified poro-mechanics formulation:

$$\sigma = \sigma' - b_0 \left(1 - \frac{\sigma}{\sigma_{dc}}\right) \pi \tag{2}$$

σ_{dc} is a material parameter considering both the stress field heterogeneity and its effect on the equivalent pressure amplitude.

The effective stress σ' can be assessed by means of a basic creep model. The viscoelatic solid skeleton is modeled by a Burger chain. The Hooke element allows instantaneous elastic response to be simulated, the Kelvin-Voigt element is associated with reversible viscous strain and the Maxwell element is associated with irreversible viscous strain. This scheme is applied separately to each component of σ' tensor.

Experimental results obtained over a long term period point out that creep rate slows down progressively with time. This "creep potential decrease" is taken into account in the model by an increase of the apparent viscosity of the Maxwell element η^M.

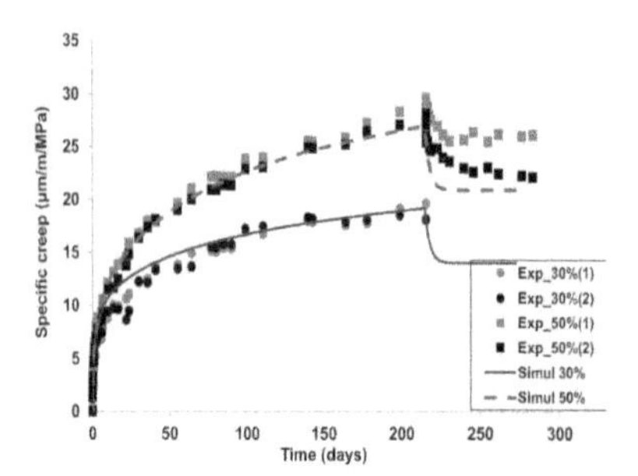

Figure 1. Simulated compressive creep curve at 30% and 50% obtained with law A.

$$\eta^M = \eta^0 C_C \tag{3}$$

where η^0 is the viscosity at the beginning of the loading and Cc the consolidation coefficient defined by:

$$C_C = \exp\left(\frac{\varepsilon^{eq}}{\varepsilon^k}\right) \tag{4}$$

In this expression, ε^{eq} is the equivalent strain which quantify the cumulated viscous strains undergone by concrete during its history and ε^k a parameter playing the role of a creep strain potential.

Therefore, the poro-mechanics formulation (2) has to be modified to take into account the coupling between creep-induced damage and poro-mechanical loading as follows:

$$\sigma = (1 - d^{bc})\sigma' - b_0\left(1 - \frac{\sigma}{\sigma_{dc}}\right)\pi \tag{6}$$

APPLICATION OF THE MODEL TO PREDICT BASIC CREEP OF CONCRETE UNDER DIFFERENT TYPES OF LOADING

A parametric study is conducted to have three different consolidation states before loading in order to analyze its impact on creep strain for the different types of loading. For that purpose, three shrinkage final amplitudes were considered by means of three water content evolution laws (called evolution laws A, B and C in the following). The shrinkage amplitude calculated after 300 days are respectively about 35 µm/m, 70 µm/m and 110 µm/m for the laws A, B and C respectively.

The compressive basic creep simulations are carried out by considering the three previous assumptions on shrinkage. As in practice, shrinkage is deduced from total strains to get creep strains. The specimens were loaded at 28 days while unloading has been carried out after 215 days of loading.

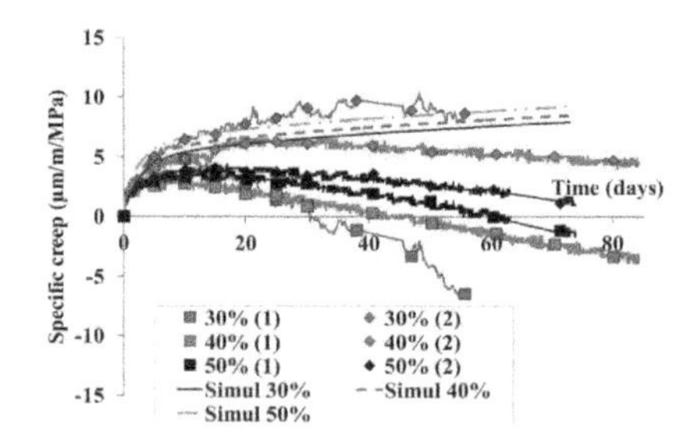

Figure 2. Simulated tensile creep curve at 30%, 40% and 50% obtained with law C.

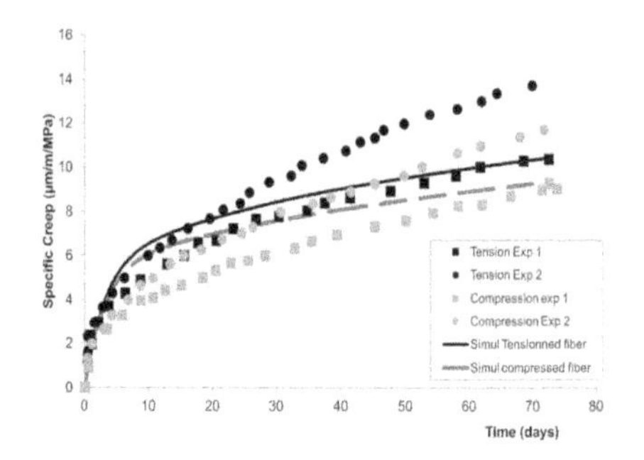

Figure 3. Experimental results and predicted ones of bending specific creep at 50%.

Simulation of basic creep tests in tension is presented in figure 2.

We have conducted simulation of bending basic creep only using law C. The strains obtained on most compressed and the most tensioned fiber of the beam are represented in figure 3.

As it can be seen, experimental results and predicted ones are in a good agreement.

CONCLUSIONS

A numerical model has been used in this study to analyze the basic creep of concrete. The reproduction of the delayed behaviour of concrete under different types of loading by means of a single formulation was possible. The originality of the model relies on physical phenomena taken into account which are consolidation allowing creep-site depletion under hydro-mechanical loading and creep-induced damage to be taken into account. This study points out the role played by shrinkage in the creep modeling. Such an approach allows thus a better understanding of the coupling between shrinkage, damage and creep in concrete. Finally, one have to note that this model is actually in progress and ongoing modifications will certainly be brought in the future to address missing aspects.

Concrete Repair, Rehabilitation and Retrofitting IV – Dehn et al. (Eds)
© 2016 Taylor & Francis Group, London, ISBN 978-1-138-02843-2

Distributions of bond stress between plain round bars and low strength concrete under cyclic loadings

H. Araki
Hiroshima Institute of Technology, Hiroshima, Japan

C. Hong
Constec Engi. Co., Tokyo, Japan

ABSTRACT

The purpose of this study is to determine the distribution and the bond characteristics of plain round bars in low strength concrete. Cyclic pull-out loadings were performed using bars embedded in concrete prisms with strength less than 13.5 N/mm^2. Cyclic loading program is shown in Figure 1. The local tensile and compressive stresses of the bar were obtained by strain gauges mounted inside the bar, as per the references. The main parameter being considered is the length of the embedded bars. By calculating the tensile force along the bar, the relative slip in the bar is also investigated.

The distributions of tensile stress and bond stress under the monotonic pull-out loading are shown in Figure 2. The distribution shape is smoothed by 3-dimensional curves. In the bond stress distribution, the peak of the bond stress shifted to the opposite side of the loading in the early stage of the pull-out loading. The resistance of the adhesion force rapidly vanished from the loaded end. Therefore, the bond stress due to the friction force near the loading side was approximately zero. When the pull-out load was close to point A bond stress in the opposite direction was observed at the loaded end because the embedded bar at this end was affected by the concrete compressed by the reaction force of the pull-out load. In previous studies (Abram 1913) concerning bond stress it was reported that the average bond stress was not only influenced by the concrete strength, but also various factors; the mix properties of concrete, curing, construction method and diameter of the bar. In this study the maximum bond stress in the distribution was approximately 2 MPa regardless of the bond length. The distributions from point A to C in Figure 1 did not change significantly. The peak of the bond stress did not change. Bond stress in the same direction as the pull-out load was observed near the loaded end. In the monotonic pull-out loading tests for the plain round bars, the transition of the distributions of the tensile stress and the bond stress were consistent with the results of the previous studies.

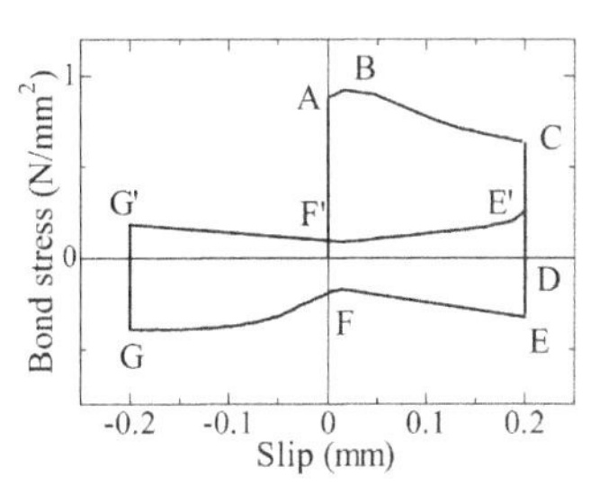

Figure 1. Loading proguram

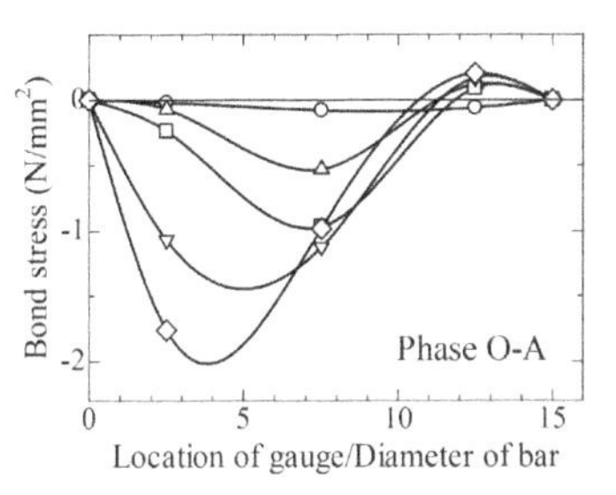

Figure 2. Distributions of bond strength.

Based on the results of the pull-out tests using the plain round bar and low strength concrete the following conclusions are made:

1. The distributions of the tensile stress and bond stress have the same characteristics regardless of the bond length.
2. The peak of bond stress shifted to the unloaded end when the pull-out load increased until the slip occurred.
3. The distributions of the tensile stress and the bond stress do not change significantly although the bond characteristics deteriorated due to the cyclic loadings.
4. The characteristics of local bond stress are approximately the same as the average bond stress along the bond length.

Concrete Repair, Rehabilitation and Retrofitting IV – Dehn et al. (Eds)
© 2016 Taylor & Francis Group, London, ISBN 978-1-138-02843-2

Damage evaluation of RC columns subjected to seismic loading by energy dissipation using 3D lattice model

M.R. Simão & T. Miki
Kobe University, Kobe, Japan

ABSTRACT

Reinforced Concrete (RC) structures exhibit a highly nonlinear behavior in the event of an earthquake. In current seismic design, large consideration is given to issues of strength and deformation capacity, but it is recognizable that in structural damage analysis energy dissipation capacity is one of the key factors in structural damage.

This study is focused on the applicability of the 3D lattice model (Miki & Niwa 2004) to perform damage evaluation of RC columns with circular cross section having consideration for the energy dissipation characteristics of columns under seismic excitation.

A new analysis concept has been developed using the 3D lattice model for circular cross section RC columns based on more realistic multi-directional polygonal discretization. The development of the 3D lattice model is based on the 2D lattice model shown in Figure.1, which based on arch and truss analogy allows the representation of elements in terms of concrete and reinforcement as well as the shear resisting mechanism.

Figure 2 illustrates a schematic representation of arch and truss analogy for a circular cross section column in 3D space.

The diagonal members which include a part of representation of truss action consist of three parts, which are Inner Diagonal Members (IDM), Surface Diagonal Members (SDM) and Diagonal Members in Transverse direction (DMT) respectively.

Longitudinal reinforcement is represented as vertical reinforcement member along the sixteen nodes per layer defining the geometry of the model and transverse reinforcement; it is represented in the form of horizontal reinforcement members uniformly distributed at intervals of $0.5D$ throughout the model as the intervals of arrangement are not taken into account. The representation of multi-directional polygonal 3D lattice model members is detailed in Figure 3.

The 3D lattice model offers some unique characteristics in terms of analytical capacity of the composing elements, because of the truss and arch analogy. It is assumed that an average stress and

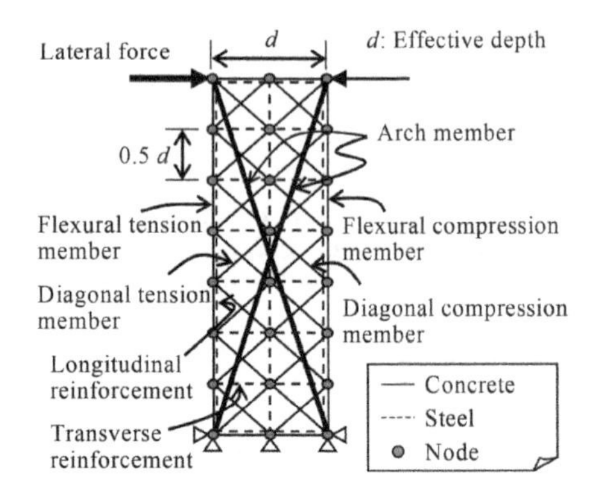

Figure 1. Schematic diagram of 2D lattice model (Miki & Niwa 2004).

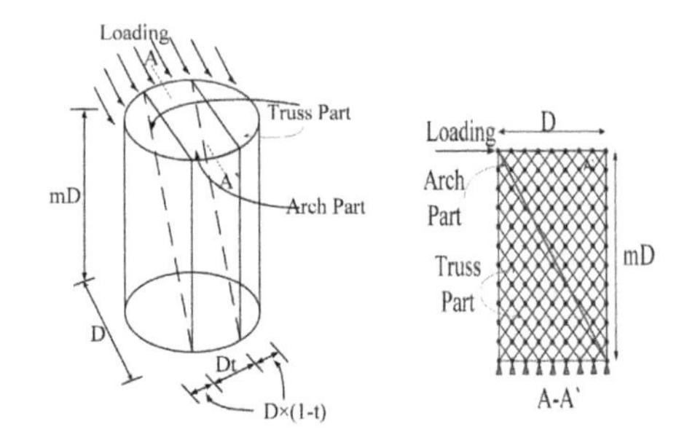

Figure 2. Arch and truss discretization in the 3D lattice model for circular column.

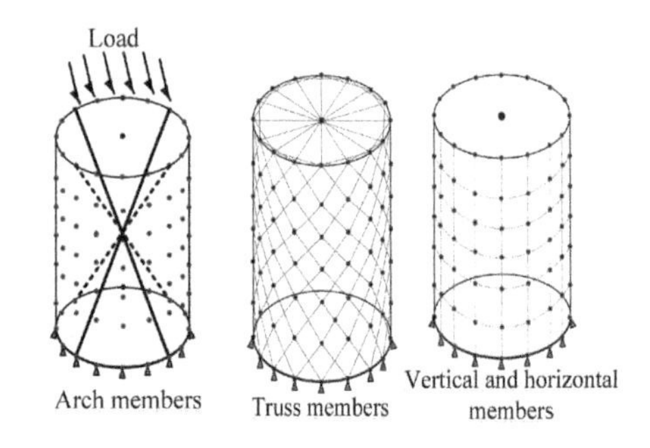

Figure 3. Schematic diagram of multi-directional polygonal 3D lattice model.

strain relationship governs every component of the system. That means that stress-strain relationships define the strain energy for each element in the lattice model.

By taking into account the energy dissipation in individual elements, the distribution of energy dissipation in a RC column can be evaluated by the dynamic lattice model. Based on this assumption the energy dissipated in each element can be calculated from the product of the strain energy dissipated in each element, where the strain energy is the area enclosed by the stress-strain relationship for the unloading and reloading curves, and the elemental volume. RC member accumulated energy dissipation will be the sum of all the elemental energy dissipation histories.

$$E_{dissip} = \sum_{i=1}^{n} \left(Es_i \times V_i\right) \tag{1}$$

In order to use the energy dissipated as an indicator of seismic damage in concrete structures, the total energy dissipated must somehow be normalized so as to compare the results of different size specimens (Inoue 1994).

In this study, the damage range evaluation is performed from the material point of view, which means that the stress-strain relationships of the RC structural members are used to determine energy dissipation capacity and ultimately damage range. Since reinforced concrete is composed of concrete and reinforcing steel, its damage can be defined by the composition of the damage sustained by concrete and by reinforcing steel.

$$D_{RC} = D_{concrete} + D_{steel} \tag{2}$$

Under defined strain conditions, energy dissipation in the material can be obtained using the constitutive models of materials and compared to analytical energy dissipation. In that manner damage index in the material will be derived as the ratio between the analytical response energy dissipation and the energy dissipation calculated using the constitutive models

Since elemental energy dissipation has been defined as the product between strain energy and elemental volume, the damage index in material can be expressed as the ratio between strain energy of analysis response and strain energy obtained using the constitutive models.

$$D_{concrete} = D_{steel} = \frac{Es_{response}}{Es_{constitutive}} \tag{3}$$

The applicability of the model is verified by performing dynamic analysis on a circular cross section RC column that has been tested by the E-Defense shown in Figure 4. The analytical target is a circular cross section column named C1-5 tested using a shake-table by E-Defense.

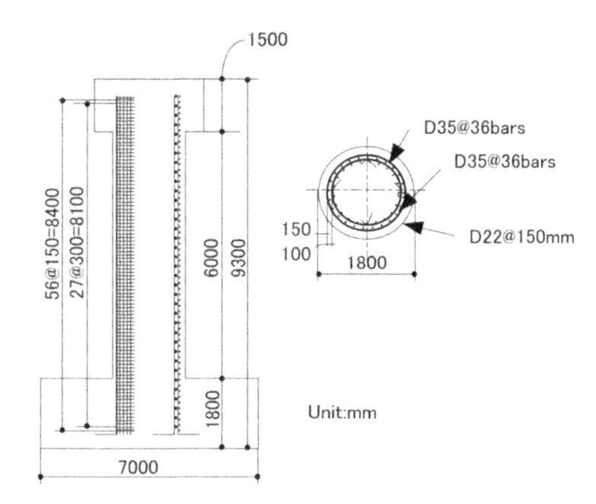

Figure 4. C1-5 specimen details (Kawashima et al. 2010).

The analytical results show that response strain energy of concrete is about a third of constitutive strain energy of concrete, while in reinforcement the analytical strain energy is sixteen times smaller than the constitutive. This reinforces the idea that energy the energy dissipation potential of steel reinforcement is by large most significant than that of concrete.

The energy dissipation capacity obtained from the analysis for concrete shows that in the first excitation about 30% of energy is dissipated at the bottom, corresponding to1000 mm of column height, while reinforcement dissipates less than 1%. This means that by large in the given conditions, the damage potential is bigger in concrete than in steel.

In conclusion, the applicability of the 3D lattice model to perform damage evaluation of reinforced concrete columns with circular cross section using the multi-directional polygonal 3D lattice model is confirmed comparing the analytical and experimental results.

Damage evaluation proposed from the material point of view considering the accumulated energy dissipation of concrete and reinforcement, which are dependent on strain energy show based on the analytical results that by large scale, steel reinforcement is the most dominant material in seismic resistance capacity.

REFERENCES

Inoue, S. 1994. Ductility and energy dissipation of concrete beam members and their damage evaluation based on hysteretic energy dissipation, Kyoto: Kyoto University Press.

Miki, T. and Niwa, J. 2004. Nonlinear analysis of RC structural members using 3D lattice model, *Journal of advanced concrete technology*, Vol.2, No.3: 343–358.

Concrete Repair, Rehabilitation and Retrofitting IV – Dehn et al. (Eds)
© 2016 Taylor & Francis Group, London, ISBN 978-1-138-02843-2

Influence of aggregate size and the effect of superplasticizer on compression strength

H. Aljewifi
Materials Research Laboratory, University of Omar Almukhtar, Albida, Libya

X.B. Zhang
Institut Pascal, Univercité Blaise Pascal de Clermont-Fd, Montluçon, France

J. Li
Univercité Paris XIII, LSPM, CNRS UPR 3407, France

ABSTRACT

This paper describes firstly, the effect of all size of aggregates on the compressive strength of the ordinary concrete used in buildings. The importance of this effect is determined by conducting laboratory experiments on ordinary concrete prepared using aggregates of different size (but same type: limestone) without varying water cement ratio in this phase. Jain et al. 2011 demonstrated that for all types of aggregates, pervious concrete mix prepared using smaller size of aggregates demonstrated higher compressive strength. The experimental results indicated that Mechanical behavior of cylindrical samples appeared that maximum stress (σ_{cyli}) tend to increasing with smaller size of aggregates; and the longitudinal deformation diminish with 1–62.5 mm size; Elastic module remain approached except with 1–62.5 mm size of aggregate. For cubic samples, maximum stress (σ_{cubic}) greater than cylindrical samples and they gave the same results as cylindrical samples, see table 1. Physical properties of aggregates classified using the initial experiments such as: sieves analysis, the water content test; coefficient of absorption, bulking of sand test, cleaning of aggregate, porosity and real volume bulk density.

In addition, Micro Deval test used to determine Los Angles coefficient. In the first phase of this research and the same as, for all size of aggregates, ordinary concrete prepared using smaller size of aggregates demonstrated higher compressive strength values. Secondly, this paper studied the effect of HAA Dynamon and Adricete BVF admixture on the compressive strength of the ordinary concrete with vary water cement ratio, in terms of reducing the water content within the normal concrete, according to the specific proportions of cement weight and classification of the mechanical properties of fresh and solid concrete at 28 days and examined the benefits of admixture in terms of improving the compression resistance. Furthermore, this paper studied the effect of superplasticizer on compression strength. Figure 1 appeared the comparison between compression strength without addition an admixture and with HAA and BVF. Compression strength increase with decreasing size of aggregates and without using an admixture. For same size (1/19 mm) we note that approached behavior with or without introducing an admixture. Consequence, introduce Dynamon HAA and Adicrete BVF does not carry great influence on mechanical strength.

Table 1. Mechanical results on hard concrete.

Size (mm)	ρ_{av} (g/cm^3)	ε (mm/mm)	$\sigma_{cyli.}$ (MPa)	E (GPa)	σ_{cubic} (MPa)
1–19	2.32	0.0338	27.71	8.10	35.06
1–25	2.42	0.0029	26.97	9.65	34.11
8–37.5	2.35	0.0031	25.49	8.40	34.18
1–62.5	2.36	0.0017	21.67	14.5	26.11

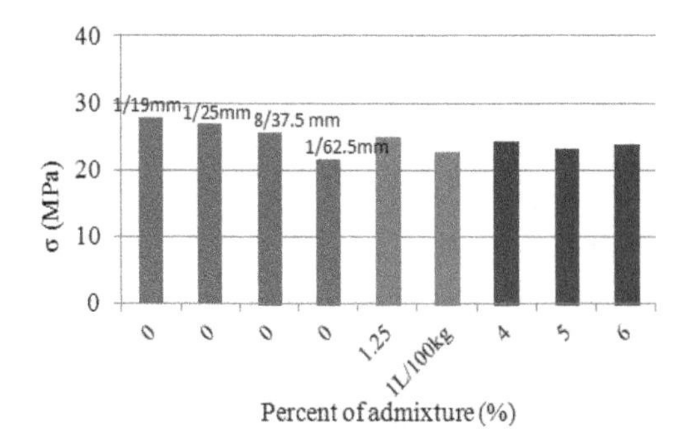

Figure 1. Influence introduce the admixture on compression stress at 28 days.

Concrete Repair, Rehabilitation and Retrofitting IV – Dehn et al. (Eds)
© 2016 Taylor & Francis Group, London, ISBN 978-1-138-02843-2

Bond of reinforcing bars in cracked concrete

P. Desnerck, J.M. Lees & C. Morley
Engineering Department, University of Cambridge, Cambridge, UK

INTRODUCTION

Concrete is an inhomogeneous material with a relatively low tensile strength. Therefore it is often used in combination with steel reinforcement so that the steel can resist tensile stresses after cracking. One of the most severe forms of cracking in reinforced concrete is the result of the corrosion of the reinforcing bars.

To investigate crack formation due to corrosion, researchers have undertaken accelerated corrosion tests on reinforced concrete specimens where impressed currents were applied.

One issue with accelerated corrosion bond tests is that they study the combined effect of cracking and the formation of a soft layer of corrosion products around the reinforcement. While these effects co-exist in practice, it leads to difficulties in analysing the processes at a fundamental level. Therefore this study aims to quantify the bond strength of reinforcing bars in cracked concrete. Rather than performing accelerated corrosion tests, it focuses on the more fundamental effect of the cracking itself.

EXPERIMENTAL PROGRAM

In this study the principles of a controlled split tensile test is applied to pre-crack specimens. In this way rough crack surfaces are formed along a predefined cracking plane running through the axis of the reinforcing bar. Of particular interest are the bond reductions during the early stages of corrosion and thus the onset of cracking. The splitting tests seek to achieve 0.03 to 0.04 mm crack widths.

In order to represent common crack patterns, the test method includes one or two pre-cracking phases followed by standard bond strength "pull-out tests".

In the pre-cracking phase(s) the specimens are subjected to a split cylinder test (Figure 1). Pull-out tests are then conducted on the cracked specimens to determine the influence of the cracks on the bond behaviour.

Cylindrical specimens with a diameter of either 107 mm or 60 mm, and a height of 100 mm are used (Figure 2). The specimens are cast in a plastic

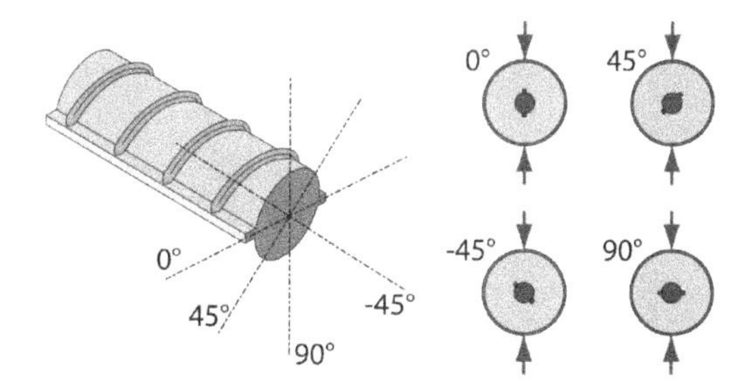

Figure 1. Applied crack orientations with respect to the reinforcing bars rib pattern.

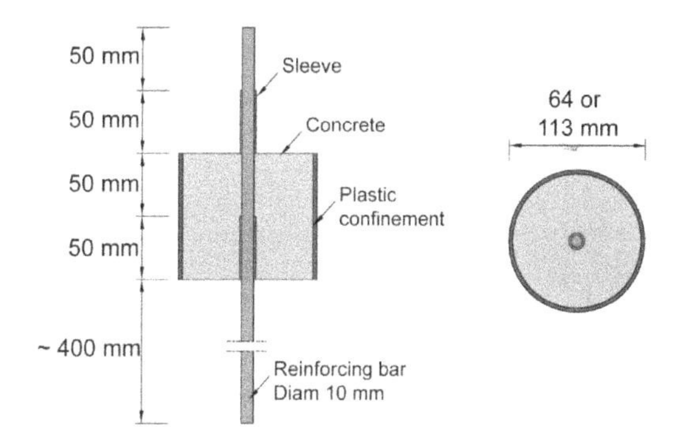

Figure 2. Specimen geometry.

cylindrical mould which was used as confinement for the specimen during the pre-cracking phases and in many cases also remained in place for the pull-out tests.

Based on previous work, the following parameters were selected for investigation: single or double cracking (number of cracks), the crack orientation relative to the reinforcing bar rib pattern, the confinement and the concrete cover.

BOND STRENGTH RESULTS

From the pull-out experimental results, values of the bond stress can be derived. The results of the bond tests are expressed in terms of a bond reduc-

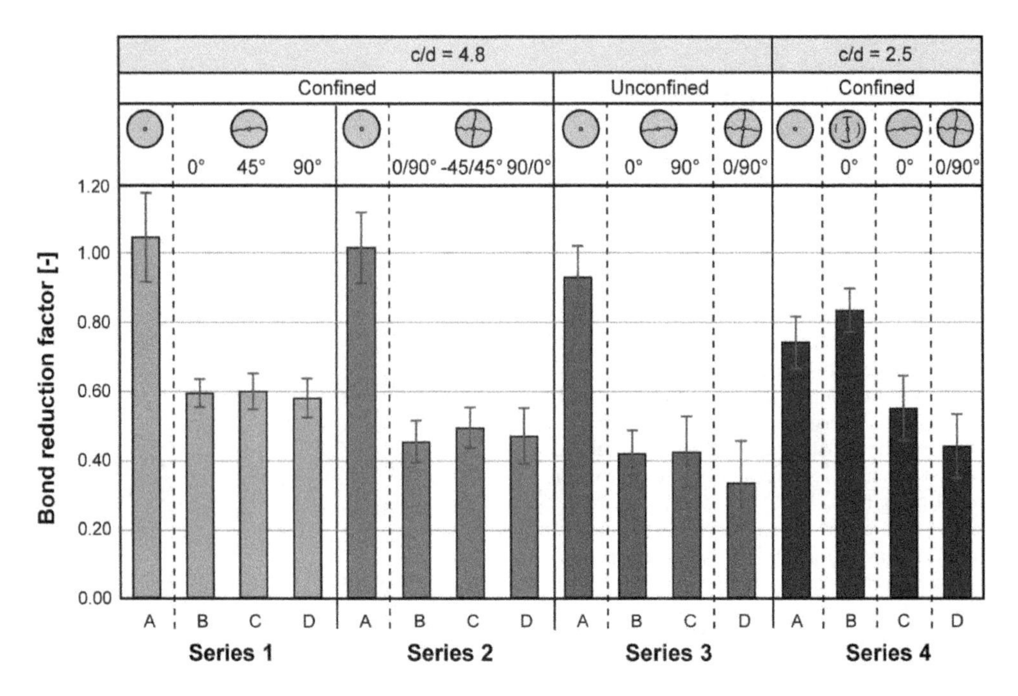

Figure 3. Bond reduction with respect to uncracked reference specimens.

tion factor. This factor is defined as the ratio of the actual strength of a specific set to the mean bond strength of the equivalent uncracked reference specimens in that series. In this way direct comparisons of the test results within and between series can be made (Figure 3).

The presence of cracks, even with minor crack widths of 0.03 to 0.04 mm, result in a significant reduction of the bond strength. For specimen with a single crack the reduction was on average 44% and for double cracked specimens the reduction was 54%. The measured values for single and double cracked specimens are within the relatively wide range of values reported by other researchers in the past.

The crack orientation with respect to the rib pattern of the reinforcing bars has little or no effect on the obtained bond properties. Three different crack orientations were tested and the results showed similar ultimate bond strengths.

For double cracked specimens the order in which the cracks are formed (linked to the test method) has no significant influence on the bond behaviour.

Confinement influences the ultimate bond strength of a pre-cracked specimen. In the absence of a restraining force, existing cracks can open up fully enabling the reinforcing bar to more easily slip out of the specimens. The reduction of bond capacity between confined and unconfined specimens was an additional 11% for a given parameter combination.

When the concrete cover is reduced, the residual bond strength after cracking is reduced as well. For smaller covers the failure mode of the uncracked concrete is shifted from a pull-out failure to a splitting failure.

The bond strength of reinforcing bars in cracked cylinders embedded in an uncracked concrete ring of 23 mm is 18% lower than the original uncracked specimens with the same total concrete cover.

The obtained test results indicate that the presence of longitudinal cracks can significantly influence the bond behaviour of ribbed reinforcing bars in concrete. This suggests that bond reduction factors are necessary for cracks that run along the reinforcement bars when undertaking load bearing capacity checks of existing reinforced concrete structures.

ACKNOWLEDGEMENTS

The authors would like to gratefully acknowledge the financial support of the UK Engineering and Physical Sciences Research Council (EPSRC) through the EPSRC Project 'Reinforced concrete half-joint structures: Structural integrity implications of reinforcement detailing and deterioration' [Grant no. EP/K016148/1].

Concrete Repair, Rehabilitation and Retrofitting IV – Dehn et al. (Eds)
© 2016 Taylor & Francis Group, London, ISBN 978-1-138-02843-2

Improved formulation for compressive fatigue strength of concrete

E.O.L. Lantsoght
Universidad San Francisco de Quito, Quito, Ecuador
Delft University of Technology, Delft, The Netherlands

C. van der Veen
Delft University of Technology, Delft, The Netherlands

A. de Boer
Ministry of Infrastructure and the Environment, Utrecht, The Netherlands

ABSTRACT

The compressive strength of concrete decreases as an element is subjected to cycles of loading. The lower and upper limits are expressed as a fraction of the concrete compressive strength, and can be written as $S_{min}f_{ck}$ and $S_{max}f_{ck}$. The result of fatigue tests on concrete cylinders in compression is the so-called Wöhler-curve, or *S-N* curve, showing S_{max} versus the number of cycles.

The expression for concrete under compression subjected to cycles of loading from NEN-EN 1992-2+C1:2011 (CEN, 2011) is more conservative than previously used expressions in the Netherlands. Therefore, different expressions are given in the National Annex NEN-EN 1992-2+C1:2011/NB:2011 (Code Committee 351 001, 2011). The *S-N* curve described by the two expressions given in the code is discontinuous at 10^6 cycles. Because of this anomaly in the current code provisions, it is necessary to propose a new expression for concrete under cycles of compressive loading. Moreover, the proposed expression should be valid, yet not overly conservative, for high strength concrete.

Currently, there seems to be a disagreement in the literature on whether or not the fatigue strength of concrete under compression decreases as the compressive strength of the concrete increases, or, in other words, if the *S-N* curve should be steeper and reducing more quickly for higher strength concrete. Therefore, it is deemed conservative to descrease the fatigue life for high strength concrete, as applied in the currently available codes.

The full paper given an overview of the current code provisions for fatigue of concrete under compression that are used in this study: NEN 6723:2009, the Eurocode for concrete structures NEN-EN 1992-1-1+C2:2011 and its Dutch National Annex NEN-EN 1992-1-1+C2:2011/NB:2011, the Eurocode for concrete bridges NEN-EN 1992-2+C1:2011 and its Dutch National Annex NEN-EN 1992-2+C1:2011/NB:2011, and the recently published *fib* Model Code 2010.

To develop a new proposal for the fatigue strength of concrete under compression, a database of fatigue tests is compiled, with an emphasis on recent test results on high strength concrete. In total, 429 experiments are in the database, of which 234 experiments do not contain fibers in the concrete mix (maximum compressive strength of 145 MPa) and of which 195 experiments contain fibers and have a maximum compressive strength of 226 MPa. An additional 165 test results of normal strength concrete are used for verification purposes of the proposed formulas. As the range of a low number of cycles to failure is interesting for applications in bridge engineering, a histogram showing the distribution of the number of cycles to failure for the 429 experiments from the database is given in Figure 1.

For the design of new structures, a method should be proposed that is easy to use and does not require iterations. The expression from NEN-EN 1992-2+C1:2011 (CEN, 2011), is used as a starting point to develop this new method, with $k_1 = 1$ and $\gamma_c = \gamma_{c,fat} = 1.5$:

$$f_{cd,fat} = k_1 \beta_{cc}(t_0) f_{cd}\left(1 - \frac{f_{ck}}{400}\right) \quad (1)$$

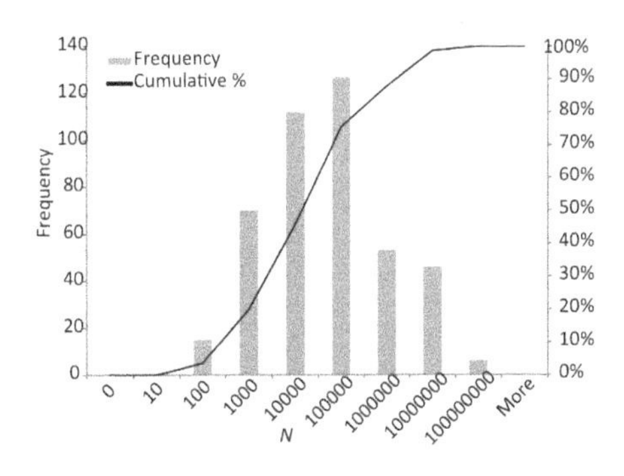

Figure 1. Histogram of number of cycles in the experiments from the database.

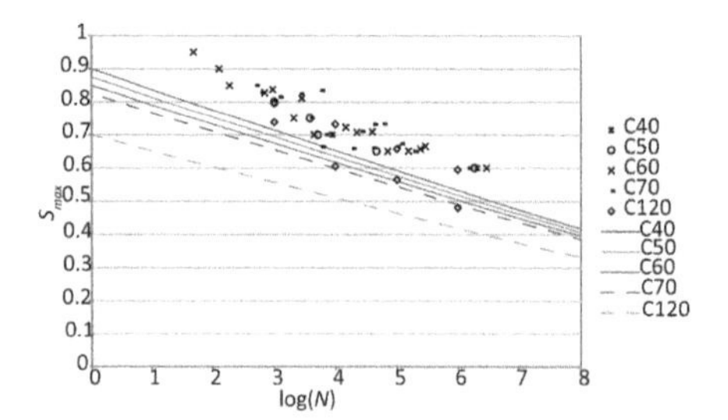

Figure 2. S-N curves from Eq. (2) and comparison to test results with $k_1 = 1$ and $S_{min} = 0.05$.

$$N_i = 10^{\left(14\frac{1-E_{cd,max,i}}{\sqrt{1-R_i}}\right)} \tag{2}$$

$$R_i = \frac{E_{cd,min,i}}{E_{cd,max,i}} \tag{3}$$

$$E_{cd,min,i} = \frac{\sigma_{cd,min,i}}{f_{cd,fat}} \tag{4}$$

$$E_{cd,max,i} = \frac{\sigma_{cd,max,i}}{f_{cd,fat}} \tag{5}$$

$$\beta_{cc}(t_0) = \exp\left\{s\left[1-\left(\frac{28}{t_0}\right)^{0.5}\right]\right\} \tag{6}$$

For existing structures a new, iterative method is developed, as follows:

$$\log N_i = \frac{6(S_{max}-1)}{S_{max,EC}-1} \text{ for } S_{max} \geq S_{max,EC} \tag{7}$$

$S_{max,EC}$ is here the value of S_{max} which is found for 10^6 cycles, and can be expressed as:

$$S_{max,EC} = \left(1-\frac{f_{ck}}{400}\right)\left(1-\frac{3}{7}\sqrt{1-R_i*}\right) \tag{8}$$

with f_{ck} in MPa and $R_i* = S_{min}/S_{max,EC}$ and

$$\log N_i = 14\frac{1-E_{cd,max,i}}{\sqrt{1-R_i}} \text{ for } N_i > 10^6 \tag{9}$$

$$R_i = \frac{E_{cd,min,i}}{E_{cd,max,i}} \tag{10}$$

$$E_{cd,max,i} = \frac{\sigma_{cd,max,i}}{f_{cd,fat}} \text{ for } N_i > 10^6 \tag{11}$$

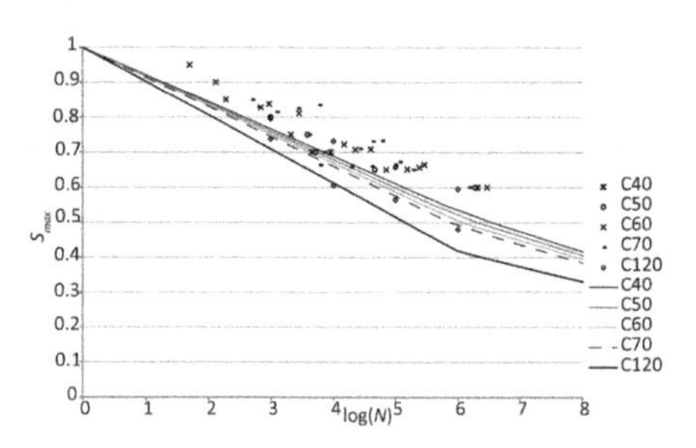

Figure 3. S-N curves from Eqs. (7) and (9) and comparison to test results with $S_{min} = 0.05$.

$$f_{cd,fat} = k_1\beta_{cc}(t_0)f_{cd}\left(1-\frac{f_{ck}}{400}\right) \tag{12}$$

with $\beta_{cc}(t_0)$ from Eq. (6), $k_1 = 1$ and $\gamma_c = \gamma_{c,fat} = 1.5$.

A comparison between the proposal for design and the test results from the database with $S_{min} = 0.05$ is given in Figure 2. In Figure 2, the S-N curves for different concrete classes are shown (lines), as well as test results from the database (datapoints) for their respective concrete classes. A similar comparison with the proposal for existing structures is shown in Figure 3. As can be seen from these figures, the best predictions are found when using the method for existing structures, which requires iterations.

REFERENCES

Cen 2011a. Eurocode 2: Design of concrete structures—Concrete bridges—Design and detailing rules. NEN-EN 1992-2+C1:2011. Brussels, Belgium: European Committee for Normalization.

Cen 2011b. Eurocode 2: Design of Concrete Structures - Part 1-1 General Rules and Rules for Buildings. NEN-EN 1992-1-1+C2:2011. Brussels, Belgium: European Committee for Normalization.

Code Committee 351 001 2009. Regulations for concrete—Bridges—Structural requirements and calculation methods, NEN 6723:2009. Delft, The Netherlands: Dutch Institute for Normalization. (in Dutch)

Code Committee 351 001 2011a. National Annex to NEN-EN 1992-2+C1, Eurocode 2: Design of concrete structures—Concrete bridges—Design and detailing rules. NEN-EN 1992-2+C1:2011/NB:2011. Delft, The Netherlands: Dutch Institute for Normalization. (in Dutch)

Code Committee 351 001 2011b. National Annex to NEN-EN 1992-1-1+C2, Eurocode 2: Design of concrete structures—Part 1-1: General rules and rules for buildings NEN-EN 1992-1-1+C2:2011/NB:2011. Delft, The Netherlands: Nederlands Normalisatie-instituut.

FIB 2012. Model code 2010: final draft, Lausanne, Switzerland: International Federation for Structural Concrete.

Concrete Repair, Rehabilitation and Retrofitting IV – Dehn et al. (Eds)
© 2016 Taylor & Francis Group, London, ISBN 978-1-138-02843-2

Determination of diffusivities of dissolved gases in saturated cement-based materials

Quoc Tri Phung
Belgian Nuclear Research Centre and Magnel Laboratory for Concrete Research, Ghent University, Belgium

Norbert Maes, Diederik Jacques, Elke Jacops & Arno Grade
Belgian Nuclear Research Centre (SCK•CEN), Institute for Environment, Health, and Safety, Belgium

Geert De Schutter
Magnel Laboratory for Concrete Research, Ghent University, Belgium

Guang Ye
Microlab, Delft University of Technology, Delft, The Netherlands

ABSTRACT

Diffusion is an important property for characterizing concrete durability because it governs the penetration of aggressive substances responsible for degradation. However, data on the diffusion of substances (other than Cl^-) in concrete are very scarce due to time and resource consuming measurements. This work describes a method to determine the diffusion coefficients of dissolved gases in saturated cement-based materials in order to study the effects of degradation on the transport properties. The method is based on a through-diffusion methodology and allows simultaneous determination of diffusivities of two dissolved gases diffusing in opposite directions. A cement plug is mounted between two water reservoirs pressurized by two different gases at equal pressure (~1 MPa) to avoid advection. The changes in the dissolved gas concentration (at the opposite sides) are measured indirectly via gaseous phases which are in equilibrium with aqueous phases according to Henry's law. Additionally, a simple 1-D diffusive transport model (based on 1st and 2nd Fick's laws) is developed to interpret the experimental data. The concentrations at outlet and inlet are used as inputs for the model, and the diffusivity is obtained by a fitting procedure.

DETERMINATION OF DIFFUSIVITIES OF DISSOLVED GASES

Sample preparation and experimental setup

The diffusion experiments were performed on two intact samples S1 (water/powder = 0.425, limestone filler/powder = 0%) and S2 (water/powder = 0.375, limestone filler/powder = 10%). Furthermore, the change in diffusivity due to carbonation was determined on carbonated sample S1 (named as S1C). Type I ordinary Portland cement (CEM I 52.5 N) and limestone fillers (Calcitec 2001S) was used. Superplasticizer (Glenium 27) was added to the mix with content of 0.5% with respect to mass of cement.

Cement paste was poured in a cylindrical PVC tube with inner diameter of 97.5 mm. The sample was then rotated during 24 hours to prevent segregation. Afterwards, the sample was cured in a sealed condition in a temperature controlled room (22 ± 2oC) for 27 days. The 28-day-cured cement paste was sawn into 25 mm thick slices. The disks were then embedded into the polycarbonate part of the permeability cells. It took 24 hours for the resin to dry before polishing to obtain the final sample thickness of 25 mm. The samples were afterwards saturated under vacuum conditions. For carbonated sample, the sample S1C was carbonated by applying a pressure gradient of 5 bar pure CO_2 for 28 days (Phung et al., Under review). Prior to diffusion test, the carbonated sample needs to be saturated after carbonation test.

The gas couple He-Xe was used for all diffusion experiments. A schematic view of the experimental setup is presented in Fig. 1. The setup is the same as the one used for clay materials (Jacops et al., 2013) except for the diffusion cell which is exactly the same of the permeability cell. Two precise pressure transducers were used to measure gas pressure in the pressurized water vessels. The water with the dissolved gas was circulated over the contact filter in the cell by means of magnetic coupled gear pumps. The diffusion cell was connected to the system through quick connectors.

Experimental procedure

The experiment was performed in a temperature controlled room at 21 ± 2°C. Prior to the diffusion measurement, the sealing of the whole system was checked

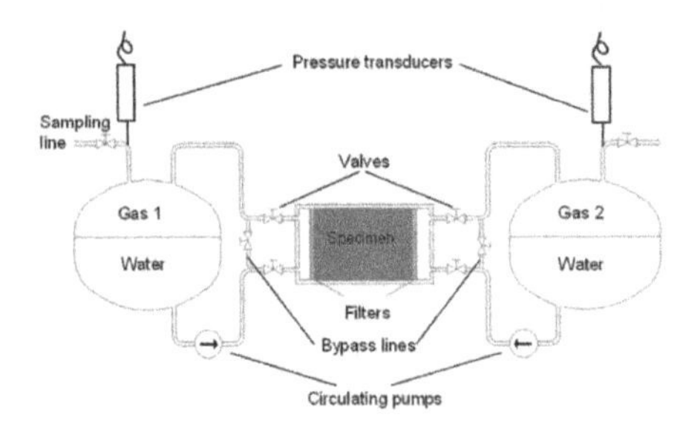

Figure 1. Schematic experimental setup of diffusion.

by applying a gas pressure of 12 bar and following the pressure evaluation over time (few weeks). The entire setup is considered to be gastight if the pressure is stable. The air or residual gas from previous tests in the system was removed as much as possible by flushing the system with the testing gases. During flushing, bypass lines were opened while all valves connected to the sample were closed to prevent diffusion proceeding at this step. Subsequently, a volume of 0.5 litre of degassed water was added to each vessel, where after they were pressurized by the respective gases to a similar pressure of about 10 bar to prevent advective transport. Prior to the measurement, a sample was taken to determine the initial gas composition of both vessels and referred as the sample at the time zero. In order to start the test, all valves connected to the sample were opened, the bypass lines were closed. At that moment, the pressures in both vessels slightly dropped due to small volume increase of the system, but still in balance. Gas samples were regularly taken via external samplers until enough data points were collected to obtain diffusivity. To avoid a too large pressure drop due to sampling, the sample volumes taken were only 6 ml. The gas composition was analysed with a CP4900 micro gas chromatograph.

Data analysis

In order to interpret the experimental data, a 1-D diffusive transport model was developed (Jacops et al., 2013). Fick's laws were used to describe the diffusion of the dissolved gases. The diffusion problem was numerically solved in COMSOL Multiphysics. The effective diffusion coefficient was obtained by using a least squares fitting procedure with the MATLAB Optimization Toolbox.

RESULTS AND DISCUSSION

Intact samples

Due to lower diffusivity of Xe compared to He (5 times lower in free water), the concentration of Xe at the outlet was not high enough to be detected reliably by gas chromatography within the experimental time (4–5 months). Therefore only the data of He diffusion are presented for the intact samples. The effective diffusion coefficients of samples S1 and S2 were 2.32×10^{-11} m²/s and 1.22×10^{-11} m²/s, respectively. The limestone filler addition improved the microstructure which resulted in a decrease of diffusivity of about 50%.

Table 1. Estimation of effective diffusivity of carbonated zone using series model.

D, m²/s	D_{un}, m²/s	D_{car}, m²/s	D_{un}/D_{car}
1.64×10^{-11}	2.32×10^{-11}	4.72×10^{-12}	4.92

Carbonated sample

The effective diffusion coefficient of S1C was 1.64×10^{-11} m²/s, which is 30% lower than the effective diffusion coefficient of the intact sample. It is worth mentioning that the effective diffusion coefficient of the carbonated sample (D) should be treated as the composite effective diffusion coefficient. Therefore, the series model (Phung, 2015) was applied for the calculation of the effective diffusion coefficient of the carbonated zone with an assumption that the effective diffusion coefficient of sound zone equals to the one of the intact sample. Estimation of the effective diffusion coefficient of the carbonated zone (D_{car}) resulted in a value of 4.72×10^{-12} m²/s, which is 5 times lower than the effective diffusion coefficient of the intact sample (D_{un}).

CONCLUSIONS

In the present work, a promising method to determine the diffusion coefficients of dissolved gases in saturated cement-based materials was described. The proposed method enabled to measure the diffusivities of two dissolved gases in a single experiment and was integrated to the other setups in order to study the effects of degradation on the transport properties. A 1-D diffusive transport model accounting for the pressure drop was used to better interpret the experimental results.

The method proves its reliability through its good correlation coefficients between the measured and modelled partial pressures. It was observed that the addition of limestone filler had significant influences on diffusivity. With a similar w/c ratio, adding 10% of limestone filler reduced the diffusivity of cement paste by 50%.

The carbonation exhibited a significant decrease in diffusivity resulting from the changes in microstructure. A series model was used to estimate the diffusivity of the carbonated zone. Estimations showed that the diffusivity of the carbonated zone were significantly decreased (factor of 5) compared to the composite diffusivity (only 30% decrease).

Concrete Repair, Rehabilitation and Retrofitting IV – Dehn et al. (Eds)
© 2016 Taylor & Francis Group, London, ISBN 978-1-138-02843-2

Comparison of residual strengths of concretes with quarzitic, limestone and slag sand constituents after cyclic high-temperature exposure

S. Anders
Construction Materials, Institute of Structural Engineering, Bergische Universität Wuppertal, Wuppertal, Germany

ABSTRACT

A series of experiments was conducted in order to investigate the effects of different types of cements, different types of aggregates, increased number of thermal cycles and different storing climates on concrete mixtures. An overview on the variations is given in Table 1

From the experiments the following conclusions can be drawn.

The residual compressive strength does not differ significantly from the compressives strength at room temperature as long as the maximum temperature does not exceed 250°C.

The flexural strength decreases even at temperatures of 250°C.

At higher temperature levels the mixture containing CEM III/A as cement and slag sand as aggregate performs significantly better than mixtures containing quazitic aggregates. Therefore, the type of aggregates plays obviously a dominant role.

Thermal cycles seem not to deteriorate the concrete structure further after some initial thermal cycles. Thus, the damage in the concrete structure seems to depend on the maximum temperature and not on the number of thermal cycles. This behaviour could be observed in terms of the compressive strength, flexural strength as well as development of stress-strain relations.

Storing the specimens after an exposure to 250°C as a maximum seems to decrease the residual compressive strength by about 20%. The type and duration of the storing does not seem to have a significant effect on the strength reduction.

If the specimens were heatd up to 900°C storing the specimens in climate seems to increase the residual compressive strength compared to an immediate testing after cooling down of the specimen.

Concerning the effects of storing in climate the tests are not finished yet. Maybe cements blended with carbonate will show a different behavior.

Table 1. Tested concrete properties and thermal properties.

Parameter	Properties
Cements	CEM I, CEM II/A-LL, CEM III/A
Aggregates	Quartz sand, limestone powder, slag sand
Temperature levels	250°C, 500°C, 900°C
Thermal cycles	0, 1, 8 and 24
Storing climates	the storing climate was only changed after exposure to one thermal cycle: "A": immediate testing after cooling down "B": 7 days in climate 20°C/65% relative moisture "C": 5 minutes submerged in water, then climate "B" "D": extension of climate "B" to 28 days

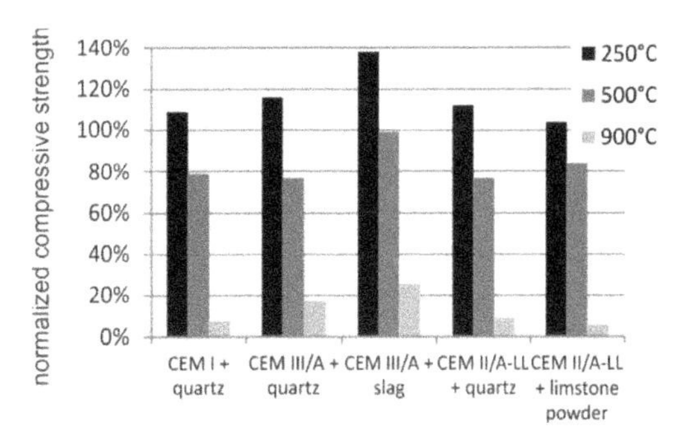

Figure 1. Effect of different types of cement on the residual compressive strength after one thermal cycle.

Concrete Repair, Rehabilitation and Retrofitting IV – Dehn et al. (Eds)
© 2016 Taylor & Francis Group, London, ISBN 978-1-138-02843-2

Author index